System Synthesis with Vhdl

System Synthesis with VHDL

Edited by

Petru Eles
Timisoara Technical University

Krzysztof Kuchcinski
Linkoping University

and

Zebo Peng
Linkoping University

KLUWER ACADEMIC PUBLISHERS
BOSTON / DORDRECHT / LONDON

A C.I.P. Catalogue record for this book is available from the Library of Congress.

ISBN 978-1-4419-5024-6

Published by Kluwer Academic Publishers,
P.O. Box 17, 3300 AA Dordrecht, The Netherlands.

Sold and distributed in the U.S.A. and Canada
by Kluwer Academic Publishers,
101 Philip Drive, Norwell, MA 02061, U.S.A.

In all other countries, sold and distributed
by Kluwer Academic Publishers,
P.O. Box 322, 3300 AH Dordrecht, The Netherlands.

Printed on acid-free paper

CONTENTS

Preface

We have witnessed during the last two decades a tremendous growth in the area of design automation for digital circuits and embedded systems. The driving force of this growth is the underlying development in microelectronics technology, which has delivered an exponential increase in computational power and storage capacity. It has been shown that the number of transistors on a single chip has increased ten times every six years and the speed of microprocessors has increased ten times every eight years. And it is expected that this exponential growth will continue in the near future.

Design automation techniques have provided an enabling technology for designing complex integrated circuits and systems. Design methodologies and tools that operate at the physical and logical level are widely available today and extensively used in the industry. More and more recent research activities have therefore been devoted to the design tasks at the behavioral and architectural level. Many high-level synthesis techniques and systems have been developed in recent years.

Embedded systems are usually composed of several interacting components such as custom or application specific processors, ASICs, memory blocks, and the associated communication infrastructure. The development of tools to support the design of such systems requires a further step from high-level synthesis towards a higher abstraction level. The lack of design tools accepting a system-level specification of a complete system, which may include both hardware and software components, is one of the major bottlenecks in the design of embedded systems. Thus, more and more research efforts have been spent on issues related to system-level synthesis.

This book addresses the two most active research areas of design automation today: high-level synthesis and system-level synthesis. In particular, a transformational approach to synthesis from VHDL specifications is described.

Features of the Book

Throughout this book, we try to provide a coherent view of system synthesis which includes the high-level and the system-level synthesis tasks. VHDL is used as a specification language and several issues concerning the use of VHDL for high-level and system-level synthesis are discussed. These include aspects from the compilation of VHDL into an internal design representation to the synthesis of systems specified as interacting VHDL processes.

Our emphasis has been placed on the use of a transformational approach to system synthesis. A Petri net based design representation is rigorously defined and used throughout the book as a basic vehicle for illustration of transformations and other design concepts. Iterative improvement heuristics, such as tabu search, simulated annealing and genetic algorithms, are discussed and illustrated as strategies which are used to guide the optimization process in a transformation-based design environment. Advanced topics, including hardware/software partitioning, test synthesis and low power synthesis are discussed from the perspective of a transformational approach to system synthesis.

Use of the Book

This book can be used for advanced undergraduate or graduate courses in the area of design automation and, more specifically, of high-level and system-level synthesis. At the same time the book is intended for CAD developers and researchers as well as industrial designers of digital systems who are interested in new algorithms and techniques supporting modern design tools and methodologies.

The first part of the book, consisting of chapters 1 to 5, gives an overview of the state-of-the-art in high-level and system-level synthesis. It also offers a short introduction to VHDL and optimization heuristics. This makes the work accessible to a broad spectrum of potential readers without a special background in digital design or design automation. The second part of the book consists of chapters 6, 7 and 8, and is dedicated to the transformational approach to system synthesis. It first presents a transformational approach to high-level synthesis. Then, several advanced issues related to the synthesis of VHDL system specifications are discussed. Finally, a transformational approach to hardware/software partitioning is presented. The third part of the book, consisting of chapters 9 and 10, is dedicated to the issues of test synthesis and low-power synthesis which currently are of increasing importance and interest.

Acknowledgements

We would like to thank all those who, over the years, contributed to our system-synthesis research which became the backbone of this book. In particular, we thank Björn Fjellborg, Marius Minea, Xinli Gu, Alexa Doboli, Jonas Hallberg, Erik Stoy and Peter Grün. We would also like to thank the Ph.D. students who attended the graduate course on Synthesis Methodologies for Digital Systems at Linköping University and helped us to improve early drafts of this book. We are especially grateful to Erik Stoy and Jonas Hallberg for their careful reading and valuable comments.

We want to acknowledge the financial support from the Swedish National Board for Industrial and Technical Development (NUTEK) for many of our research projects in the system synthesis area and from Linköping University which helped to sponsor several visits by Petru Eles to Linköping.

Finally we want to thank our families for their patience and love which constituted the most important support of this work.

Petru Eles
Krzysztof Kuchcinski
Zebo Peng

PART I   PRELIMINARIES

1

Introduction

The rapid development of microelectronics technology in the last thirty years has made it possible now to integrate millions of gates in a single chip. It was obvious from the beginning that the design of such chips and complex systems based on them will not be possible without advanced design techniques and computer-aided design (CAD) tools. We have therefore seen an equally rapid development of design methods and CAD tools, which are widely used for specification, simulation, verification, placement, routing, and, to some extent, synthesis.

This book concentrates on methods and tools for system synthesis. By system synthesis we mean the synthesis of an implementation from a system-level specification of a complete system, which may include both hardware and software components. We will present the methods and tools developed recently for system synthesis and, in particular, a transformational approach to system synthesis from VHDL specifications. We will also identify related problems in the area and challenging issues for ongoing and future research.

1.1 Design Specification and VHDL

A typical *design process* consists of a sequence of design steps. The input to a design step is given as a *specification* which defines the functionality of the designed system together with some design requirements. The specification can be given using a natural language, a programming or hardware-description language, or formal methods. Given a specification, a synthesis technique can be used to create an implementation at a lower level of abstraction.

The *implementation* is given in terms of lower level primitives, usually in the structural domain. This means that it contains additional implementation-related information added during the design process. It can be noted that the implementation, obtained from a given design step, can usually be interpreted as the specification of the next abstraction level. For example, an RT-level design is an implementation of a system specification but it is also a specification for the logic design step.

The current development of design techniques for digital systems aims at automating the different steps of the design process by providing specialized CAD tools. And there is strong motivation for this development. First, automatic tools can shorten the design cycle and reduce time-to-market for new products. Second, automation of routine and tedious design steps can improve the quality of a design by reducing the number of design errors. Finally, good CAD tools make it possible to explore many implementation alternatives for a given specification in a short time.

Automation of the design steps requires a precise representation of the design. A *design representation* is used to model an implementation of the design by capturing some of its important aspects. It is usually used throughout the synthesis process to represent the results of design decisions.

Digital hardware can be specified and represented at different levels of abstraction. Different representations can also be used to capture the same design in different domains. Traditionally the different views of digital hardware have been illustrated using the so called Y-chart originally introduced by Gajski [GaKu83]. The axes of the Y-chart represent different domains: *behavioral*, *structural* and *physical* (see Figure 1.1). In every domain, a given design can be represented at different levels of abstraction. The points closer to the center of the Y-chart correspond to the lower levels of abstraction while the points at the periphery of the Y-chart represent the most abstract representations. The most commonly used levels of abstraction are *circuit*, *logic*, *register-transfer* (RT), and *system* level.

The system level comprises, in the behavioral domain, algorithmic specifications usually given in terms of interacting concurrent processes. These specifications can be written in hardware-description languages (e.g., behavioral VHDL) or using related formalisms (e.g., SpecCharts, SDL, StateCharts). In the structural domain a design at the system level can be represented by basic building blocks such as processors, memories and buses described by, for example, structural VHDL.

At the RT level, the implementation of the system-level specification is captured in terms of register-transfer operations in the behavioral domain and ALUs, registers, multiplexers in the structural domain. Both can be described in VHDL.

At the logic level, digital hardware is represented in terms of Boolean logic

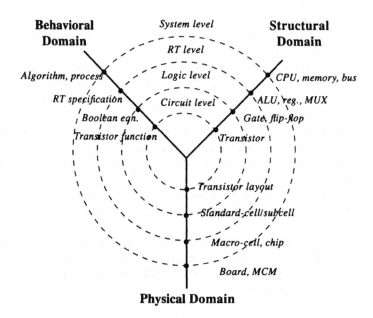

Figure 1.1: The Y-chart.

in the behavioral domain, which makes use of the well-known logic values "0" and "1" (or "true" and "false") and the basic logic operators, such as "and", "or" and "not". In the structural domain, simple logic gates and flip-flops are used to implement the given set of Boolean equations.

The circuit-level representations capture the lowest-abstraction-level primitives of a design. In the structural domain they use, for example, transistors, resistors and capacitors to form the simple logic gates and flip-flops. In the behavioral domain, differential equations are used to define relations between, for example, voltage and current of the transistors.

1.1.1   Hardware Description Languages

Hardware description languages (HDLs) are used to describe different aspects of hardware designs. They usually cover several levels of abstraction in both the behavioral and the structural domain. A typical HDL contains basic features of programming languages, such as data types and program statements. It also has many features which are not available in programming languages and are specific to hardware description. The semantics of a HDL is often defined in terms of the functions of the implemented hardware. It can, however, be also defined in terms of a simulation mechanism.

An important feature of HDLs, which is not present in common programming languages, is the concept of *time*. Time is explicitly represented in HDLs usually as some kind of pre-defined type and different mechanisms are provided which allow direct reference to time. They are used in order to specify time-dependent behavior of the modeled system.

Another important feature of HDLs is the ability to describe the *structure* of designs. Current HDLs can be used to specify components and their interfaces, the way they are interconnected, as well as different configurations (or versions) of their implementation.

Digital hardware is inherently parallel. Different parts of a piece of hardware can operate in parallel independently or they can be involved in different communication and synchronization activities. Many HDLs allow specification of parallel execution by using features normally available in concurrent programming languages, such as concurrent processes. This makes it possible to specify very complex parallel behaviors, such as pipelining or communication protocols.

It should be noticed that some important features of programming languages, such as dynamic data types and recursion, are not directly useful for hardware description. They are very often supported by HDLs but are not directly used to model hardware implementations.

1.1.2   VHDL

VHDL stands for Very High Speed Integrated Circuits Hardware Description Language. It was developed by the Very High Speed Integrated Circuits (VHSIC) Program Office of the Department of Defense and standardized by IEEE [IEEE87, IEEE93]. Being a standard, VHDL is not a propriety language of any CAD vendor and can be used by all CAD companies on the same basis. It provides therefore a common platform for different CAD tools. Moreover, the language is not defined for any particular CAD environment or design technology. It can be used to support different design methodologies as well as different fabrication technologies.

VHDL is a very rich language which covers several levels of abstraction in the behavioral and the structural domain of description. It can be efficiently used to describe digital hardware at logic, RT and system level. It is basically an algorithmic language which incorporates features of the modern programming language Ada. In extension to these features it also defines specific hardware description facilities, such as signals, components, configurations and interface definition.

The semantics of VHDL is defined in terms of simulation. It is also defined independently of any particular simulator, which has many advantages. A simulation-based semantic definition creates, however, problems for the use of the language for other design activities than simulation. If used for synthesis the standard simulation semantics must be taken into account by the synthesis tools. This may create several complex problems which have to be solved efficiently to avoid large overhead in the designed hardware. Moreover, VHDL contains many features which either cannot be synthesized at all or would result in huge hardware overhead. These problems have to be carefully considered when synthesizing VHDL constructs and defining a subset of the language for synthesis.

1.1.3   Design Representations

A design specified by a HDL needs to be translated into a design representation to be efficiently analyzed and synthesized. Existing design representations for RT-level designs use typically graphs as the basic means for representing the functionality and structure of a design. *Data-flow graphs* (DFG) are widely used for this purpose. A DFG represents operations as nodes and data transfers as arcs in a graph. If the DFGs are acyclic, loop constructs must be captured in some special ways, which can be done by introducing hierarchy into the graph [DeMi94]. In this case, every graph is acyclic and a loop is represented as a lower-level graph.

Control/data-flow graphs (CDFGs) are another method of representing RT-level designs. In this representation operations on data are captured in a data-flow graph while control related information, such as operation sequencing, branching or looping, is represented in the control-flow graph. Each node of the control-flow graph can have several associated data-flow nodes which represent the operations controlled by it.

In this book, we will use a design representation which is called Extended Timed Petri Net (ETPN). This representation captures a design as two separate but related parts: the control part and the data path. The data path is represented as a directed graph with nodes and arcs. The nodes are used to capture data manipulation and storage units. The arcs represent the connections of these units. The control part, on the other hand, is represented as a timed Petri net with restricted transition-firing rules [PeKu94].

1.2   Synthesis

Synthesis is usually defined as the translation of a behavioral representation of a design into a structural one [GDWL92]. In some cases, however, the translation process does not necessarily mean the creation of a purely structural representation. For example, high-level synthesis usually produces a symbolic state machine to model the control function and a data path structure at RT level. The state machine description has to be further synthesized in order to produce a structural design.

The whole synthesis process consists of several consecutive steps performed at different abstraction levels. The different steps usually make use of different basic implementation primitives and employ different synthesis methods. The following synthesis steps can usually be identified:

- *System level* — Accepts an input specification in the form of communicating concurrent processes. The synthesis task is to generate the general system structure defined by processors, ASICs, buses, etc. System-level synthesis operates at the highest level of abstraction where fundamental decisions are taken which have great influence on the structure, cost and performance of the final product.

- *High level* — The input specification is given as a behavioral description which captures the functionality of the designed system. High-level synthesis is, therefore, also called behavioral synthesis. Using functional units and memory elements as basic primitives, a high-level synthesis tool generates an implementation at the RT level. The basic high-level synthesis steps are scheduling, allocation and binding. RT-level synthesis is typically considered as a subset of high-level synthesis where allocation and binding are done automatically while scheduling is carried out by the designer.

- *Logic level* — Can be divided into combinational and sequential logic synthesis. The combinational logic synthesis accepts as input Boolean equations while the sequential logic synthesis accepts some kind of finite state machines. Logic synthesis produces gate-level netlists as output.

- *Physical design* — Accepts a gate-level netlist and produces the final implementation of the design in a given technology. This synthesis step depends on the implementation technology (e.g, FPGA or full custom). However, the common main tasks are placement and routing.

In this book, we will concentrate on system synthesis which covers the two highest synthesis levels: system-level and high-level synthesis.

1.2.1   System Synthesis

System synthesis accepts as input a specification of a complete system and performs system-level synthesis and high-level synthesis. The final implementation usually consists of several heterogeneous components. Parts of the system can be implemented in hardware and parts in software. The communication structure between different parts has to be designed and included in the final implementation.

System-Level Synthesis

System-level synthesis is the first step of system synthesis. During this step several decisions regarding the architecture of the system are made. For example, the allocation of system components is carried out to determine a system level view of the architecture by selecting a set of processing, storage, and communication components on which one can implement the specified system functionality. The components can be, for example, microprocessors, microcontrollers, DSPs, ASIPs, ASICs, FPGAs, memories, and buses.

Partitioning at this level divides the specification into modules which can be implemented using the allocated system components.

Another task of system-level synthesis is the generation of the communication structure between different parts, which is called *communication synthesis*. It can be done by an interface synthesis and/or by a protocol selection technique.

Finally, *special design styles* can be suggested for performance-critical parts of the system. For example, a pipeline implementation can be suggested for a component which is the bottleneck of the system.

A specific area of the system-level synthesis research, called *hardware/ software co-design*, has gained a big interest recently. It concentrates on the synthesis problems of complex systems which have to be implemented by interacting hardware and software parts. Identification of the hardware and software partitions is a specific system-level partitioning task. Behavioral descriptions of the modules assigned to the hardware partition are submitted for further synthesis to high-level synthesis tools. Software can be compiled using commercial compilers or application specific software-synthesis packages. In addition to that, the communication structure between software and hardware partitions has to be generated during this design step.

Design decisions taken in system-level synthesis have to be supported by an analysis of the basic design parameters. To partition a system, for example, we would like to know the size of the different parts when implemented in different technology and the traffic through the interfaces. Some parameters, such as pin count, can usually be directly extracted from the design specification. There are, however, parameters which have to be estimated. Hence *estimation* techniques play an important role during early decisions of system design. The estimation accuracy also influences the final design quality.

High-Level Synthesis (HLS)

High-level synthesis (HLS) accepts a behavioral specification of a digital system and produces an RT-level implementation. Logic synthesis will then synthesize the RT-level design into an implementation at the gate level. Usually high-level synthesis comprises three main tasks: *scheduling, allocation* and *binding*. Scheduling deals with the assignment of each operation to a time slot corresponding to a clock cycle or time interval. Allocation and binding carry out selection and assignment of hardware resources for a given design. Allocation selects a number and types of hardware resources for a given design while binding assigns the particular instance of a selected hardware component to a given data path node. Sometimes the term allocation is used to denote both the allocation and binding tasks.

A typical HLS system, depicted in Figure 1.2, performs the basic synthesis steps in sequence. First, the behavioral specification is compiled into an internal design representation. During this step compiler optimization techniques can be used. After that, allocation of components for the internal design representation is carried out. It determines the number and type of components used to implement the given specification. The next step, scheduling, assigns operations to clock steps. Finally, the binding step assigns physical components, such as functional units, registers and buses, to data-path elements of the design representation.

The above high-level synthesis tasks are not independent of each other. Scheduling assigns operations to time slots and limits therefore the allocation freedom. For example, the assignment of two "add" operations to the same clock step means that they are executed in parallel and they cannot share the same adder. It limits the freedom of binding since these two operators have to be bound to different adders or other functional units. On the other hand, performing allocation first will limit scheduling in a similar way.

Scheduling, allocation and binding can be implemented using optimization algorithms. There exists a wide spectrum of different approaches ranging from classical formulations, such as integer linear programming, to optimization heuristics specially proposed for synthesis purpose, such as list scheduling, force-directed scheduling or the left-edge algorithm for allocation and binding. In some cases, rule-based schemes are also used. The main problem is how to reach the optimal result, which is not always possible to be achieved since it has been proved that the problems of finding the optimal solutions both for

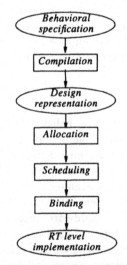

Figure 1.2: A typical HLS system.

scheduling and allocation are NP-complete (see, for example, [MiLD92]). This means that these problems are intractable, for practical designs, by algorithms that guarantee optimal solutions. This explains why many systems use heuristics which involve domain-specific knowledge to come as close as possible to the optimum.

1.2.2   Optimization

Synthesis entails the translation of one representation into another. This translation is not a simple one-to-one mapping between two design descriptions. There exist many different and correct mappings that can be applied for a given description and that satisfy functional correctness criteria. Synthesis entails therefore the selection of the optimal mapping. Such *optimization* should take into account several design quality constraints. Two of the quality measures, *area* and *performance*, are most often considered. Area cost can be defined in several ways, for example, as the VLSI silicon area or the number of gates. Performance is usually defined differently for different classes of circuits. For combinational circuits it is measured as the input/output propagation delay. For sequential circuits it can be measured as cycle time. At the RT level performance is measured as the latency which defines the time for executing a given sequence of operations.

Digital-system optimization can be defined in different ways. It can be defined as the problem of maximizing performance while fulfilling a given upper-bound constraint on the area or that of minimizing area while fulfilling a given lower-bound on performance. The selection of the method depends very much on the optimization strategy used and goals of the synthesis process.

One usually-used optimization strategy is to search the *design space* to find the optimal solution. One view of the design space can be defined as a multidimensional space spanned by different characteristics of the design, such as performance, area, and power consumption. Points in this space represent possible implementations (or sets of implementations) for a given design with some characteristics. In the simplest case, we consider a two-dimensional design space spanned by the performance and area. Figure 1.3 depicts an example of this two-dimensional space and different design implementations representing different trade-offs between performance and area. It can be noted that Design 1 is the biggest one in terms of silicon area and at the same time it offers the best performance (the lowest latency). On the other hand, Design 5 is the smallest but it is also the slowest one. There exist three other designs which offer other performance/area trade-offs. Based on the requirements the designer will select the solution which fulfils his/her needs. The process of finding and evaluating different design implementations is called *design space exploration*, which is usually supported by some CAD tools. It is impossible to visit all solutions during design space exploration and strategies have been developed to

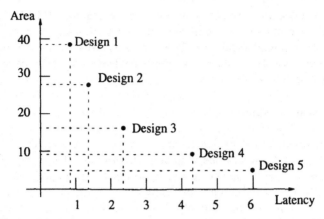

Figure 1.3: A design space example and different design trade-offs.

guide the exploration so that the search space is reduced but good solutions are still reached.

During design space exploration the designer, supported by a CAD tool, performs trade-offs between different design characteristics. In Figure 1.3, for example, the design trade-offs include only two parameters: performance and area. In practical situations there exist much more parameters to consider, for example, testability, power consumption and pin count. The design methodology and CAD tools have to take all of these into account.

1.2.3   Related Issues

When the complexity and the size of designs grow, *testability* and *power consumption* become very important features. Testing large systems with limited accessibility to their internal structure is a complex task. It usually involves two main activities: test-pattern generation and test application. The time spent on these activities depends very much on the testability characteristics of the design. The test generation and the test application activities are easier for designs equipped with special design-for-testability (DFT) features.

Traditionally DFT improvements of designs were performed at very late stages of the design process. Usually, when a design was nearly ready some DFT techniques were applied to it to improve its testability. Unfortunately, in many cases this is too late since it can cause the violation of other design constraints, such as performance, area or pin count. Recently some research has been done to improve this situation and consider DFT earlier in the design process. Test synthesis approaches which integrate testability considerations together with other synthesis goals have been proposed.

The test synthesis approaches try to combine several methods to improve the testability of a design. First, well-known techniques are used, such as scan path or built-in-self-test (BIST), to improve the testability of different parts of the design. These techniques are usually used selectively based on the global analysis of the design. Second, new testability-improvement features are incorporated into the synthesis algorithms, which optimize, in this case, not only performance and area but also testability. This is achieved very often without inclusion of special DFT techniques; instead the flexibility of scheduling and allocation is exploited to avoid generating difficult-to-test designs.

With the current trend in mobile electronics, such as cellular telephones, portable computers, portable CD players and cameras, the power consumption issue is becoming an extremely important parameter of designs. A product that consumes much power has a limited operation time and it therefore has problems to compete with similar products which consume less power. The power consumption of a design is therefore an important parameter to be considered in the design process.

It can be noted that there exist many techniques for low-power design which are applicable at the logic and the physical level. For example, reducing voltage affects power consumption very much due to the quadratic relation between voltage and power consumption. Another method often used in practice is to minimize switching activities in the designed circuit. This technique can be very successful since transistor switching is one of the main factors responsible for power consumption.

There exist, however, a new trend to consider power consumption at the system and RT levels. The methods applied at these levels usually concentrate on special features of the design which are important from the power-consumption point of view. They can be treated as complementary to techniques used at the logic and the physical level. For example, the selection of number representation can influence the switching activities in a circuit and thus can reduce or increase its power consumption. Binding of several operations into one physical component or into several components creates not only area/performance trade-offs but also power consumption trade-offs. Such trade-offs can be considered in the synthesis process by using an optimization algorithm which takes power-consumption into account.

1.3 Transformational Approach

A design can be improved by applying transformations which change its representation while keeping its semantics unchanged. We usually call such transformations semantic-preserving. A design transformation can be applied to a design at any stage of the design process. It can be applied to designs at different abstraction levels as well as in different design domains. Usually, transformations are grouped into classes, such as compiler oriented transformations and graph transformations. Examples of compiler oriented transformations include well-known transformations, such as common sub-expression elimination, loop unrolling/folding, dead-code elimination and tree-height reduction. The graph transformations are dependent on the design representation. Typical transformations are, for example, operator-node merging/ splitting and control-node splitting/folding.

In this book, we will concentrate on graph transformations applied at the system and the RT level. Only the behavioral and structural domains will be considered. A transformation which merges two graph nodes representing operations is an example of such a transformation. This transformation is used to captured the result of a design decision to share a functional unit by two operations, which can lead to a design with improved area cost. The fundamental idea of a transformational approach is to make use of a sequence of such transformations to transform a design from its initial specification into its final implementation.

The transformational approach is well known by designers who use it very often when redesigning, improving or adding new features to existing designs. For example, converting a register to a scan register for testing purpose is a testability-improving transformation. Transforming an arithmetic unit to a pipeline implementation is a performance-enhancing transformation.

Transformations can also be used by automatic synthesis tools. Such tools select transformations based on certain selection criteria in order to improve some aspects of the design. Usually some kind of optimization algorithm is used for the selection of transformations. The optimization algorithm uses a cost function and given constraints to evaluate improvements of the design after transformations. It tries to select the transformations in such a way that the final design will be as close to the optimum as possible.

The optimization problem can be solved by using different techniques but usually some kind of neighborhood search heuristics is used. For example, simulated annealing and tabu search heuristics are widely used, which improve a cost function, defined for the design representation, by applying local transformations. Since these transformations perform only local changes in the design, a more general view of the design process is captured by the optimization algorithm. It evaluates the cost function and selects transformations with respect to the goal of reaching the global optimum. This does not mean that, at a certain moment, the design transformation with the best improvement of the cost function has to be selected. Optimization algorithms very often involve hill-climbing heuristics which also accept, under special conditions, transformations leading to a larger cost, even if the final goal is to minimize the cost function.

1.4 Current Trends

System synthesis is an area which currently is in a very dynamic evolution. This book will discuss several aspects of system synthesis which are currently studied by the research community with an emphasis on a transformational approach. In this section, we try to outline some of the current trends and future directions in system synthesis, as can be predicted today.

The recent development in microelectronics technology, such as sub-micron technology, demonstrates that a complete digital system can be integrated on a single chip, thus becoming a system-on-a-chip (SOC). This development can be compared to the introduction of the single-chip microprocessor about two decades ago which led to a drastic change of the design methods. The intro-duction of SOCs requires new design methodologies which will make it possible to keep a continuous increase in the designers' productivity and decrease in time-to-market. In particular, efficient design methodologies at system-level and high-level have to be developed to make it possible to design SOCs in reasonable time and budget. In the rest of this section, we identify some important topics which, in our opinion, require more research and devel-opment.

Concerning the specific area of hardware design, it is expected that VHDL will consolidate its position as one of the mostly used specification languages. There are, however, several problems to be solved with VHDL as a synthesis language. Even though there exist standards aiming to define a synthesis semantics for VHDL constructs at the logic level, several problems remain to be solved concerning the interpretation of VHDL for high-level and system-level synthesis.

Classical high-level synthesis algorithms and optimization techniques will be further developed to cover particularities of specific application domains and to satisfy the requirements of special high-performance designs like those involving pipelined, superscalar, or very large instruction word architectures.

The development of efficient estimation techniques will remain of crucial importance for system synthesis. As we move towards higher abstraction levels it becomes more and more difficult to estimate layout, memory, power consumption, testability or performance implications of certain design decisions. Thus, better integration of system synthesis with logical and physical synthesis is one way to provide sufficiently accurate estimations in the future.

As already mentioned, testability improvement and power optimization are and will remain for a while among the hot topics of high-level synthesis research.

Evolution towards the system-level produces a continuous change in the level of granularity at which design trade-offs and design transformations are performed. Module libraries will consist of more and more complex pre-designed components like processor cores or memory blocks. These compo-nents will be parametrized and different versions can be automatically generated on request. It is very likely that methods developed for high-level synthesis can be used for this purpose. In this context, design of glue logic and generation of the hardware and software components of the interconnection sub-system becomes of critical importance. The use of CPU cores or DSP processors as instantiable modules in system design is more and more encouraged by the need for flexibility and short design times as well as performance considerations. The advent of SOC will shift the main design effort from the design of processors or similar components to the design of systems consisting of several existing processing elements as the main building blocks. Further research is needed in the directions of system architectures, architecture modeling, system-level partitioning, and interface generation.

Synthesis of embedded applications implies the concurrent development of both the hardware and the software components of the system. In this context, more and more attention will be focused on the software synthesis process. Embedded software usually runs in a dedicated and specific environment. Very often it does not run on standard microprocessors but on micro-controllers, specialized DSPs or highly optimized application-specific processors. The development process of such software differs very much from that of standard, general-purpose applications. Thus, great effort has to be invested in order to develop sophisticated analysis and optimization methods for embedded software synthesis. The large variety of processors used in embedded systems, many of which are strictly application specific, and the need of highly optimized software to run on this processors, triggered new research and development efforts in the area of retargetable compiling. At the same time, several real-time aspects, like performance prediction and scheduling, are also in the focus of the research community working in the area of software synthesis for embedded systems.

Specification at the system level is another critical aspect of any system design methodology. Thus, it is no wonder that the problem of a system specification language is and will remain the subject of intensive debate in the research community. Which will be the language, if any, that allows us to produce a clear, concise, executable and implementation-independent specification of a system? This will probably remain an open question for several years to come. Many researchers argue that there will not exist any unique language to satisfy all requirements but several languages will be used to specify systems in different application areas or to describe different parts of the same system.

Even more important than the specification language problem is that of the computational models used as design representations of a system. The heterogeneity of components interacting inside a system, from dataflow elements to pure control, make it very difficult to produce a clear specification of a whole system and to efficiently verify and synthesize it. The development of clear computational models as a basis of new specification, verification and synthesis tools still represents an open research area for the future.

Verification is another crucial bottleneck of any system design process.

Simulation is widely used today by both hardware and system designers for the verification of hardware and mixed hardware/software systems. But even with the rapid progresses in simulation techniques and the availability of very fast simulation platforms, extensive verification of large systems can not be performed by simulation. Formal methods will lead to reliable system verification tools. There are already important results in the area of hardware verification using formal methods. The extension of these methods to the system level, including hardware and software components (and possibly also analog, optical or mechanical elements) is an actual research area.

Many research and development efforts have been directed towards the development of new algorithms and tools to support different design tasks. However, what the designer really needs is a user-friendly environment to support his or her efforts during the whole design cycle. Substantial efforts are still needed to come up with efficient design methodologies and to realize integrated environments which support the concurrent specification, validation and synthesis of all components of a system. Such an environment will have to integrate and coordinate tools working at different levels of abstraction and to support the interaction of design components based on different design representations. User-friendly graphical front-ends will have to be integrated together with large databases and the WWW. Design reuse has to be supported by library tools to catalogue, distribute, and search for modules which range from gates and transistors to custom and application specific processors, ASICs, memory blocks, software drivers and real-time kernels.

1.5   Outline of the Book

This book is divided into three parts. Part I, consisting of Chapters 1 to 5, provides an overview of the system synthesis area. In particular, it defines the basic terminology of high-level and system-level synthesis, introduces the basic features of the VHDL'92 language which are relevant for system synthesis, and presents some basic techniques and optimization heuristics used for system synthesis.

Chapter 1 gives a general introduction to the typical design process and defines the basic terms in system synthesis. It discusses the issues related to system specification and the use of VHDL in system synthesis. It defines the different synthesis levels and describes the main tasks of each of these levels. It then introduces the basics of the transformational approach as well as its advantages.

In Chapter 2, a basic introduction to the VHDL'92 language is given. This includes the basic constructs, data types, objects, expressions, processes and sequential statements. It then describes the simulation mechanism in details and discusses the use of subprograms and concurrent statements. The chapter focuses on those features of VHDL which are relevant for system synthesis and required in order to understand the rest of the book.

Chapter 3 summarizes the fundamental issues of high-level synthesis. It discusses the basic high-level synthesis tasks, including scheduling, data path allocation and binding, and controller synthesis. Several well-established techniques to perform these tasks are presented.

Chapter 4 presents the main problems in system-level synthesis. These include allocation of system components (architecture selection), system partitioning, communication synthesis, and hardware/software co-design. Since many of the techniques used to solve these problems are still being researched, we have provided a brief overview of several important approaches recently developed by the research community.

In Chapter 5, it is pointed out that many system synthesis tasks can be formulated as combinatorial optimization problems. Due to the complexity of these problems, the development of efficient heuristics becomes an important issue. We present then several general heuristics which are based on neighborhood search techniques and are directly applicable to the transformational approaches to system synthesis. These include simulated annealing, tabu search, and genetic algorithms. This chapter provides a basic knowledge for the readers to understand many of the techniques discussed in the other chapters of the book.

Part II of the book consists of Chapters 6, 7 and 8 and is dedicated to the transformational approach to system synthesis. It presents first a transformational approach to high-level synthesis. It discusses then several advanced issues related to the synthesis of VHDL system specifications. Finally a transformational approach to hardware/software partitioning is presented.

Chapter 6 presents the basics for our transformational approach. This includes the formal definition of a design representation, called ETPN (Extended Timed Petri Net), the mapping of VHDL specifications into ETPN, and the hardware implementation of ETPN as well as the ETPN transformations. The problem of how to select transformations during the synthesis process is also discussed and two basic techniques have been described for the selection problem.

In Chapter 7, several advanced issues related to our transformational approach to system synthesis are described. These are the synthesis of VHDL subprograms, synthesis strategies for interacting VHDL processes, synthesis with signal level interaction, and synthesis with system-level interaction. Finally, the issues of specification and synthesis with timing requirements are also presented.

Chapter 8 presents problems and techniques concerning hardware/software partitioning. In particular, it describes a transformational approach to hardware/software partitioning, which is based on a graph partitioning formulation and makes use of iterative improvement heuristics including simulated annealing and tabu search algorithms.

Finally, Part III of the book, consisting of Chapters 9 and 10, presents several advanced issues for system synthesis.

Chapter 9 discusses problems and techniques related to test synthesis, namely the integration of testability consideration in the synthesis process. It first gives an overview of the digital system testing area, focusing on test pattern generation and design for testability. It then describes an approach to high-level test synthesis which includes two integrated techniques, one used for testability analysis and the other for testability improvement transformations.

Chapter 10 addresses the issues related to low-power synthesis. It first discusses the sources of power consumption and its reduction in general. Several techniques for power estimation are then presented as well as techniques for high-level synthesis transformation for low power. Finally, several low-power synthesis systems and algorithms are described.

Figure 1.4 depicts the dependencies of the different chapters. After Chapter 1, which gives a general introduction, Chapters 2, 3, 4 and 5 are relatively independent. A reader with knowledge of VHDL, high-level and system-level synthesis, and optimization heuristics can skip them. The chapters in Part II

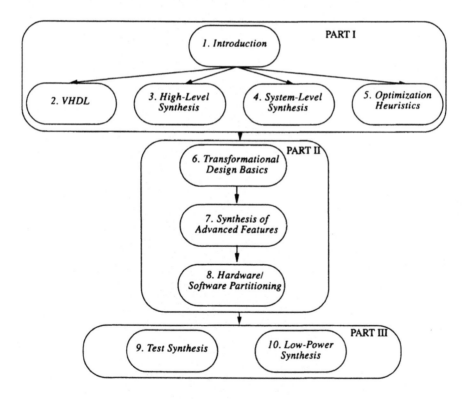

Figure 1.4: Chapter dependencies.

should be read sequentially. To understand a given chapter, knowledge of its previous chapters is assumed. Part III discusses two topics which are relatively unconnected and can be studied independently.

2

VHDL

VHDL is an acronym for VHSIC Hardware Description Language (VHSIC stands for Very High Speed Integrated Circuits). The language was intended to be a standardized HDL to support design, documentation, and verification of digital systems. The development of VHDL was sponsored by the US Department of Defence during the 1980s. In 1987 the language was adopted by the IEEE as a standard; this version of VHDL is known as the IEEE Std. 1076-1987 [IEEE87]. A new version of the language, VHDL'92 (IEEE Std. 1076-1993) [IEEE93, BFMR93], resulted after revision of the initial standard in 1993. The discussion in the present chapter is based on this new standard.

The language was defined to support the whole design process from the specification at the highest abstraction level to the physical level. Thus both system-level specifications and specifications at the register-transfer, logic, or (to some extent) circuit level can be formulated in VHDL.

As some of our examples show, VHDL includes facilities which make it suitable both for specifications in the behavioral style and for structural specifications. The behavior of a design can be described using programming language constructs. At the same time specific language constructs are available for describing the design as a structure consisting of several interconnected components.

A precise simulation semantics is associated with the language constructs and therefore models corresponding to any abstraction level or description style can be verified using a VHDL simulator. Although not explicitly considered during the initial language definition and standardization process, synthesis quickly became one of the most important application domains of VHDL. Being a *standard* HDL it is natural that VHDL is commonly used as an input language for synthesis systems. This is why, throughout this book, we are considering VHDL as the specification language for system synthesis.

The input specifications we are considering for synthesis are at the system level, and consequently they are formulated in a purely behavioral style. This is why our short introduction to VHDL will mostly concentrate on the behavioral aspects of the language. Nevertheless, the structural aspect of VHDL is still important in this context, as we generate as output of the synthesis process a mixed description in which the netlist is represented as a structure and only the controller is specified in a behavioral style.

In this chapter a short introduction to some aspects of VHDL is provided. Our main goal is to introduce those concepts which, in the following chapters, will be discussed from the synthesis point of view. In order to keep the overview accurate and succinct we will present the (sometimes simplified) syntax of some constructs using a notation similar to that introduced in the Language Reference Manual [IEEE93][1]. We also assume that the reader is familiar with a programming language such as Pascal, C, or Ada.

2.1   The Basic Constructs

A digital system can be modeled in VHDL as an *entity*. The entity is the most basic building block in a design. An entity may be used as a component in another design (which itself is described as an entity) or may be the top-level module of the design.

An entity is described in VHDL as a set of *design units*. There are five different types of design units, each of which can be compiled separately:
- Entity declaration
- Architecture body
- Package declaration
- Package body
- Configuration declaration

In this chapter we introduce the first four units. Configuration declarations are an advanced facility for structural specification and will not be discussed in this book.

---

1. Reserved words are written in **boldface**.
   The vertical bar (|) separates alternative items.
   Square brackets ([...]) denote optional items.
   Curly braces ({...}) identify an item that is repeated zero or more times.
   A name in *italics* prefixed to a non-terminal name represents the semantic meaning associated with that non-terminal.

2.1.1   Entity Declaration

An *entity declaration* specifies the external view of the entity. This view
basically consists of a name associated with the entity and a list of ports
through which the entity communicates with its external environment.

The syntax of an entity declaration is[1]:

        entity_declaration ::=
                entity identifier **is**
                        entity_header
                        [entity_declarative_part]
                [begin
                        entity_statement_part]
                end [entity] [*entity*_simple_name];

        entity_header ::=
                [port_clause] [generic_clause]

The *port clause* in the *entity header* specifies the signals on which information
is passed into and out of the entity. In Figure 2.1 we show both the block representations and the entity declarations corresponding to three circuits:

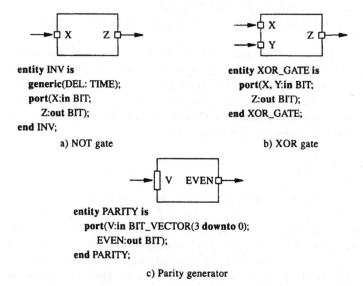

entity INV is
   generic(DEL: TIME);
   port(X:in BIT;
      Z:out BIT);
end INV;

    a) NOT gate

entity XOR_GATE is
   port(X, Y:in BIT;
      Z:out BIT);
end XOR_GATE;

    b) XOR gate

entity PARITY is
   port(V:in BIT_VECTOR(3 **downto** 0);
      EVEN:out BIT);
end PARITY;

    c) Parity generator

Figure 2.1: NOT gate, XOR gate, and Parity generator: external view.

---

1. Throughout this presentation we will ignore the optional *entity_declarative_part* and
   entity_statement_part.

a) a NOT gate with input port X and output port Z;

b) an XOR gate with input ports X and Y, and output port Z;

c) a four-bit parity generator with input port V (a four-bit vector) and output port EVEN;

The header of entity INV in Figure 2.1a also contains a *generic clause* defining the generic constant DEL. The actual value of this constant, representing the delay time on the gate, will be specified later when a certain instance of the gate will be used. Therefore, we actually define a class of entities INV. Instantiations of this class, having different delay times, can be created later in other entities using the INV gates (Figure 2.4).

2.1.2   Architectures

The entity declaration describes the external interface of an entity. A separate part, associated with the same entity, describes how the entity is implemented. This part is called *architecture body* and has the following syntax:

```
architecture_body ::=
        architecture identifier of entity_name is
                architecture_declarative_part
        begin
                architecture_statement_part
        end [architecture] [architecture_simple_name];
```

The *architecture statement part* consists of one or several concurrent statements such as processes, concurrent signal assignments, component instantiations, etc. (see Subsection 2.3.1 and Section 2.6). All these statements are executed in parallel so the order in which they are specified is not important. They describe the internal details of the entity using the behavioral or structural modeling style or a combination of the two.

Behavioral Modelling

In Figure 2.2 we show an architecture body corresponding to the parity generator declared as an entity in Figure 2.1c. This is a purely *behavioral specification* consisting of a single process statement. It describes the functionality of the design without specifying anything about the actual structure of a circuit implementation. The value '1' is produced on the output port EVEN if an even number of bits in the input vector V have value '1'; otherwise '0' is assigned to the output port.

In the behavioral style, an architecture is usually modeled as one or several processes executing in parallel. Each process itself consists of a set of sequential statements that are executed in the specified order. A process can

```
architecture PARITY_BEHAVIOURAL of PARITY is
begin
  process(V)
    variable NR_1: NATURAL;
  begin
    NR_1:=0;
    for I in 3 downto 0 loop
      if V(I)='1' then
        NR_1:=NR_1+1;
      end if;
    end loop;
    if NR_1 mod 2 = 0 then
      EVEN<='1' after 2.5 ns;
    else
      EVEN<='0' after 2.5 ns;
    end if;
  end process;
end PARITY_BEHAVIOURAL;
```

Figure 2.2: Parity generator: behavioral specification.

have an associated *sensitivity list*. It consists of one or several signals specified within parentheses after the reserved word **process** (in Figure 2.2 the sensitivity list consists of the single signal V). The process is activated each time any signal in the list changes value, i.e., there is an event on the signal.

Structural Modelling

A *structural model* of a system specifies the components of the design and their interconnections. The behavior of the design is not explicitly described in such a model but it is implied by the actual structure. The architecture body in Figure 2.4 presents a structural model of the four-bit parity-generator circuit depicted in Figure 2.3. It is an implementation of the entity declared in Figure 2.1c.

The declarative part of the architecture body PARITY_STRUCTURAL contains two *component declarations* declaring components XOR_GATE and

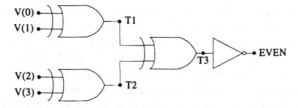

Figure 2.3: Parity generator circuit.

```
use WORK.all; -- makes all units in the current library visible
architecture PARITY_STRUCTURAL of PARITY is
   component XOR_GATE
      port(X,Y: in BIT; Z: out BIT);
   end component;
   component INV
      generic(DEL: TIME);
      port(X: in BIT; Z: out BIT);
   end component;
   signal T1, T2, T3: BIT;
begin
   XOR1: XOR_GATE port map (V(0), V(1), T1);
   XOR2: XOR_GATE port map (V(2), V(3), T2);
   XOR3: XOR_GATE port map (T1, T2, T3);
   INV1: INV
      generic map (0.5 ns)
      port map (T3, EVEN);
end PARITY_STRUCTURAL;
```

Figure 2.4: Parity generator: structural specification.

INV. These components are not active elements of the entity but merely templates for models that will be used inside the architecture. An instance of a declared component is created by a *component instantiation* statement. Three instances of component XOR_GATE and one of component INV are created inside the architecture PARITY_STRUCTURAL. The four instantiated components are interconnected by the three signals declared inside the architecture: T1, T2, and T3. The actual connections between the ports of the sub-components, the ports of the specified entity and the signals are specified by the *port map* in the component instantiation statement. Once activated, the four components will work in parallel and interchange information through their ports, according to the specified interconnections.

At the instantiation of component INV1 a *generic map* has been also provided. It specifies that the instance of component INV, labelled INV1, will have an actual value of 0.5 ns for the generic parameter DEL (representing in this case the internal delay on the gate).

The system sub-components utilized inside a structural model have to be described using distinct VHDL design entities. The implementations of the XOR and INV gates used inside the parity checker are given in Figure 2.5. They correspond to the entity declarations in Figure 2.1a and b respectively.

In order to simulate a structural model, like PARITY_STRUCTURAL in Figure 2.4, entity declaration/architecture pairs have to be associated to component instances. In VHDL, this binding is called *component configuration*. VHDL provides a sophisticated mechanism for performing this associ-

```
architecture ARCH_INV of INV is
begin
   Z<=not X after DEL;
end ARCH_INV;

architecture ARCH_XOR of XOR_GATE is
begin
   Z<=X xor Y after 1 ns;
end ARCH_XOR;
```

Figure 2.5: An implementation of entities INV and XOR_GATE.

ation in a flexible manner. In our example, we use the so called *default binding* which solves configuration of the components in absence of any explicit binding indication. As the result of the default binding, those entity declarations will be associated to each instance of components XOR_GATE and INV (the instances labelled XOR1, XOR2, XOR3, and INV1 in PARITY_STRUCTURAL) which have the same name as the respective component (these are, in our case, the entity declarations in Figure 2.1a and b).

For the binding of an architecture body to an entity declaration the following default rule is applied: if a single architecture body has been defined for a given entity (as is the situation in our example for XOR_GATE and INV) that architecture will be associated to the entity; if there are several architecture bodies for a given entity, the most recently analyzed will be used.

Now that we have specified two models of the parity generator we can verify them by simulation. In order to do this, a test bench has to be provided. In Figure 2.6 we give a very simple, minimal[1], test bench that produces a set of test stimuli, which are directed to the input port of an instance of the parity generator. There are three points to mention in connection with the model in Figure 2.6:

- The entity BENCH has an empty interface which is the typical situation for a top-level module like a test bench;
- The architecture body is a mixed behavioral/structural one, consisting of a concurrent signal assignment (see Subsection 2.6.1) and a component instantiation;
- We have defined two architecture bodies corresponding to the entity declaration PARITY. In order to select the one we want to test (regardless to the order in which the architectures are analyzed) we place a *configuration specification* in the declarative part of the architecture body. It determines which architecture body should be bound to entity declaration PARITY for the instance PARITY_GENERATOR of component PARITY.

---

1. We assume that output values, directed to signal E, are checked "manually" through the user interface of the simulation environment.

```
entity BENCH is
end BENCH;

use WORK.all; -- makes all units in the current library visible
architecture ARCH_BENCH of BENCH is
  component PARITY
    port(V: in BIT_VECTOR (3 downto 0); EVEN: out BIT);
  end component;
  for PARITY_GENERATOR:PARITY use              -- configuration specification
    entity PARITY(PARITY_STRUCTURAL);
  signal VECTOR: BIT_VECTOR (3 downto 0);
  signal E: bit;
begin
  VECTOR <= "0010",
       "0000" after 3 ns,
       "1001" after 5.8 ns,
       "0110" after 9 ns,
       "1011" after 13 ns,
       "1111" after 15.8 ns,
       "0001" after 18.5 ns,
       "1110" after 21 ns,
       "0100" after 25.7 ns,
       "1010" after 30 ns,
       "0011" after 32.5 ns,
       "1000" after 35.5 ns,
       "0101" after 39 ns,
       "1100" after 42 ns,
       "0111" after 44.5 ns,
       "1101" after 50 ns;
  PARITY_GENERATOR: PARITY port map (VECTOR, E);
end ARCH_BENCH;
```

Figure 2.6: Test bench for the parity generator model.

2.1.3   Packages

A package is a collection of declarations such as subprograms, types, constants, components, and possibly others, which are grouped in a way that allows different design units to share them. The interface to the package, consisting of those declarations which are intended to be seen from the outside, is defined in the *package declaration*, according to the following syntax:

```
package_declaration ::=
        package identifier is
            package_declarative_part
        end [package] [package_simple_name];
```

The *package body*, on the other hand, contains the hidden details which are not visible from the outside. The following is the syntax of a package body specification:

```
package_body ::=
        package body package_simple_name is
            package_body_declarative_part
        end [package body] [package_simple_name];
```

An example of a package declaration is the following:

```
package ILLUSTRATE is
    type TAB is array(NATURAL range <>, NATURAL range <>) of BIT;
    constant PATTERN: BIT_VECTOR (3 downto 0);
    function INDEX (T: TAB; V: BIT_VECTOR) return NATURAL;
end ILLUSTRATE;
```

The package body corresponding to the package declaration above has to define the body of function INDEX and the value of constant PATTERN (This is a *deferred constant*: it is visible to the outside by its identifier, but its value is hidden inside the package body):

```
package body ILLUSTRATE is
    constant PATTERN: BIT_VECTOR (3 downto 0):= "0101";
    function INDEX (T: TAB; V: BIT_VECTOR) return NATURAL is
    begin
    -- implementation of the function
    end INDEX;
end ILLUSTRATE;
```

Items declared inside a package declaration can be used within other design units by selection:

```
    X:= ILLUSTRATE.PATTERN;
    Y:= ILLUSTRATE.INDEX(TABLE, VECTOR);
```

These items can be made directly visible by an appropriate *use clause*:

```
    use ILLUSTRATE.PATTERN;
```

This use clause makes identifier PATTERN visible so that it can be subsequently used without qualification with the package name. Using the reserved word **all**, all names declared inside a package declaration can be made directly visible:

```
    use ILLUSTRATE.all;
```

Packages are a very efficient way for creating a standard or vendor-specific VHDL environment. The VHDL language standard includes two predefined packages: STANDARD and TEXTIO. The package STD_LOGIC_1164, which is an IEEE standard, defines a nine-value logic type with associated operators. Synthesis oriented packages for logic and arithmetic are also available.

2.2 Data Types, Objects, and Expressions

VHDL is a strongly typed language. This means that each object has an associated type specifying which values the object can store and the operations that can be performed on it. During compilation, consistency of data types as enforced by the language definition is statically checked. A set of predefined types is exported by the package STANDARD. Other types can be defined by the designer.

2.2.1   Data types

There are four classes of data types in VHDL: scalar types, composite types, access types, and file types. Access and file types are not relevant from the point of view of hardware synthesis and will not be discussed here.

<u>Scalar Types</u>

Scalar types have simple, single values that are ordered, i.e., relational operators can be used on these values:
1. *Enumeration types*
   The values of an enumeration type are defined by listing them in an ordered list:
   > **type** STATE **is** (READY, WAITING, RUN);
   > **type** STD_ULOGIC **is** ('U','X','0','1','Z','W','L','H','-');
   Some of the predefined enumeration types are CHARACTER, BIT, BOOLEAN, SEVERITY_LEVEL.
2. *Integer types*
   The definition of an integer type defines a set of values representing the integers inside a specified range:
   > **type** INDEX **is range** 0 **to** 64;
   INTEGER is a predefined integer type.
3. *Floating point types*
   Floating point types provide approximation to real numbers. The precision of the representation is implementation dependent. A definition of a floating point type defines the range that bounds the set of values corresponding to that type:
   > **type** LEVEL **is range** -1.00 **to** 1.00;
   REAL is a predefined floating point type.
4. *Physical types*
   A physical type is a numeric type representing some physical quantity, such as time, length, voltage, etc. The declaration of a physical type includes the

specification of a primary (base) unit, and possibly a number of secondary units which have to be integral multiples of the primary unit. The only physical type we are interested in here is the predefined type TIME, which is defined as follows:

```
type TIME is range -2147483647 to 2147483647
    units                       -- the range is implementation dependent
        fs;                     -- femtosecond (this is the primary unit)
        ps = 1000 fs;           -- picosecond
        ns = 1000 ps;           -- nanosecond
        us = 1000 ns;           -- microsecond
        ms = 1000 us;           -- millisecond
        sec= 1000 ms;           -- second
        min= 60 sec;            -- minute
        hr = 60 min;            -- hour
    end units;
```

Composite Types

Composite types are used to define collections of values. These can be array types or record types:

1. *Array types*

    An array object is a composite object consisting of elements that have the same type (subtype). The name for an element of an array object uses one or more indices with values belonging to a specified index type. Both the type of the elements and that of the index value(s) are specified in the definition of the array type:

    ```
    type TAB is array (200 downto 0) of INTEGER;
    type MATRIX is array (0 to 99, 0 to 99) of REAL;
    type TAB_CHAR is array (INTEGER range <>) of CHARACTER;
    ```

    Type TAB_CHAR is an example of an *unconstrained array type*. The notation <> means that the range of indices is unconstrained. The user specifies the index range when declaring an object of type TAB_CHAR:

    ```
    variable T: TAB_CHAR (15 downto 0);
    ```

    There are two predefined array types, defined as follows:

    ```
    type STRING is array (POSITIVE range <>) of CHARACTER;
    type BIT_VECTOR is array (NATURAL range <>) of BIT;
    ```

2. *Record types*

    A record object is a composite object consisting of named elements, which may have different types. Names and types of the components, called fields, are specified in the definition of the record type:

    ```
    type FEATURES is record
        SIZE: INTEGER range 100 to 20000;
        PRICE: REAL range 20.00 to 2000.00;
        DELAY: TIME;
    ```

end record;

Record fields are referred to by selection: if F is an object of type FEATURES then F.DELAY refers to field DELAY of that object.

Subtypes

A *subtype* is derived from a type, called its *base type*, by specifying a certain constraint. All the values of the base type that satisfy the constraint belong to the subtype. The operations defined for the subtype are the same as those associated with the base type. A subtype may be defined as a constraint on a scalar type or on an unconstrained array:

     subtype SMALL **is** INTEGER **range** -9 **to** 9;

     subtype BIT_8 **is** BIT_VECTOR (7 **downto** 0);

There are two predefined subtypes:

     subtype NATURAL **is** INTEGER **range** 0 **to** INTEGER'HIGH;

     subtype POSITIVE **is** INTEGER **range** 1 **to** INTEGER'HIGH;

HIGH is a *predefined attribute* which, applied to a scalar type, returns the upper bound of this type; thus, INTEGER'HIGH is the upper limit for INTEGER values in the given implementation. Similarly, the predefined attribute LOW returns the lower bound of the type. There are two other related predefined attributes, RIGHT and LEFT, which return the right and the left bound of a type, respectively.

2.2.2   Objects

A VHDL object is a named entity that contains a value of a given type (subtype). There are four *classes* of objects in VHDL: constants, signals, variables and files[1].

Constants

A *constant* is initialized to a specified value when it is created and this value can not be modified later. Constant values are specified in *constant declarations*:

     constant PI: REAL:=3.14159;

An exception to this rule are deferred constants which are declared, inside a package declaration, without specifying their value (see Subsection 2.1.3); this value is given only inside the corresponding package body.

A generic is implicitly considered as an object of class constant.

---

1. We will not elaborate here on objects of class *file*.

Variables

The value of a *variable* can be changed after this has been created. An initial
value can be explicitly specified in the *variable declaration*:

variable FLAG: BOOLEAN := TRUE;

variable REG: BIT_VECTOR(31 **downto** 0);

An important feature of variables, with differentiates them from objects of class
signal, is that their value is immediately updated after an assignment. This
makes them similar to variables in most programming languages.

Variables are normally declared only inside a process or a subprogram. This
allows no shared access from several processes to a variable, which is a
condition for the deterministic behavior of a VHDL model. As a new feature of
VHDL'92, so-called *shared variables* can be declared outside a process or
subprogram. Given the potential of non-determinism implied by the use of
these variables, they are not considered for synthesis.

Signals

Signals are used to connect different parts of a design. Thus, signals are the
objects through which information is propagated between processes and
between sub-components of an entity. A *signal declaration* is similar to the
declaration of a variable. Optionally an initial value can be specified:

signal READY: BIT:='1';

signal HEADER: STRING(0 **to** 64);

The semantics of signals is strongly connected to the notion of time in VHDL.
A signal has not only a current value but also a projected waveform with determines its future values at certain moments of simulation time. According to this
semantics the result of a signal assignment is not an instantaneous change of
the signal value but a change of the projected waveform which will be effective
only at some future time. We will discuss signal assignments and the VHDL
simulation mechanism in Section 2.4.

Signals may not be declared within processes or subprograms. Ports are
implicitly objects of class signal.

2.2.3   Expressions

Expressions in VHDL, like those in most programming languages, are formulas
that define the computation of a value. They combine *primaries* with *operators*.
Primaries include names of objects, literals, function calls and parenthesized
expressions. The following operators may be used inside an expression:

Logical operators

and   or   nand   nor   xor   xnor   not
These operators are defined for the predefined types BIT and BOOLEAN, as well as for one-dimensional arrays of BIT and BOOLEAN.

Relational operators

=   /=   <   <=   >   >=
Equality (=) and inequality (/=) are predefined on any type except file types. The other relational operators are predefined on any scalar type or on discrete array types (one-dimensional arrays with elements of enumeration or integer type). When the operands are arrays, comparison is performed one element at a time from left to right.

Shift operators

sll   srl   sla   sra   rol   ror
Each of the operators takes an array of BIT or BOOLEAN as the left operand and an integer value as the right one.

Adding operators

+   -   &
The operands for + and - have to be of numeric type (integer, floating point, or physical) with the result being of the same numeric type. The concatenation operator & is predefined for any one-dimensional array type.

Sign operators

+   -
Signs + and - are predefined for any numeric type and have their conventional mathematical meaning: identity and negation function respectively.

Multiplying operators

*   /   mod   rem
The * and / operators are predefined so that both operands are of the same integer or floating point type. The result is of the same type. Both are also defined, with certain restrictions, for physical types. The **rem** (remainder) and **mod** (modulus) operators are predefined for any integer type, and the result is of the same type.

Miscellaneous operators

abs **
The **abs** (absolute) operator is defined for any numeric type. The ** (exponentiation) operator is defined for the left operand to be of integer or floating point type, and for the right operand to be of integer type only.

2.3 Processes and Sequential Statements

The major modeling element for behavioral specifications in VHDL is the process. The process is essentially a sequential body of code which is activated in response to certain events. The *process statement* is a concurrent statement, thus several processes can be specified inside an architecture which are executed in parallel. They also execute in parallel with the processes specified inside the other architectures that are part of the design. The statement body of a process consists of *sequential statements* which are: variable assignment, if statement, case statement, loop statement, next statement, exit statement, assertion statement, report statement, wait statement, signal assignment, procedure call, return statement, and null statement. Most of these statements will be introduced in this section. Signal assignment and wait statement will be discussed in Section 2.4 while procedure call and return statement are presented in Section 2.5.

2.3.1   Process Statement

Processes are described according to the following syntax[1]:

```
process_statement ::=
        [postponed] process [(sensitivity list)] [is]
                process_declarative_part
        begin
                {sequential_statement}
        end process;
```

Items declared inside a process are typically types, variables, constants, subprograms. No signals can be declared inside a process, as they serve only for communication *between* processes.

A process is an implicit loop. The simulator executes the sequential statements in the prescribed order and after the last statement execution continues with the first one. After being created at the start of simulation, the process is

---

1. Every (concurrent and sequential) statement can be labeled according to VHDL'92; we will not specify the label in the syntax descriptions. Postponed processes are discussed later (see Subsection 2.4.3).

either in an active state or it is suspended and waiting for a certain event to occur.

Suspension of a process can result after execution of a wait statement which has been explicitly specified by the designer. If there is a sensitivity list in the process header, then no wait statement is allowed inside the process, but a wait is introduced automatically at the end of the sequential statement list. Thus, after executing its last sequential statement such a process will be suspended. It will resume and execute, starting from the first statement, after an event has been produced on one of the signals in its sensitivity list. The following two processes are equivalent:

```
process (A, B, C)                    process
begin                                begin
    . . .                                . . .
end process;                         wait on A,B,C;
                                     end process;
```

Wait statements will be discussed in Subsection 2.4.5.

2.3.2   Variable Assignment Statement

The syntax of a *variable assignment* is the following:
```
variable_assignment_statement ::=
        target := expression;
```

The target is typically an object of class variable and must have the same base type as the expression. The expression is evaluated when the statement is executed and the computed value is assigned to the variable object instantaneously. Thus, the variable changes value at the current simulation time. This is similar to variable assignments in usual programming languages but differs essentially from the way signal assignments are treated in VHDL (see Subsection 2.4.4).

2.3.3   If Statement

The *if statement* allows selection between different alternatives depending on one or more conditions:
```
if_statement ::=
        if condition then
                sequence_of_statements
        {elsif condition then
                sequence_of_statements}
        [else
                sequence_of_statements]
        end if;
```

Conditions are evaluated successively until one is found that is true; the corresponding sequence of statements is executed. If no condition is true and the else clause is present the corresponding sequence is executed; otherwise none of the sequences is executed.

In the following example one of three variables is incremented depending on the value of X:

```
if X<0 then
   COUNT_NEG:=COUNT_NEG+1;
elsif X=0 then
   COUNT_ZERO:=COUNT_ZERO+1;
else
   COUNT_POS:=COUNT_POS +1;
end if;
```

2.3.4 Case Statement

The *case statement* allows selection between different alternatives depending on the value of an expression:

```
case_statement ::=
      case expression is
            when choices => sequence_of_statements
            {when choices => sequence_of_statements}
      end case;
choices ::=
      choice {Ichoice}
choice ::=
      simple_expression I discrete_range I others
```

The expression must be of a discrete type (enumeration or integer), or of an one-dimensional array type with elements of a character type[1]. The execution of the case statement consists of the evaluation of the expression followed by the execution of the sequence of statements whose choices include the resulting value. All the choices must be distinct and all possible values of the expression must be covered. If at least one possible value is not covered explicitly in any choice, an **others** choice is required as the last choice of the case statement. It will be selected whenever the value of the expression is not given in the choices of the previous alternatives.

The following statement increments the variables SMALL, MULTIPLE, and OTHER depending on the value of X:

---

1. A *character type* in VHDL is any enumeration type that has at least one element which is a character literal. Thus, the predefined type BIT, for instance, is also a character type.

```
case X is
  when 0 to 9 => SMALL:=SMALL+1;
  when 10|20|30|40|50|60|70|80|90 => MULTIPLE:=MULTIPLE + 1;
  when others => OTHER:=OTHER+1;
end case;
```

2.3.5   Loop Statement

The *loop statement* executes a given sequence repeatedly, zero or more times.
The syntax of the loop statement is the following:

```
loop_statement ::=
       [iteration_scheme] loop
              sequence_of_statements
       end loop;
iteration_scheme ::=
       while condition | for identifier in discrete_range
```

Using one of the two iteration schemes the usual *while* or *for* loops will be
created. For the *while* scheme, the condition is evaluated before each iteration,
and execution proceeds if the condition is true. The following sequence
computes the number of bits in vector V which have value '1':

```
I:=3;
while I >= 0 loop
  if V(I)='1' then
    NR_1:=NR_1+1;
  end if;
  I:=I-1;
end loop;
```

The following sequence, from the example in Figure 2.2, executes the same
computation using a *for* scheme:

```
for I in 3 downto 0 loop
  if V(I)='1' then
    NR_1:=NR_1+1;
  end if;
end loop;
```

The loop parameter, in our example I, is implicitly declared within the *for* loop
and is incremented at the end of each iteration.

When the iteration scheme is omitted the loop will be executed indefinitely. An
exit or **return** statement can be used to leave such a loop (see Subsection 2.3.6
and Section 2.5).

2.3.6   Next and Exit Statements

The *next statement* is used to complete the execution of one of the iterations of an enclosing loop statement. The remaining statements in the current iteration are skipped and execution continues with the next iteration.

next_statement ::=
     next [loop_label] [**when** condition];

An *exit statement* completes the execution of an enclosed loop statement. It is specified according to the following syntax:

exit_statement ::=
     exit [loop_label] [**when** condition];

If a loop label has been specified the statements apply to the nominated loop. If there is no explicit loop label they apply to the innermost loop. Both statements have a conditional form with a *when clause*. If a condition has been specified it is first evaluated and if the result is *false* the iteration continues normally.

2.3.7   Assertion and Report Statements

An *assertion statement* checks that a specified condition is true and reports an error if it is not. It is mainly used to check certain constraints, modeled by the condition, and to report if a constraint is not fulfilled which means that an *assertion violation* has occurred. The syntax is the following:

assertion_statement ::=
     assert condition
         [**report** expression]
         [**severity** expression];

The expression in the report clause has to be of type STRING and specifies the error message generated if the condition is false; if no report clause is specified the default message is "Assertion violation".

The expression in the severity clause has to be of the predefined type SEVERITY_LEVEL which is defined as follows:

type SEVERITY_LEVEL **is** (NOTE, WARNING, ERROR, FAILURE);

In the absence of the severity clause the default value for severity is ERROR. The report generated for an assertion violation contains both the message string and the severity level. In addition, the simulator can initiate appropriate actions depending on the severity level (for instance to abort simulation if severity level is greater than some implementation dependent threshold).

VHDL'92 provides the *report statement* which can be used to generate a message without the assertion check:

        report_statement ::=
                report expression
                        [**severity** expression];

2.4   The Simulation Mechanism

As mentioned earlier, VHDL has been defined primarily as a language for the modeling and simulation of digital circuits. The Language Reference Manual defines several aspects of the language semantics in terms of their simulation. Understanding VHDL requires some knowledge about the way a VHDL model is executed by the simulation kernel.

The first step during execution is the so called *elaboration* of the model. In this phase components are bound to entities, the hierarchy of the model is flattened and what results is a set of processes connected through signals. This structure will be simulated under the control of an event-driven simulation kernel (the *simulator*).

The simulator is responsible for updating signal values, as result of which certain events may occur. Processes waiting on these events will then resume and are executed until they suspend on a wait statement. When all of them have been suspended the simulator again updates signals which results in a new set of processes being activated. Simulation is thus a cyclic process where each *simulation cycle* consists of a signal-update and a process-execution phase. The basic principle behind this simulation philosophy is that it models how functional components react to changes on their inputs and respond through changes on outputs which again will produce some further reaction of other or of the same units.

As part of the simulation cycle a global clock, holding the *current simulation time*, is incremented. Simulation time in this way advances with discrete values taking into account only instances at which signals are scheduled to get values or a process has to be reactivated after waiting for a certain amount of time.

An essential feature of this simulation mechanism is that current signal values are only updated by the simulator and only at certain moments during simulation of the model. A signal assignment statement executed inside a process has no direct effect on the current value of that signal. It only schedules a new value to be placed on the signal (by the simulator) at some later time which is specified by the designer as part of the signal assignment.

One basic data structure which underlies this execution mechanism is the *signal driver*. A process that assigns values to a signal will automatically create a driver for that signal. This driver contains the *projected output waveform* of the signal, which is a set of transactions, each *transaction* being a pair consisting of a value and a time. The signal is planned to take the indicated

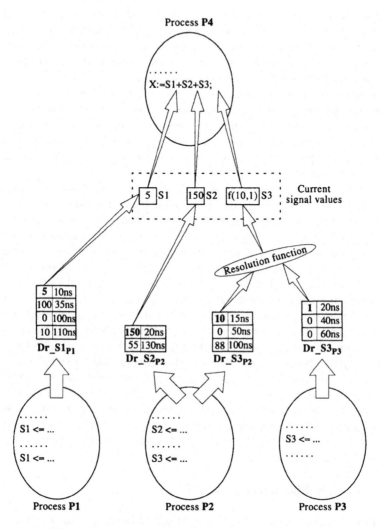

Figure 2.7: Processes, drivers, and resolution functions.

value at the moment indicated by the respective time. A signal assignment executed by a process only affects the projected output waveform by placing one or more transactions into the driver corresponding to the signal and possibly by deleting other transactions.

Figure 2.7 shows a group of four processes. Driver Dr_S1_{P1} corresponds to signal S1 and belongs to process P1 (which assigns values to S1). Process P2 assigns to both S2 and S3 and thus has two drivers associated: Dr_S2_{P2} and Dr_S3_{P2}. Process P3, which assigns to S3, has an associated driver called

Dr_S3$_{P3}$. As we can see, there are two drivers corresponding to signal S3, Dr_S3$_{P2}$ and Dr_S3$_{P3}$, as two processes are assigning values to this signal.

In each driver the transactions are placed in ascending order of the time. The value part of the first transaction is the *current value of the driver*. This first transaction in the driver is the only one which has a time component that is not greater than the current simulation time. As simulation time advances and the current time becomes equal to the time component of the next transaction, the first transaction is deleted and the next becomes the current value of the driver. In this way the driver gets a new value and, regardless if this value is different or not from the previous one, the driver and the signal is said to be *active* during that simulation cycle.

The current value of each signal is kept by the simulator in a structure automatically created at the start of the simulation. The value stored here is used by any process referring to the signal. In Figure 2.7 we showed this for process P4 which is accessing signals S1, S2, and S3. During each simulation cycle the value of all signals which are active during that cycle is updated by the simulator. If as result of this update the *current value of the signal* has changed to a different value, we say that an *event* has occurred on that signal.

Updating the current signal value typically consists of moving the actual value of the driver into the location that stores the current signal value. What if there are several drivers for a given signal? Which value should be moved by the simulator? We illustrate this situation in Figure 2.7 with signal S3 which is assigned by two processes and thus has two drivers. Such a signal is called a *resolved signal*. For each resolved signal the designer has to specify an associated *resolution function* which will be called by the simulation kernel every time the signal has to be updated. Given the current values of the drivers corresponding to the signal, the resolution function returns the value which will be used for updating the current signal value. Resolved signals will be discussed in Subsection 2.4.6.

2.4.1   The Simulation Cycle

Now that we have introduced the basic elements implied in the execution of a VHDL model we can give a more exact description of the simulation process which consists of two parts: the initialization phase, which is executed after the elaboration of the model, and the simulation cycle, which is run repeatedly.

During the *initialization phase*, the current simulation time T_c is set to 0 ns and each signal is set to its initial value[1]. Then each process is executed until it suspends. Finally the time T_n for the next (which initially is the first) simulation cycle is calculated as in the last step of the simulation cycle

---

1. This initial value is given explicitly in the signal declaration (Subsection 2.2.2) or it is the implicit initial value which is equal to T'LEFT, where T is the type of the signal.

described below.

Each *simulation cycle* consists of the following steps:

1. The current time T_c is set to T_n;
2. Each active signal is updated; we have shown above how signals are activated as time advances and how they are updated by the simulator; as a result of signal updates events are generated.
3. Each process that was suspended waiting on signal events that occurred in this simulation cycle resumes; processes which were waiting for a certain, completed, time to elapse[1] also resume;
4. Each resumed process executes until it suspends;
5. The time T_n of the next simulation cycle is determined as the earliest of the following three time values:
   1. TIME'HIGH;
   2. The next time at which a driver becomes active;
   3. The next time at which a process resumes[1];

Simulation is completed when time arrives at TIME'HIGH or, depending on the implementation, as the result of some assertion violation.

2.4.2   Delta Delay and Delta Cycle

The simulation mechanism we presented in the previous section (together with the rest of the language definition given in the reference manual [IEEE93]) precisely defines the semantics of any VHDL description and guarantees a deterministic simulation result, regardless of the particular features of any VHDL platform used by the designer. It is based on the ordering of events in time in which any event is executed in response to another event that has been scheduled for a previous simulation time.

But not only events scheduled for successive simulation times are to be ordered deterministically but also different events that occur at the same simulation time. This has been solved in VHDL by introducing the so-called *delta delay*, which is an infinitesimally small delay that separates events occurring in successive simulation cycles but at the same simulation time.

Let us consider the two processes in Figure 2.9 which model the functionality of the circuit shown in Figure 2.8. If, at a certain simulation time T_c, an event has occurred at one of the input ports X or Y (i.e. the value on the corre-

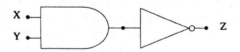

Figure 2.8: Circuit modelled in VHDL in Figure 2.9.

---

1. See Subsection 2.4.5 for the *time clause* in the **wait** statement.

```
entity DELTA_DELAY_EXAMPLE is
  port(X, Y:in BIT;
        Z: out BIT);
end DELTA_DELAY_EXAMPLE;

architecture DELTA of DELTA_DELAY_EXAMPLE is
  signal S: BIT;
begin
  AND_GATE: process(X,Y)
  begin
    S<=X and Y;
  end process;
  INVERTER: process(S)
  begin
    Z<=not S;
  end process;
end DELTA;
```

Figure 2.9: Example to illustrate delta delay.

sponding port has changed) the process AND_GATE will resume and execute until it suspends on the (implicit) wait corresponding to its sensitivity list. During its execution the process performs an assignment to signal S, which means that a new transaction will be scheduled on the corresponding signal driver; the time component for this transaction is equal to T_c, the value for the current simulation time, as no delay has been specified in the signal assignment statement (which, by default, means a delay of 0 ns, as will be discussed in Subsection 2.4.4). In the last step of the simulation cycle, the time T_n for the next cycle will be set to the time of the next transaction to activate a driver, and this time is in our case equal to T_c. Thus, the new simulation cycle will be executed with the same simulation time as the previous one. If as a result of updating signal S its value changes, the resulting event triggers the process INVERTER to resume. This process executes an assignment to signal Z and the resulting transaction will again be scheduled for the same simulation time T_c. Hence, a third simulation cycle will be performed without any change of the current simulation time. During this cycle the current value of Z will be updated.

To summarize the example in Figure 2.9, the change of an input port, the change of signal S, and that of output port Z are performed in three successive simulation cycles but all at the same simulation time. We say that the events in these three cycles are separated by a delta delay. A simulation cycle that is performed at the same simulation time as the previous one is called a *delta cycle*.

The delta delay mechanism allows a correct simulation of models where the delay of some components is ignored and so there is no difference in simulation time between the events on input end output of these components. This is very important in the context of system synthesis. VHDL models developed for synthesis at the system level usually describe only functionality of the system without considering timing aspects like circuit delays. Simulation of such models is based on delta cycles.

2.4.3   Postponed Processes

According to the syntax given in Subsection 2.3.1 a process can be optionally declared as **postponed**. As follows from our description of the VHDL simulation mechanism a process is executed during the simulation cycle in which it was resumed as consequence of a certain event. This rule does not apply to *postponed processes*[1]. A postponed process will be executed in the cycle in which it was resumed *only if* the next simulation cycle is not a delta cycle. If the next cycle is a delta cycle the process is not executed but is "postponed"; it will be executed by the first simulation cycle for which the next cycle is not a delta cycle.

Postponed processes are executed only at the end of all deltas of the simulation time in which they have been resumed. In order to cover postponed processes, the description we gave of the simulation cycle in Subsection 2.4.1 has to be changed slightly. In the fourth step only non-postponed processes are executed. As an additional last step, if T_n /= T_c (which means that the next cycle is not a delta cycle) each postponed process that has been resumed and not executed is executed until it suspends. T_n is then recalculated and simulation continues with a new cycle.

In order to be followed by a simulation cycle which is not a delta, postponed processes must not produce events which generate new delta cycles. It is an error, for instance, if a postponed process contains signal assignments without an explicit (non-zero) delay.

2.4.4   Signal Assignment Statement

The projected output waveform stored in the driver of a signal can be modified by a *signal assignment statement*. This statement schedules one or more trans-actions on the signal driver and, possibly, deletes some of the transactions that have been placed there earlier. The following syntax is used:

signal_assignment_statement ::=

    target <= [**transport** I [**reject** *time*_expression] **inertial**] waveform;

The target is typically an object of class signal and must have the same base type as the value component of each transaction produced by the *waveform* on

---

1. Postponed processes were introduced as a new feature in VHDL'92 [IEEE93].

the right-hand side. The waveform is described as a set of *waveform elements*, each one specifying a transaction:

```
waveform ::=
        waveform_element {, waveform_element}
waveform_element ::=
        value_expression [after time_expression]
```

Evaluation of one waveform element produces a single transaction. The time component of this transaction is determined by the current simulation time plus the value of the time expression in the waveform element. The value component of the transaction is determined by the value expression of the waveform element.

A signal assignment statement directly affects only the signal driver and no signal value will be changed in the simulation cycle in which the signal assignment itself is executed. This is true even for an assignment statement with no explicitly specified delay. The following sequence, which is part of a process, illustrates this important difference between variable and signal assignments:

```
  . . .
  X<=1;
  if X=1 then
    statement_sequence_1
  else
    statement_sequence_2
  end if;
  . . .
```

Since the waveform element in the assignment to X has no explicit *after* clause, an implicit 0 ns (or delta) delay is assumed. If we assume that in the sequence above the current value of X is different from 1 before the flow of the process reaches the signal assignment to X, then in the following if statement *statement_sequence_2* will be executed. The condition of this if statement will not be satisfied as the current value of signal X is still the old one. It will be updated to 1, only after suspension of the process, as the result of executing a wait statement, and will be available during the next simulation cycle (which will be a delta cycle in this case). Obviously, a different behavior would have been obtained had X been a variable. The new value would then have been available immediately after the assignment and *statement_sequence_1* would have been executed inside the if statement.

The evaluation of the waveform during the execution of a signal assignment statement results in a sequence of transactions, each one corresponding to one waveform element. These transactions are called *new transactions*, and their sequence has to be in ascending order of the time component. This sequence of new transactions is used to update the *old transactions* which constitute the projected output waveform stored in the signal driver. The concrete way a driver is updated as result of a signal assignment depends on the *delay mechanism*. This mechanism can be explicitly specified as part of the signal assignment using the reserved words **transport** and **inertial** respectively. If no delay mechanism is specified then inertial delay is considered by default.

Transport Delay

The *transport delay* models devices that exhibit nearly infinite frequency response: any pulse is transmitted, no matter how short its duration. This is typical, for instance, for transmission lines. Updating the projected waveform according to the transport delay model is performed in the following two steps:

1. All old transactions scheduled to occur at the same time or after the first new transaction are deleted from the projected waveform.
2. The new transactions are appended to the end of the driver.

Let us consider the following sequence of assignments to signal S:

    S<=transport 100 after 20 ns, 15 after 35 ns;
    S<=transport 10 after 40 ns;
    S<=transport 25 after 38 ns;

We assume that the statements are executed at simulation time 100 ns and the projected waveform consists of a single transaction with value 0. After executing the first two assignments the driver for S will look like this:

0	100	15	10
100 ns	120 ns	135 ns	140 ns

The new transaction to be added as result of the third assignment causes the last transaction to be deleted before the new one is added. After executing the last assignment the new projected output waveform is:

0	100	15	25
100 ns	120 ns	135 ns	138 ns

As we can see no transaction scheduled to be executed before a new one is affected by a signal assignment with transport delay. Thus, considering a device modeled with a certain transport delay, every change at the input will be processed, regardless of how short the time interval between this change and the next one is. We illustrate this in Figure 2.10, with a buffer element having a delay of 15 ns. The waveform X shows how events are generated on the input X while waveform Z illustrates the events on the output. Times on the waveforms are given in ns. It is easy to observe how spikes are propagated by such a circuit.

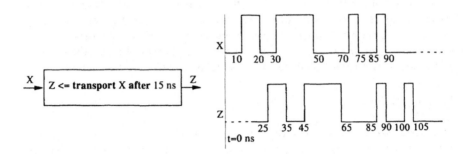

Figure 2.10: Buffer with transport delay.

Inertial delay

The *inertial delay* models the timing behavior of current switching circuits: an input value must be stable for a specified *pulse rejection limit* duration before the value propagates to the output. If the input is not stable for the specified limit no output change occurs. The following signal assignment propagates, with a delay of 15 ns, pulses on X that are not shorter than 5 ns:

S <= **reject** 5 ns **inertial** X **after** 15 ns;

If no reject clause has been provided, the pulse rejection limit is equal to the time value in the first waveform element[1]. The following assignment propagates, with a delay of 15 ns, pulses that are not shorter than 15 ns:

S <= **inertial** X **after** 15 ns;

As already mentioned, the inertial delay model is considered by default. Thus, the following assignment is equivalent to the previous one:

S <= X **after** 15 ns;

Updating the projected waveform according to the inertial delay model implies first the deletion of all old transactions scheduled to occur at the same time or after the first new transaction and the addition of the new transactions to the driver. This is performed exactly as was the case for the transport delay model. For the inertial delay an additional step is performed in order to filter out transactions corresponding to pulses which are shorter than the prescribed pulse rejection limit:

All the old transactions that are scheduled to occur at times between the time of the first new transaction and this time minus thepulse rejection limit are deleted; excepted are those transactions which are immediately preceding the first new

1. In the VHDL'87 standard no explicit specification of the pulse rejection limit was provided; it was considered automatically equal with the value of the inertial delay.

transaction and have the same value as it has.

Let us consider the following sequence of signal assignments:

S<=1 **after** 20 ns,15 **after** 35 ns;

S<=8 **after** 40 ns,2 **after** 60 ns,5 **after** 80 ns,10 **after** 100 ns;

S<=**reject** 35 ns **inertial** 5 **after** 90 ns;

We assume that the statements are executed at simulation time 100 ns and that the projected waveform consists of a single transaction with value 0. After executing the first assignment the driver for S will look like this:

0	1	15
100 ns	120 ns	135 ns

The second assignment will cause the old transactions to be deleted after the new transactions have been added to the driver (the implicit pulse rejection limit of this assignment is 40 ns, as no explicit limit has been specified):

0	8	2	5	10
100 ns	140 ns	160 ns	180 ns	200 ns

The third assignment will first delete the last transaction from the driver as it is scheduled after the first new transaction (this is similar to transport delay). The transactions scheduled for the time interval between 100+90 ns and 100+90-35 ns are then deleted after adding the new transaction. An exception is the transaction with value 5, as this is equal to the value of the new transaction:

0	8	5	5
100 ns	140 ns	180 ns	190 ns

Figure 2.11 shows how an input waveform is filtered by a buffer element with inertial delay. We have used the same waveform for input on X as in Figure 2.10. Times on the waveforms are given in ns. Only events which are stable for at least 15 ns are propagated to output Z. In the assignment to Z1 a pulse rejection limit of 8 ns is specified. Thus, the events occurring on X at 10 ns and 20 ns respectively have been also propagated.

2.4.5   Wait Statement

A process may suspend itself by executing a *wait statement*:

    wait_statement ::=
        wait [sensitivity_clause] [condition_clause] [timeout_clause];

The three clauses define the signals on which the process is sensitive, the conditions which have to be satisfied in order for the process to resume, and a maximal time for the process to wait:

    sensitivity_clause ::=

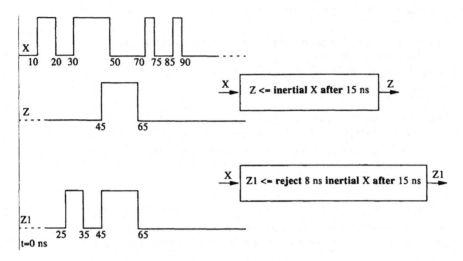

Figure 2.11: Buffer with inertial delay.

> **on** *signal*_name {, *signal*_name}
> condition_clause ::=
>   **until** condition
> timeout_clause ::=
>   **for** *time*_expression

All three clauses can be combined in a single wait statement:

> **wait on** A,B,C **until** A<2*B **for** 100 ns;

A process suspended after execution of a wait statement will resume, at the latest, immediately after the timeout interval specified in the time clause has expired. It may resume earlier as the result of an event occurring on any signal it is sensitive on. If such an event occurs, the condition in the condition clause is evaluated. If the condition is true the process will resume otherwise it resuspends.

If no condition clause has been specified in the wait statement, an implicitly true clause is assumed. If the sensitivity clause is omitted, then the process is sensitive to all of the signals mentioned in the condition expression. If no time clause has been specified the process will wait indefinitely which means that it will only resume as the result of an expected event and then only if the condition clause is true.

A process executing the following wait statement will be sensitive on signals A and B and it will resume execution after an event has occurred on one of these signals and the specified condition is true:

> **wait until** A<2*B;

After executing the following wait a process will resume and continue execution after an event occurred on A or B:

    wait on A,B;

The following statement will cause a process to suspend for *exactly* 100 ns:

    wait for 100 ns;

A process suspended on a signal, like on A or B in the examples above, is waiting for an event on that signal to occur. This means that the value of the signal has to change as result of an update. Under certain circumstances it can happen that a process has to be activated every time a signal is updated, regardless if its value changed or not. The predefined attribute TRANS-ACTION prefixed with the signal name can be used in such circumstances:

    wait on A'TRANSACTION;

TRANSACTION is a predefined signal attribute. Such an attribute creates an implicit signal that can be used like any explicit (declared) signal in, for instance, the sensitivity lists of wait statements or processes. A'TRANS-ACTION creates a signal of type BIT that toggles its value every time signal A becomes active. Thus, each time A is active (regardless if it changes value or not) an event is generated on A'TRANSACTION (since it changes from '0' to '1' or back).

In Subsection 2.3.1 we saw that a process statement can have a sensitivity list included in its header. Such a process is assumed to have an implicit wait statement at the end of its statement part. This wait statement has the sensitivity list consisting of the signals specified in the process header. No explicit wait statement is allowed inside such a process.

The execution mechanism of a VHDL model is presented at the beginning of this section. We remember that *after all processes have been executed until their suspension* (on wait statements) the simulation kernel determines the next simulation time and starts a new simulation cycle. Thus, after being resumed, a process will always be executed during another simulation cycle than before suspension, with updated values for signals and at a later simulation time (simulation time can be the same if a delta cycle is executed). Between two wait statements, on the other hand, a process always executes in *zero time* (the simulation time never changes, regardless how sophisticated the statement sequence is) and with unchanged values of signals. This simulation-based particularities of VHDL semantics have to be considered carefully in the context of system synthesis.

2.4.6   Resolved Signals and Resolution Functions

Signals in VHDL normally have a single driver which is created by the process that assigns values to the signal. We mentioned already, in our discussion about the simulation mechanism, that signals with multiple drivers are allowed but

have to be handled in a particular way. Such signals are called *resolved signals*
and are illustrated in Figure 2.7 with signal S3 which is assigned both by
process P2 and P3 and thus has two drivers connected. A *resolution function*
computes the actual value which is used to update the signal, starting from the
actual values of the drivers. This function is automatically called by the
simulation kernel, whenever needed.

The resolution function corresponding to a resolved signal is explicitly
specified by the designer. This can be done in two ways: by specifying the
name of the function in the signal declaration or by including the function name
into the declaration of the subtype of the signal.

We consider the signal RESOLVED_SIGNAL of a four-level logic type,
which models a bus that connects the output of several processes:

```
type LOGIC_4 is ('X','0','1','Z');
type L_VECTOR is array(INTEGER range <>) of LOGIC_4;
function WIRED_OR(INPUT: L_VECTOR) return LOGIC_4;
signal RESOLVED_SIGNAL: WIRED_OR LOGIC_4;
```

The same signal can also be declared by introducing a new subtype:

```
subtype RESOLVED_WIRE is WIRED_OR LOGIC_4;
signal RESOLVED_SIGNAL: RESOLVED_WIRE;
```

In both cases a resolved signal has been declared, specifying that resolution
function WIRED_OR has to be called by the simulation kernel in order to
produce the value for signal RESOLVED_SIGNAL. Function body WIRED_OR is
described in Figure 2.12 (functions are further discussed in Section 2.5).

A resolution function has only one parameter, a constant one-dimensional
unconstrained array. The array element type and the return type are the same as
the type of the signal. Each time a resolution function is invoked by the
simulation kernel, it is passed an array value, each element of which is determined by a driver of the corresponding resolved signal. The returned value will

```
function WIRED_OR(INPUT: L_VECTOR) return LOGIC_4 is
   variable RESULT: LOGIC_4:='0';
begin
   for I in INPUT'RANGE loop-- the predefined attribute RANGE
      if INPUT(I)='1' then              -- returns the index range of the
         RESULT:='1';                   -- actual vector;
         exit;
      elsif INPUT(I)='X' then
         RESULT:='X';
      end if;
   end loop;
   return RESULT;
end WIRED_OR;
```

Figure 2.12: Resolution function.

be used for the signal.

2.5   Subprograms

There are two forms of subprograms in VHDL: *functions* and *procedures*. They define algorithms for computing a value (functions) or exhibiting certain behavior (procedures). A subprogram is defined by describing a *subprogram body*:

```
subprogram_body ::=
      subprogram_specification is
            subprogram_declarative_part
      begin
            sequence_of_statements
      end [function] [procedure] [subprogram_name];
```

Items declared inside a subprogram are types, constants, variables, and subprograms, which are accessible for use within the given subprogram. The statement part contains sequential statements which define the computation to be performed by the subprogram.

The *subprogram specification* defines the kind of the subprogram (function or procedure), its name and interface. The interface describes name, class and mode of the formal parameters, and for functions also the type of the returned result:

```
subprogram_specification ::=
      procedure identifier [(formal_parameter_list)] |
      [pure] [impure] function designator [(formal_parameter_list)]
                                          return type_name
```

Formal parameters may be of *class* **variable**, **constant**, **signal**, or **file**. Each parameter is considered inside the subprogram as an object of the respective class. The *mode* of a formal parameter can be **in** (can be read but not updated within the subprogram), **out** (cannot be read but can be updated inside the subprogram), or **inout** (can be both read and updated within the subprogram).

The following is an example of a procedure-body specification:

```
procedure SEARCH (constant PATTERN:in BIT_8;
                  variable FLAG:out BOOLEAN;
                  variable INDEX:out INTEGER) is
   -- declarations
begin
   -- statements
end SEARCH;
```

A *subprogram call* contains the name of the subprogram and a list of actual parameters which are used to pass values to and from the subprogram. The association between formal and actual parameters can be by position, name, or a combination of the two. Here is a call to procedure SEARCH with association by position:

```
SEARCH("00110101",FOUND,I);
```

The same call, using named association, can be like this:

```
SEARCH(FLAG=>FOUND,INDEX=>I,PATTERN=>"00110101");
```

The types of corresponding actual and formal parameters have to match. Only actual signals, variables, and files may be associated with formals of class **signal, variable**, and **file** respectively. Expressions may be used to pass values for formals of class **constant**.

Once a subprogram is called its statements are executed until either the end of the statement list is reached or a *return statement* is executed. In both cases after leaving the subprogram a return is made to the caller. The following is the syntax of a return statement:

```
return_statement ::=
        return [expression];
```

Only return statements placed within a function include an expression. Functions must always be left by executing a return statement (which also specifies the return value).

In Subsection 2.1.3 we showed how subprograms are declared inside a package declaration without specification of the subprogram body. A *subprogram declaration* only describes the interface of the subprogram (name, formal parameter list, and type of returned value for functions) without its body. The subprogram body, in this case, is specified separately inside the body of the package. Separate declaration is also useful in other contexts like, for instance, when subprograms are calling each other recursively.

2.5.1   Functions

A function is called as part of an expression and returns a single value. Functions are allowed to have only parameters of mode **in** (which is also considered by default) and of class **constant** or **signal**; the default class is **constant**.

A typical VHDL function produces no side effect. This means that no variable or signal declared outside the function body is assigned to or altered by the function and that the same value is returned each time the function is called with the same arguments. Such a function is a **pure** one. Functions which do not correspond to these restrictions are declared as **impure**.

While procedure designators may be only identifiers, a function designator may also be an operator symbol. This is an example of a function declaration that implements addition of bit vectors:

function **"+"(LEFT,RIGHT:BIT_VECTOR) return BIT_VECTOR;**

Functions are not allowed to execute a wait statement. Thus, a function always executes in zero simulation time and returns the value at the same time it has been called.

2.5.2   Procedures

Procedures are not limited to return a single value. Using parameters of mode **out** or **inout** any number of values can be returned. A procedure call is a statement.

Procedures are allowed to have formal parameters of mode **in, out,** or **inout**; the default mode is **in.** The parameters may be of class **constant, variable, signal,** or **file.** If the class is not specified it is considered **constant** for parameters of mode **in** and **variable** for those of mode **out** or **inout.**

Procedures, unlike functions, are allowed to execute wait statements.

2.5.3   Overloading

Several subprograms in VHDL are allowed to have the same name (designator). The subprogram name and the respective subprograms are then said to be *overloaded.* Functions can also overload an operator.

When a call to an overloaded subprogram is compiled or an overloaded operator is used the following information is relevant for determining the actual subprogram (operator): number, order, name (if named association is used) and type of parameters and, for functions, type of the result. Thus, an expression like

A+B

will perform the standard integer addition if A and B are of type INTEGER, but will call the function "+", declared in Subsection 2.5.1, if A and B are vectors of bits.

2.6 Concurrent Statements

As discussed in Subsection 2.1.2, the statement part of an architecture body consists of several *concurrent statements.* They are all executed in parallel and therefore the order in which they are specified is not relevant. One of the concurrent statements, the *component instantiation statement,* has been discussed in Subsection 2.1.2. It is the basic mechanism used for structural specification. Behavioral specification is based on another concurrent statement, the process statement. It was presented in Subsection 2.3.1. The *concurrent signal assignment, concurrent procedure call,* and *concurrent assertion statement* merely represent convenient short-hand notations for expressing processes that contain only a signal assignment, a procedure call, or an assertion statement respectively, together with a wait statement.

2.6.1 Concurrent Signal Assignment Statement

A signal assignment statement appearing outside a process or a subprogram, as part of an architecture body, is interpreted as a concurrent statement and is equivalent to a process containing only that signal assignment followed by a wait statement. The following concurrent signal assignment will be executed whenever there is an event on signals X or Y:

```
S<=X+Y+2;
```

It is equivalent to the following process which is sensitive on signals X and Y:

```
process
begin
  S<=X+Y+2;
  wait on X,Y;
end process;
```

The functionality of an architecture may be specified by primarily using concurrent signal assignments. Such models reflect the flow of information through the design and thus are often called *dataflow models*. Dataflow models can be considered as being mainly behavioral as they express functionality by specifying arithmetic and logic operations which are executed in parallel.

Let us come back to the four-bit parity generator we used as an example in Subsection 2.1.1 and Subsection 2.1.2. In Figure 2.2 and Figure 2.4 we gave behavioral and structural specifications of the circuit. In Figure 2.13 we show a dataflow model of the same parity generator.

In its general form the concurrent signal assignment allows the selection of different values for the target signal based on a condition or on the value of a select expression:

```
architecture PARITY_DATAFLOW of PARITY is
  signal T1, T2, T3: BIT;
begin
  T1<=V(0) xor V(1) after 1 ns;
  T2<=V(2) xor V(3) after 1 ns;
  T3<=T1 xor T2 after 1 ns;
  EVEN <= not T3 after 0.5 ns;
end PARITY_DATAFLOW;
```

Figure 2.13: Parity generator: dataflow specification.

```
concurrent_signal_assignment_statement ::=
    [postponed] conditional_signal_assignment |
    [postponed] selected_signal_assignment
```

Conditional Signal Assignment Statement

The *conditional signal assignment* selects the value for the target signal based on the specified conditions. It is a shorthand for a process containing signal assignments in an if statement:

```
conditional_signal_assignment ::=
    target <= [transport | [reject time_expression] inertial] conditional_waveforms;
conditional_waveforms ::=
    {waveform when condition else}
    waveform [when condition]
```

Waveforms are the same as for sequential signal assignments and their syntax is given in Subsection 2.4.4.

The conditional signal assignment is executed whenever an event occurs on one of the signals in any waveform expression or condition. The conditions are then evaluated one at a time in order. For the first true condition the corresponding waveform is scheduled to be assigned to the target signal. Let us consider, for example, the following conditional assignment:

```
S <= X+Y+2 after 10 ns when I1='0' and I2='0' else
     X+Y-2 after 10 ns when I1='0' and I2='1' else
     X after 10 ns when I1='1' and I2='0' else
     Y after 10 ns;
```

The equivalent process looks like this:

```
process
begin
  if I1='0' and I2='0' then
    S <= X+Y+2 after 10 ns;
  elsif I1='0' and I2='1' then
    S <= X+Y-2 after 10 ns;
  elsif I1='1' and I2='0' then
    S <= X after 10 ns;
  else
    S <= Y after 10 ns;
  end if;
  wait on X,Y,I1,I2;
end process;
```

Selected Signal Assignment Statement

The *selected signal assignment* selects the value for the target signal based on the value of a select expression. It is a shorthand for a process containing signal assignments in a case statement:

```
selected_signal_assignment ::=
    with expression select
        target <= [transport | [reject time_expression] inertial] selected_waveforms;
selected_waveforms ::=
    {waveform when choices,}
    waveform when choices
```

The syntax for *choices* is the same as specified in Subsection 2.3.4 for the case statement. Waveforms are defined in Subsection 2.4.4.

The selected signal assignment is executed whenever an event occurs on one of the signals in any waveform expression or in the select expression. The select expression is then evaluated. For the choice value that matches the value of the select expression the corresponding waveform is scheduled to be assigned to the target signal. Here is an example of a selected signal assignment:

```
with A+B select
    S <= X+Y after 5 ns when 0,
            X-Y after 5 ns when 1 to 10,
            0 after 1 ns when others;
```

The equivalent process looks like this:

```
process
begin
  case A+B is
    when 0 => S <= X+Y after 5 ns;
    when 1 to 10 => S <= X-Y after 5 ns;
    when others => S <= 0 after 1 ns;
  end case;
  wait on X,Y,A,B;
end process;
```

2.6.2   Concurrent Procedure Call Statement

Subprogram calls have been introduced in Section 2.5. If such a call appears inside a process or another subprogram it is interpreted as a sequential statement. Otherwise it is a concurrent statement and is equivalent to a process containing only that procedure call followed by a wait statement. The procedure will be activated whenever an event occurs on one of the signal parameters of mode **in** or **inout**. A procedure called by a concurrent call is not allowed to have parameters of class **variable**. We consider the following procedure:

```
procedure COMPUTE(signal A,B:in INTEGER;
                           signal RES:out INTEGER) is
begin
    . . .
end COMPUTE;
```

If it appears as a statement inside an architecture statement part, the following is interpreted as a concurrent procedure call:

```
COMPUTE(INSIG1,INSIG2,OUTSIG);
```

It is equivalent to the following process:

```
process
begin
    COMPUTE(INSIG1,INSIG2,OUTSIG);
    wait on INSIG1,INSIG2;
end process;
```

2.6.3   Concurrent Assertion Statement

Assertion statements have been discussed, as sequential statements, in Subsection 2.3.7. The same statement will be interpreted as a concurrent statement when it is used outside a process or a subprogram. In this case it is equivalent with a process containing a (sequential) assertion statement followed by a wait statement. The assertion statement will be active each time an event occurs on one of the signals which are referred to in the assertion condition. The following assertion statement will produce a report whenever the values of signals S1 and S2 become equal:

```
architecture CHECKED_DESIGN of DESIGN is
    -- declarations
begin
    assert S1/=S2
        report "error on S1 and S2"
        severity ERROR;
    -- other concurrent statements
end CHECKED_DESIGN;
```

The following process is equivalent to the concurrent assertion statement above:

```
process
begin
    assert S1/=S2
        report "error on S1 and S2"
```

```
        severity ERROR;
      wait on S1,S2;
   end process;
```

2.7   VHDL for System Synthesis

Our presentation of VHDL in this chapter was not meant to be an exhaustive
one, but has been kept inside the limits spanned by the subject of this book. We
hope that even with this limitation, the reader now has a coherent image of a
very powerful, rich, and expressive language. As already mentioned, VHDL has
been defined to support development, verification, synthesis and testing of
hardware designs. The language allows specification of systems at various
levels of abstraction and covers the design process from the system-level down
to gate-level implementation.

Even if the intended area of application includes all design phases of
electronic systems, VHDL is defined as a simulation language. The simulation
based semantics of the language has to be carefully considered when VHDL is
used for synthesis purposes [CST91, Pos91, WoMa93].

Synthesis very quickly became one of the basic application areas of VHDL.
Now, that the language has gained wide acceptance, most of the synthesis tools,
both commercial and academic ones, accept VHDL specifications as their
input. This, however, generates several problems when implementing these
tools. Two aspects of the language definition are at the origin of most of the
difficulties:

1. VHDL has the rich capabilities and features of a modern programming
   language. This makes it, for instance, suitable for specification at the system
   level when part of the design will be, possibly, implemented as software. If
   specifications are intended to be synthesized to hardware, language facilities which are only relevant in software or which have no realistic hardware
   implementation, have to be excluded. Examples are file types, access types
   or recursive subprograms.

2. Some of the VHDL features are semantically defined in terms of simulation.
   This primarily affects process interaction facilities and the timing model.
   This creates considerable difficulties when the specification is interpreted
   for synthesis, as the functionality of the resulting implementation has to
   accurately reflect the simulation behavior of the source description.

Designers of synthesis tools have solved the problems highlighted above by
defining *subsets of VHDL* which are accepted for synthesis and by imposing
certain *modeling guidelines* which have to be followed by the system users.

Synthesis tools at the logic or even RT level are commonly available and
efficiently used today. Nevertheless, there is no general agreement yet on the
VHDL subset to be accepted by these systems or on the modeling rules which

the input specifications have to follow. Currently, work is in progress which aims to define IEEE standards in order to unify the interpretation and use of VHDL for logic synthesis.

High-level synthesis is still an academic research topic even if the first commercial tools are now available [KLMM95] and successful industrial use has already been reported. Hence, it is not surprising that the extent to which VHDL should be restricted when used for high-level synthesis is still widely discussed. There exists, however, some convergence concerning certain features which should be excluded from a VHDL subset for high-level synthesis and also on some modeling rules for specifications submitted to high-level synthesis [LiGa89, HKL90, CST91, Pos91, NBD92, BeKu93, Bie93, KLMM95]:

1. Features which are not relevant from the high-level synthesis point of view are excluded from input specifications: structural specifications, guarded blocks, resolution functions, certain signal attributes.
2. The input specification is purely sequential, formulated as a single VHDL process. Thus, VHDL specifications for high-level synthesis consist of a single process describing the intended behavior in terms of VHDL sequential statements.
3. All synchronization is restricted to a clock signal which has to be explicitly recognized. Wait statements are allowed only on this explicit clock. Signal assignments are usually allowed only on output ports.
4. The scheduling of certain operations is fixed in terms of clock cycles.
5. Strict timing, as specified by VHDL simulation models, is not considered. Thus, *after clauses* are ignored in signal assignment statements and *time clauses* in wait statements are not allowed.

As models specified according to these restrictions practically exclude process interaction aspects, most of the synthesis problems resulting from VHDL simulation semantics are avoided. At the same time these models can be synthesized according to state-of-the-art techniques in high-level synthesis and can be easily verified by the simulation of the resulting RT-level implementation. Efficient verification is facilitated by the fact that I/O operations are very often scheduled in terms of clock cycles in the input specification. This, however, reduces the optimizing capacity of the high-level synthesis tool very much. Under these circumstances, synthesis algorithms are very restricted in improving the quality of the implementation by deciding on the schedule of operations.

Some of these restrictions are however not acceptable for specifications at the system level. The most important aspect is that specifications consisting of interacting VHDL processes have to be accepted for system synthesis. Thus, no restrictions concerning the number of processes nor their interaction through signals should be introduced. The explicit use of a clock signal and pre-scheduling of operations in terms of clock cycles can not be imposed as requirement for a system-level design. Especially since specifications very often are formulated in a way which is very close to software design techniques and does not consider any hardware specific aspects. Very often, at this level, it is not yet decided which part of the design will be implemented in hardware and which will be compiled to software.

As a conclusion, rules 2, 3, and 4 given above are not acceptable in the context of system-level synthesis. Concerning the timing aspects of VHDL, the main problem is that, according to the language standard, strict timing is specified by after clauses in signal assignment statements and by wait statements with time clauses. The hardware synthesis of strict timing, as well as the implementation in hardware of the *zero time assumption* (Subsection 2.4.5), is obviously impossible. On the other hand, a large freedom in deciding when a certain operation has to be executed is left to the synthesis tool and should not to be specified by the designer. This is the reason why the restriction 5 is still to be considered for system synthesis. However, in chapter 7 we will discuss how VHDL timing facilities can be used for the specification of time constraints for synthesis.

One of the central subjects discussed in some of the following chapters is how to accept for synthesis specifications containing interacting VHDL processes while keeping at the same time simulation semantics during the synthesis process. The main task to be solved in this context is the synthesis of the interacting processes and of their underlying communication structures with a low additional implementation overhead. In order to improve the quality of system specifications and to produce efficient implementations, specific synthesis strategies have to be developed and adequate system-level mechanisms for specification have to be proposed.

3

High-Level Synthesis

3.1 Introduction

In this book, the term *synthesis* is used to denote the process of transforming a digital system from a behavioral specification into an implementation structure. Generally speaking, the specification includes some form of abstractions, i.e., some of the design decisions are not bound. The implementation, on the other hand, has to describe in detail the complete design at a given level of abstraction. Thus, synthesis can be seen as a process of creating implementation details which are left out of the specification [Pen87]. For example, a purely behavioral specification of a microprocessor may specify only what should be done in a typical instruction cycle and leaves it to the synthesis procedure to decide whether a centralized bus should be used, which technique should be employed to implement the control function, and how much parallelism should be supported.

Due to the complexity of digital systems, especially those implemented in VLSI technology, the synthesis process is usually divided into several steps. These steps usually include system-level synthesis, high-level synthesis, logic synthesis and physical design.

System-level synthesis deals with the formulation of the basic architecture of the implementation. The input to this synthesis step is a system-level specification which describes the behavior of the entire system in terms of a set of interacting processes. Such a system can be implemented by a set of cooperating processors, such as ASICs, dedicated controllers, FPGAs, and DSP processors. The allocation of the set of physical processors and the mapping of processes in the behavioral specification onto these processors are the most critical design decisions to be made during the system-level synthesis step. An important feature of the system level is that the synthesis techniques and design requirements are highly application dependent. For example, system-level synthesis techniques used in the real-time embedded-controller application area will be very different from those used in DSP systems. This feature results in also the different definitions of what design tasks constitute a system-level synthesis step. In this book, we will address a few fundamental issues related to system-level synthesis which can be considered as the common denominator for a system-level synthesis technique. These issues are system partitioning, hardware/software co-design, and interconnection-structure design. The output of the system-level synthesis step is a set of processes with well-defined interfaces. Each of the processes is specified by a behavioral specification.

High-level synthesis will then translate the behavioral specification of a process into a structural description that is still technology independent. This structural description is usually given in terms of a netlist at the register-transfer level.

System-level synthesis and high-level synthesis form the front-end of a synthesis approach to digital system design, and are together called *system synthesis*. This chapter will present the main issues related to high-level synthesis, while the next chapter will address the system-level synthesis problems. The issue of controller synthesis which is related to both high-level synthesis and system-level synthesis will be presented at the end of this chapter.

After system-level synthesis and high-level synthesis have been performed, logic synthesis and physical design are used to map the structural implementation at register-transfer level into a layout description which is the final implementation. Logic synthesis and physical design form the back-end of the synthesis approach to digital system design.

One important reason for the separation of the synthesis process into front-end (system) synthesis and back-end synthesis can be attributed to the good general property possessed by front-end synthesis and the short-lived nature of the target semiconductor process associated with the back-end synthesis. As front-end synthesis is not bound to a particular technology, it can be used in several different design environments or adopted quickly to new technologies as needed. The back-end, however, has a very short life cycle, because technologies change.

In recent years, there has been a clear trend toward automating the system-synthesis process and there are several reasons for this:
- Shorter design cycle. The use of automation in the synthesis process reduces the design time, and provides better chances for a company to hit the market

window for the products. Automation reduces also the cost of the products significantly since in many cases the design cost dominates the product development cost.

- The ability to explore a much larger design space. An efficient synthesis technique can produce several designs from the same specification in a short period of time. This allows the designer to explore different trade-offs between cost, performance, power consumption, testability, etc.

- Support for design verification. One prerequisite for automating the hardware/software co-design process, for example, is to start the synthesis process with a joint specification of both the hardware and software. This makes it possible to verify the complete design consisting of both hardware components and software procedures.

- Fewer errors. The reduction of manual design activities means that the number of human errors will be decreased. If the synthesis algorithms can be validated, we can also be more confident that the final design will correctly implement the given specification.

- Increased availability of IC technology. As more design knowledge is captured in the synthesis algorithms, it is much easier for people who are not IC-technology experts to design chips.

3.1.1   The High-Level Synthesis Tasks

The input to the high-level synthesis process is given in an algorithmic-level specification, such as behavioral VHDL. This type of specification gives the required mapping from sequences of inputs to sequences of outputs. The specification should constrain the internal structure of the system to be designed as little as possible. From the input specification, a synthesis system produces a description of a data path, that is, a network of registers, functional units, multiplexers, and buses. A control part should also be produced if it is not integrated into the data path. In a synchronous design, the control part can be given as microcode, PLA profiles or random logic.

The basic components of the data path will eventually be implemented by some physical modules available in a given technology. The technological parameters, such as silicon area, operation delay and power consumption of the physical modules are usually stored in a module library and made available to the high-level synthesis algorithms. In this way, the same high-level synthesis algorithm can be used to synthesize design based on different technologies by using different module libraries.

The basic tasks to perform in high-level synthesis are behavioral analysis, design-style selection, operation scheduling, data-path allocation, control allocation, module binding, and optimization.

3.1.2   Basic Synthesis Techniques

To carry out a synthesis task means to make design choices. There is usually a set of alternative structures that can be used to realize a given behavior. For example, an addition operation in a given behavioral specification can be implemented either by a dedicated adder or share an ALU with several other operations.

The function of a synthesis algorithm is to analyze all or a subset of these alternatives and to choose the best structure which meets given design constraints, such as limitations on cycle time, area, or power, while minimizing a cost function. The difficulty of synthesis is that trade-offs of different design aspects are highly dependent on each other. Thus, when making design decisions for a synthesis task, other tasks have to be taken into account in order to optimize some design criterion, for example, the implemetation cost. Consequently, each of the synthesis tasks cannot be carried out independently without reducing the possibility of global optimization of the design.

Another difficult factor of synthesis is the immense gap between the input specifications and the implementation results. Many design decisions are not bound in the specification and are left for the synthesis algorithms to decide. For example the synthesizer has to consider whether to multiplex a set of operations onto a single ALU or to implement several dedicated operators and the consequences of such a decision in terms of device timing, power consumption, chip size, pin-out count, and other low-level parameters.

The final source of difficulty of automated system synthesis is the parallelism inherent in a digital system. To organize the available hardware resources to efficiently and reliably perform the desired operation, the synthesis algorithms have to automatically generate parallel structures as well as their synchronization/communication schemes. They are also responsible for the scheduling of operations so as to ensure sufficient parallelism in the implementation.

It is obvious that system synthesis is a very complex problem. Many of the synthesis tasks have also been proved to be NP-complete [Gajo75, DeMi93]. Therefore, it is often necessary to divide a design into several modules and apply synthesis algorithms to one module at a time. To further reduce the complexity, a synthesis approach either partitions the synthesis task into several sub-tasks and perform one sub-task at a time, or partition the task into a sequence of transformation steps each of which makes a small change to the intermediate result of the earlier steps. The later approach is called the transformational approach to synthesis. The basic idea of the transformational approach can be illustrated with Figure 3.1, where the traditional approach to high-level synthesis divides the main synthesis task into three sub-tasks: allocation, scheduling, and binding. The transformational approach first moves the design to a structural implementation in one single step by a relatively naive

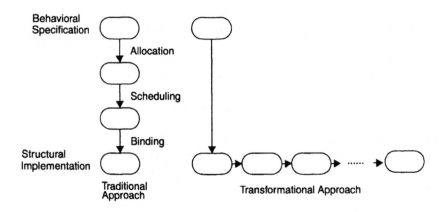

Figure 3.1: Comparison between the traditional and the transformational approaches to high-level synthesis.

mapping. The implementation produced by this step is then transformed using the iterative application of semantic-preserving transformations. These transformations do not change the level of abstraction, rather they explore the implementation space, looking for good solutions.

The use of the transformational approach has two main advantages. The first one is that it facilitates the correctness-by-construction method. Since the synthesis process is divided into a sequence of small transformations, if the transformations can be proved to be semantics-preserving, the synthesis will produce a design which correctly implements the specification [Pen88].

The second advantage of the transformational approach is that it makes it more efficient to employ optimization heuristics in the synthesis process. A transformational approach can be viewed as a neighborhood-search optimization method. The synthesis process starts with an initial implementation which is generated by a naive mapping and similar to an initial solution of an optimization problem. The synthesis process then makes small changes to transform the current solution to a neighborhood solution, which is the same as the neighborhood moves. The objective of the transformations is to reach the optimal design, which is the same as finding the optimal solution in the solution space. Therefore, we can employ existing neighborhood search techniques in the transformational approach. Techniques such as simulated annealing, tabu search, and genetic algorithms have been reported to perform well in different synthesis processes [EPKD97, HaPe96].

An important component of the transformational approach is the design representation which is used to capture the intermediate results of the synthesis process. The representation model must be able to represent the design with different degrees of completion. That is, it should be able to describe a very abstract design with a lot of unspecified information, for example, a purely behavioral specification. At the same time, it should also be able to describe a very detailed implementation with physical parameters which is, for example, produced by the synthesis process. Thus, it is not necessary or always possible to have just one design representation model; a lot of synthesis systems actually use several different representation models during different stages of the synthesis process.

In most design environments, however, it is very useful to have a unified design representation which can be used to represent the design at different levels of abstraction. The main application of a design representation is to explicitly capture the intermediate result of a design so as to allow the design algorithm to make *appropriate* design decisions. This is very important in the transformational approach to synthesis.

3.2   Scheduling

Operation scheduling, or in short scheduling, deals with the assignment of each operation to a time slot corresponding to a clock cycle or time interval. Typically, the input to this task consists of a control and data flow graph (CDFG), a set of available hardware resources and a performance constraint. A schedule will be generated such that the data/control dependency defined by the CDFG will not be violated and the performance constraint is satisfied.

Since scheduling determines which operations can be assigned to the same time slot, it affects the degree of concurrency of the resulting design and thus its performance. Further, the maximum number of concurrent operations of a given type in a schedule is a lower bound on the number of required hardware resources for that operation. Therefore, the choice of a schedule affects the cost of the implementation and consequently scheduling plays an important role in high-level synthesis.

The scheduling problem can be formulated in several ways depending on the basic assumptions made. One straightforward way is to assume that the behavioral descriptions do not contain conditional or loop constructs, that each operation takes exactly one control step to execute, and that each type of operation can be performed by one and only one type of functional unit [GDWL92]. In this case, we have the following problem:

Given: a set O of operations with a partial ordering which determines the precedence relations, a set K of functional unit types, a type function, $\tau: O \rightarrow K$, to map the operations into the functional unit types, and resource constraints m_k for each functional unit type.

Find: a (optimal) schedule for the set of operations that obeys the partial ordering and utilizes only the available functional units.

The above stated problem is called resource-constrained scheduling, since a set of constraints regarding the number of functional units (hardware resources) of each type are explicitly given. For example, in a design where only two multipliers can fit within the available chip area, it is necessary to impose a resource constraint to limit the number of multipliers to two. The other scheduling problems are time-constrained or time- and resource-constrained, which will be discussed later in this section.

Example 3.1: Figure 3.2 gives a simple behavioral specification and its corresponding data flow graph (DFG). The DFG is generated by an algorithm which analyzes the data dependency relation of the different operations. The data dependency analysis is performed based on the following principle: Let O be the set of all operations in the DFG. If the result of operation $o_i \in O$ is used by operation $o_j \in O$, then operation o_i must finish its execution before operation o_j can begin, and we say that there is a data dependency between these two operations. This data dependency is represented during the synthesis process as a precedence constraint (partial ordering) between the operations, which must be satisfied by the schedule. Figure 3.2b illustrates the data flow graph which is generated from the VHDL code shown in Figure 3.2a. We have assumed that all data flows downwards and have therefore drawn the directed edges only as edges without arrows to indicate directions.

□

The simplest scheduling technique is a greedy heuristics based on the "as soon as possible" (ASAP) principle. To schedule the DFG given in Figure 3.2b, using the ASAP algorithm, the operations are first sorted topologically according to their partial ordering; that is, if there is a partial order from o_i to o_j, o_i will be sorted before o_j. The algorithm then schedules operations one by one in the topologically sorted order by placing them in the earliest possible control step [MPC90].

For the DFG example given in Figure 3.2b, the topologically sorted order is illustrated in Figure 3.3a by the number used to label the operations. Let us assume that the available functional units include one adder and one multiplier. Both the addition and substraction operations are mapped into the adder, and

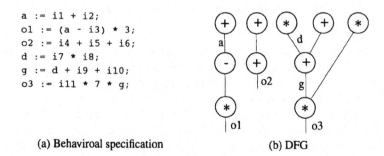

```
a  := i1 + i2;
o1 := (a - i3) * 3;
o2 := i4 + i5 + i6;
d  := i7 * i8;
g  := d + i9 + i10;
o3 := i11 * 7 * g;
```

(a) Behaviroal specification   (b) DFG

Figure 3.2: A behavioral specification and its data flow graph.

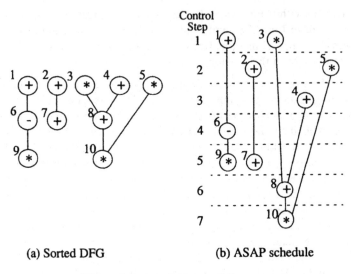

(a) Sorted DFG   (b) ASAP schedule

Figure 3.3: An ASAP scheduling example.

the multiplication operations are mapped into the multiplier. When operation 1 (the operation labeled with 1) is considered for scheduling, it is scheduled in control step 1, since the addition operation is mapped to the adder and there is an adder available. When operation 2 is considered, however, since the adder is already occupied, it cannot be scheduled in control step 1. Operation 2 will be scheduled in control step 2 instead. The algorithm will then examine operation 3. Since the multiplication operation is mapped to the multiplier and the multiplier is available, it is scheduled in control step 1. This process will continue until each operation is assigned to a control step.

> **Example 3.2:** Figure 3.3b illustrates the result of applying the ASAP scheduling algorithm to the DFG graph of Example 3.1, which is shown in Figure 3.3a. In this case, seven control steps are needed to execute the given behavioral specification, provided that one adder and one multiplier are used.
>
> □

A similar approach to the simple scheduling problem is to use the "as late as possible" (ALAP) principle. With an ALAP algorithm, all the operations are also first sorted topologically according to their data/control dependencies, as in the case of ASAP. However, the operations will be scheduled backwards by placing them in the latest possible control step. To schedule the DFG graph of Example 3.1, operation 10 is considered first. Since there is a multiplier available, operation 10 is scheduled in the last control step, which is illustrated in Figure 3.4. The next operation to be considered is operation 9, another multiplication. It cannot be scheduled in the last control step, since there is no more multiplier available. Operation 9 will therefore be scheduled in the last but one control step. When operation 8, an addition, is considered, even though there is an adder available in the last control step, it cannot be scheduled there, since there is a data dependence between operation 8 and 10. Operation 8 must be completed before operation 10 can start. Since operation 10 has been scheduled in the last control step. The latest possible control step to perform operation 8 is the last but one control step. The algorithm will then exam operation 7, which has no data dependence with any operation already scheduled. Operation 7 can therefore be scheduled in the latest possible control step where the needed hardware resource to perform it is available, which is the last control step.

> **Example 3.3:** Figure 3.4b illustrates the results of applying the ALAP algorithm to the DFG graph of Example 3.1, which is also depicted in Figure 3.4a. The ALAP algorithm generates a schedule which consists of six control steps, which is one step less than the schedule generated by the ASAP algorithm. Since the two designs need to have the same amount of functional units, the ALAP algorithm generates a better schedule in this particular case.
>
> □

Usually, we would like the generated schedule to be an optimal one, i.e., it takes the minimal number of control steps to execute the specified behavior. The general scheduling problem is however NP-complete [GaJo75], and heuristics that do not guarantee optimal results are widely used to generate satisfactory solutions. In the following sections, a few of the most widely used algorithms will be described. For a complete treatment of scheduling techniques, please refer to [DeMi93].

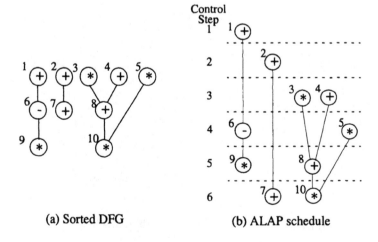

(a) Sorted DFG   (b) ALAP schedule

Figure 3.4: An ALAP scheduling example.

3.2.1   List Scheduling

The ASAP and ALAP algorithms are examples of *constructive* techniques for solving the scheduling problem. In such techniques, a schedule is constructed step by step until all operations are scheduled.

Another constructive approach is *list scheduling*, which proceeds from control step to control step. In each control step, the operations that are available to be scheduled are kept in a list, ordered by some priority function. Each operation on the list is then scheduled if the resources needed are free; otherwise it is deferred to the next control step.

An important decision for a list-scheduling approach is therefore to select the priority function, since it has a strong impact on the final results. Some systems give higher priority to operations with low mobility; here mobility of a operation is defined as the number of control steps from the earliest up to the latest feasible control step in which the operation can be scheduled. Others give higher priority to operations with more immediate successors, arguing that scheduling them in the current control step would make the largest number of operations ready, thereby allowing the earliest possible consideration of each operation. Unfortunately, there is no agreement on which priority function is best, and the selection of such a function usually depends on the application.

Let us consider the DFG of Example 3.1, which is illustrated in Figure 3.5a. Assume that the priority function is defined as the length of the path from the operation to the end of the block; an operation associated with a high number has high priority. In the first step of the list scheduling algorithm, operation 1, 2, 3, 4, and 5 are ready to be scheduled since none of them depend on any other operation to be completed. These ready operations are ordered by the priority

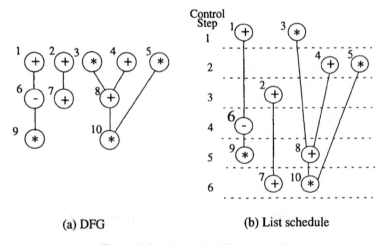

(a) DFG   (b) List schedule

Figure 3.5: A list-scheduling example.

function as follows:

Priority(o_1) = 2;
Priority(o_3) = 2;
Priority(o_4) = 2;
Priority(o_2) = 1;
Priority(o_5) = 1.

These operation will be scheduled one by one in the above order. Assume again that there are only one adder and one multiplier available. Operation 1 (o_1) and 3 will be scheduled in the first control step. The algorithm will then proceed to the second control step. Now we have operation 2, 4, 5, and 6 ready to be scheduled, since operation 2, 4, and 5 do not have any predecessors and the only predecessor of operation 6 has been scheduled in the previous step. Based on the priority function, we have the following order:

Priority(o_4) = 2;
Priority(o_2) = 1;
Priority(o_5) = 1;
Priority(o_6) = 1;

Therefore operation 4 and 5 will be scheduled in control step 2 and the algorithm proceed to control step 3. Operation 2, 6, and 8 are now ready to be scheduled and they will be ordered in the following way:

Priority(o_2) = 1;
Priority(o_6) = 1;
Priority(o_8) = 1.

Since all the three ready operations are of type addition and there is only one adder available, only one of them can be scheduled. The three operations have also the same priority value; this provides room for a secondary priority function to be used to select one of them to be scheduled. We assume however that only one priority scheme is used. Therefore, the selection is based on the topological order and operation 2 is scheduled. This process continues until all operations are scheduled. The final schedule for this example is illustrated in Figure 3.5b. This gives a schedule with six control steps, which is the optimal solution for this example.

3.2.2 Force-Directed Scheduling

The scheduling problems discussed so far are called resource-constrained scheduling (RCS). A RCS algorithm tries to find a schedule with the smallest number of control steps under some resource constraints. The resource constraints are typically given in the form of the number of functional units of each type, as in the case of the previously discussed examples. In the more general form, the resource constraints can be given in terms of a complex cost function such as the estimated area of the silicon implementation of the whole design.

There is another type of scheduling problems which are called time-constrained scheduling (TCS). A TCS algorithm tries to minimize the resources required to meet a specified global time constraint. The time constraint is typically given in terms of the number of control steps allowed for the execution of the specified behavior. The TCS problems occur very often in digital signal processing applications in which the system throughput is fixed and the silicon area must be minimized.

The force-directed scheduling algorithm [PaKn89] is one of the most widely used techniques for the TCS problem. The basic strategy is to place similar operations in different control steps so as to balance the concurrency of the operations assigned to the functional units without increasing the total execution time. By balancing the concurrency of operations, it is ensured that each functional unit has a high utilization and therefore the total number of units required is decreased.

The forced-directed scheduling algorithm consists of three main steps: determine the time frame of each operation, create a distribution graph, and calculate the force associated with each assignment.

The first step of the force-directed scheduling algorithm is to determine the time frame of each operation. A good approximation of the time frame can be determined by constructing the ASAP and ALAP schedules without any resource constraints and then combining the result of both schedules.

Example 3.4: Let us consider the example given in Figure 3.2 again. Assume that the time constraint indicates that four control steps are allowed for this example. Figure 3.6 illustrates the ASAP and ALAP schedule of the DFG given in Figure 3.2b, assuming no resource constraints. From the two schedules, it is easy to identify the time frame

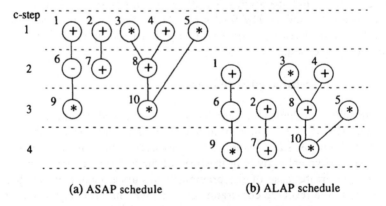

(a) ASAP schedule   (b) ALAP schedule

Figure 3.6: ASAP and ALAP schedules of Example 3.1.

for each operation. For example, operation 1 can be either scheduled in control step (c-step) 1 or 2 and therefore its time frame spans from c-step 1 to c-step 2. Operation 2, on the other hand, has a time frame from c-step 1 to c-step 3. The question that remains to be answered is in which c-step of the time frame the operation should be scheduled in order to reduce the number of functional units needed. It can be observed that if the ASAP schedule is directly used, 3 adders and two multipliers are needed, which is definitely not optimal for this example.

□

The time frames for all the operations can be collected in a graph, such as the one given in Figure 3.7 for Example 3.1. The width of the box in Figure 3.7 containing an operation in a c-step represents the probability that the operation will be eventually placed in that c-step (time slot). It is assumed that the probability distribution for each operation is uniform. The width of an operation box therefore equals 1 divided by the number of c-steps of its time frame.

The next step of the force-directed scheduling algorithm is to add the probabilities of each type of operations for each c-step and build a *distribution graph* for it. The distribution graph shows, for each c-step, how heavily loaded that step is, provided that all possible schedules are equally likely. If an operation could be done in any of the k steps in a time frame, $1/k$ is added to each of the c-steps in the graph. Using the information captured in Figure 3.7, we can calculate the value of the distribution graph, or DG, for the multiplication operations. The results are $DG_{mult}(1) = 1/2 + 1/3 = 0.833$, $DG_{mult}(2) = 1/2 + 1/3 = 0.833$, $DG_{mult}(3) = 1/2 + 1/2 + 1/3 = 1.333$, and $DG_{mult}(4) = 1/2 + 1/2 = 1$, as illustrated in Figure 3.8. The results for the addition/subtraction operation DG are $DG_{a/s}(1) = 1/2 + 1/3 + 1/2 = 1.333$, $DG_{a/s}(2) = 1/2 + 1/2 + 1/3 + 1/3 + 1/2 + 1/2 = 2.667$, $DG_{a/s}(3) = 1/2 + 1/3 + 1/3 + 1/2 = 1.667$, and $DG_{a/s}(4) = 1/3 = 0.333$..

The third step of the force-directed scheduling algorithm is to calculate the force associated with every feasible c-step assignment of each operation. For an operation with a time frame that spans from c-steps f to t, the force associated

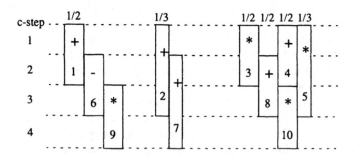

Figure 3.7: Time frames for Example 3.1.

c-step   Multiplication DG   Addition/subtraction DG

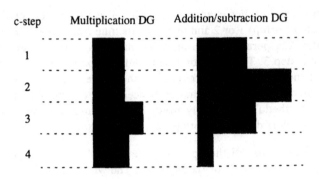

Figure 3.8: Distribution graphs for Example 3.1.

with its assignment to c-step j $(f \le j \le t)$ is

$$Force(j) = DG(j) - \sum_{i=f}^{t} \left[\frac{DG(i)}{(t-f+1)} \right]$$

In other words, the force associated with the tentative assignment of an operation to c-step j is equal to the difference between the distribution value in that c-step and the average of the distribution values for the c-steps bounded by the operation's time frame.

For example, with the assignment of operation 10 to c-step 3, we have

$Force(3) = DG_{mult}(3)$ - average DG_{mult} value over time frame of operation 10
          $= 1.333 - (1.333 + 1)/2 = 0.167.$

On the other hand, the assignment of operation 10 to c-step 4 yields

$Force(4) = DG_{mult}(4)$ - average DG_{mult} value over time frame of operation 10
          $= 1 - (1.333 + 1)/2 = -0.167.$

As the DG shows, if operation 10 is assigned to c-step 3, the distribution is not very well balanced, while the assignment of operation 10 to c-step 4 will generate a better result. This is reflected by the negative value of the force associated with the latter assignment.

We must also calculate the force for all predecessors and successors of the current operation whenever their time frames are affected. These additional forces are called indirect forces. The total force is the sum of the direct and indirect forces [PaKn89].

In Example 3.1, operation 10 has four predecessors, operations 3, 4, 5, and 8. The assignment of operation 10 to c-step 4 will not affect the time frame of any of its predecessors. Therefore the total force of assigning operation 10 to c-step 4 equals the direct force calculated above, namely -0.167. The assignment of operation 10 to c-step 3, on the other hand, will affect the time frames of operations 3, 4, 5, and 8. For example, operation 8 will now only be able to be performed in c-step 2 instead of both c-steps 2 and 3. Therefore, the assignment of operation 10 to c-step 3 implies that operation 8 will be assigned to c-step 2 and the indirect force of the latter assignment must be calculated and added to the total force of the former assignment. The indirect force of assigning operation 8 to c-step 2 equals $DG_{a/s}(2)$ - average $DG_{a/s}$ value over the time frame of operation 8 (c-steps 2 and 3), i.e., 2.667 - (2.667 + 1.667)/2, which is 0.5. The same calculation should be performed for operations 3, 4, and 5. All the indirect forces will then be added to the direct force to yield the total force of assigmning operation 10 to c-step 3.

Once all the forces are calculated, the operation-control step pair with the largest negative force (or least positive force) is scheduled. The distribution graphs and forces are then updated and the above process is repeated until all operations are scheduled.

Since the force-directed scheduling algorithm schedules one operation in each iteration, it is also constructive. Different from other constructive approaches, however, force-directed scheduling makes global analysis of the operations and control steps when selecting the next operation to be scheduled. Therefore force-directed scheduling is more expensive computationally than, for example, list scheduling. Force-directed scheduling has complexity $O(cN^2)$, while list scheduling $O(cN \log N)$, where c is the number of control steps and N the number of operations [MPC90].

3.2.3   Transformation-Based Scheduling

The scheduling techniques discussed up till now are all of constructive type. Another basic class of scheduling algorithms is based on transformations. A transformation-based algorithm begins with an initial schedule, usually either maximally serial or maximally parallel, and applies transformations to it to obtain other schedules. The basic transformations are converting serial operations, or blocks of operations, into parallel ones and the inverse, converting parallel operations into series ones. Transformation-based algorithms differ in how they choose what transformations to apply and in which order these are applied.

One extreme technique is to use exhaustive search. That is, all possible combinations of serial and parallel conversions are tried and the best design will be chosen. This method has the advantage that it looks through all possible designs and guarantee the optimal solution. However, it is computationally very expensive and not practical for large designs. Exhaustive search can be improved, to reduce the computation time needed, somewhat by using branch-and-bound techniques, which cut off the search along any path that can be recognized to be suboptimal.

Another approach to transformation-based scheduling is to use heuristics to guide the process. Transformations are chosen that promise to move the design closer to the optimal design. Both the CAMAD system [Pen86, PKL89, PeKu94] and the Yorktown Silicon Compiler [Cam90] use this approach.

One important advantage of the transformation-based approach is that in each iteration, a complete schedule exists and accurate estimation of the design in terms of different criteria can be made. It is therefore straightforward to extend this type of algorithms to handle many advanced issues related to scheduling to be discussed in the next section. Further combination of scheduling and allocation can also be achieved using the transformation-based approach, as in the case of the CAMAD system, which is discussed in Chapter 6.

3.2.4   Advanced Scheduling Topics

The scheduling problems we have discussed so far are simplified versions of the real problems. We will now discuss several advanced topics which must be considered by any practical scheduling algorithm.

Control constructs

Until now we have assumed that a CDFG corresponds to a single basic block, i.e., one section of straight-line code with only one entry and one exit point. However, most hardware description languages, such as VHDL, support conditionals, loops and other control structures. The scheduling algorithm must consider these constructs during the scheduling process.

When scheduling conditional branches, the scheduling algorithm should make use of the possibility to share functional units between mutually exclusive branches as much as possible. For example, the same adder can be used in both the "then" and "else" clauses of an "if" statement [WaCh95].

Chaining and multicycling

We have up till now assumed that all operations requires the same amount of time to execute, and this time is the control-step length or clock cycle time. In practice, different operation types may have different execution times and the above assumption results in the situation that the clock cycle time is dictated by the most time consuming operation.

> **Example 3.5:** Figure 3.9a illustrates a design where the addition and multiplication operations are mapped into an adder and a multiplier with 50ns and 100ns delay, respectively [WaCh95]. With the single operation per cycle assumption, the clock cycle time is determined to be 100ns and the overall schedule length is 200ns.
>
> □

In order to avoid the above problem, chaining and multicycling techniques

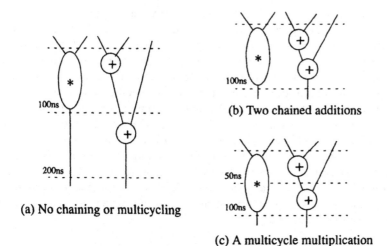

(a) No chaining or multicycling

(b) Two chained additions

(c) A multicycle multiplication

Figure 3.9: Chained and multicycle operations.

can be used.

Chaining is the task of combining more than one operation in a control step. Figure 3.9 illustrates the case when chaining is used in Example 3.5, where the two addition operations are chained and scheduled in the same control step. In this way, the overall schedule length becomes 100ns. However, an extra adder is needed in this example since the two addition operations can not longer be mapped into the same adder because they are performed in the same control step.

Chaining can be performed before scheduling, and then the combined operations can be considered as a single entity in the scheduling process. Chaining can also be performed together with scheduling.

Another alternative to improve the schedule of Example 3.5 is to set the clock period to 50ns, and to execute the multiplication over two control steps, as illustrated in Figure 3.9c. An operation which is to be executed continuously over several clock cycles is called a *multicycle* operation. In Example 3.5, multicycling decreases the schedule length to 100ns, just like chaining, but without the cost of another adder. However, multicycling uses twice as many control steps as chaining, which may result in a larger controller. Further, multicycling usually needs additional registers to latch the results from one cycle to the other. For example, the result produced by the first addition operation must be latched while this is not the case in the chaining solution.

Scheduling with timing constraints

Most scheduling techniques try to find a schedule which has the shortest schedule length while satisfying a set of constraints or one which requires the least resources and has a shorter schedule length than a given constraint. In both cases the global performance of the design, in terms of the overall schedule length, is used as one design criterion. However, few digital systems work in isolation, so there may also be a need to specify more detailed timing constraints on certain operations, or sets of operations. For example, we might want to give a minimum timing constraint, which specifies that one operation must be executed less than a specified amount of time after another operation, or a maximum timing constraint, which specifies that one operation must be executed at least a specified amount of time after another operation.

Most scheduling techniques handle these timing constraints by adding additional constraint edges to the CDFG, and then treating those additional edges in much the same way as other constraints. Special techniques based on heuristics which deal with local timing constraints in a systematic manner have also been developed [HaPe96].

Discussion

In the general sense, scheduling is to assign operations to control steps so as to minimize a given objective function while meeting a set of design constraints. The object function may include the number of control steps, delay, power, hardware resource, and testability. The general scheduling problem can be formulated as an Integer Linear Programming (ILP) problem. Using the ILP formulation, for example, the time-constrained scheduling problem with a minimal total silicon area can be easily formulated. We need only to let the cost function be the estimated total area of the final implementation. The concept of ILP will be further elaborated later in this chapter when data path allocation and binding are discussed as well as in Chapter 5.

3.3   Data Path Allocation and Binding

In general, data path allocation and binding deal with the problem of which resources are used to realize in the physical implementation. Such resources include registers, memory units and different functional units as well as their communication channels. The basic principle is to share resources as much as possible provided that the performance and other design criteria can be satisfied.

Allocation and binding carry out selection and assignment of hardware resources for a given design. Allocation determines the type and number of hardware resources for a given design. Binding assigns the instance of an allocated hardware resource to a given data path node. Different data path operations can share the same hardware resource if they are not executed at the same time. For example, an adder can be shared by two additions if they are not executed during the same clock cycle. A register can also be used to store the values of two variables if the life times of these variables do not overlap.

As pointed out earlier, the term allocation is sometimes used to denote both the allocation and binding tasks.

Example 3.6:   An allocation for the example depicted in Figure 3.2 selects two adders and one multiplier. A possible schedule for this allocation is depicted in Figure 3.10. Note that the scheduling assumes, in this case, that at most two adders and one multiplier in every clock cycle can be used, as determined by the allocation. The binding step assigns every operation of the scheduled graph to a physical hardware component. A possible binding for this example is also depicted in Figure 3.10. Addition/substraction nodes 1, 2, 6 and 7 are assigned to adder #1, addition nodes 4 and 8 are assigned to adder #2, and multiplication nodes 3, 5, 9 and 10 are assigned to the multiplier.

□

The selection of the type and number of hardware resources during the allocation step is usually formulated as an optimization problem. The main goal is to find the minimum number of resources while fulfilling given area/performance constraints.

The basic assumption made by many high-level synthesis systems regarding binding is that each data path node has at least one module in a module library which implements the function of the data path node. For example, a node which performs an addition operation may correspond to an adder or an ALU in the module library. The different modules can have different areas and/or latencies. This gives the possibility to make trade-offs between different implementations.

The binding problem is also an optimization problem and can be formulated using existing optimization methods. For example, an Integer Linear Programming method or a graph clustering technique can be used to solve it. It can also be solved using heuristic methods.

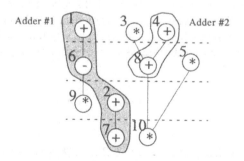

Figure 3.10: Binding for Example 3.6 with 2 adders and 1 multiplier.

3.3.1   Integer Linear Programming

Integer linear programming (ILP) is a subclass of the linear programming problems where the decision variables are of integer values. If we assume that decision variables are represented by a vector X the ILP problem can be formally stated as:

$$\text{Maximize (or Minimize)} \quad C^T X \tag{3.1}$$
$$\text{Subject to} \quad AX = B \tag{3.2}$$
$$X \geq 0 \tag{3.3}$$
$$X \quad \text{integer} \tag{3.4}$$

In this formulation, the maximization/minimization criterion is defined as (3.1) and the constraints for the optimization problem are defined as equations (3.2) and inequalities (3.3). In addition all decision variables are constrained to be integers (3.4). If binary decision variables are used instead of integer ones, the problem is called 0/1 linear programming problem. Once our problem is defined as an ILP problem known methods for solving it can be directly applied.

As an example, we will define the problem of finding a binding, for a given schedule, as a 0/1 linear programming problem. We will concentrate on the formulation of binding constraints. The minimization criterion will not be included in this discussion. The reader can imagine different types of cost functions which can be used to minimize different aspects of the design, for example, interconnection cost or power consumption. For the purpose of this presentation we also make the simplifying assumption that every operation is executed in exactly one clock cycle, i.e., no chaining or multicycle operations are possible.

We define the following decision variables and constants:

- $binding_{ij}$ where $i = 1, 2,..., ops$ and denotes operation number, and $j = 1, 2,..., r$ and denotes component number; the decision variable $binding_{ij}$ is 1 iff operation i is bound to hardware resource j.
- $schedule_{ik}$ where $i = 1, 2,..., ops$ and denotes operation number, and $k = 1, 2,..., max\_step$ and denotes the number of the steps in a given schedule. Since we assume that the schedule is already decided, the constant $schedule_{ik}$ is 1 iff operation number i is scheduled in step number k.

The binding problem can be defined as a solution satisfying the following constraints:

$$\sum_{j=1}^{r} binding_{ij} = 1, \qquad\qquad i = 1, 2,..., ops; \tag{3.5}$$

$$\sum_{i=1}^{ops} binding_{ij} \cdot schedule_{ik} \leq 1, \qquad j = 1, 2,..., r; \ k = 1,..., max\_step; \tag{3.6}$$

$$binding_{ij} \in \{0, 1\}, \quad i = 1, 2,..., ops; \; j = 1, 2,..., r. \tag{3.7}$$

The constraint (3.5) denotes that an operation can only be assigned to one resource while the constraint (3.6) requires that at most one operation can be executed on a hardware recourse during an execution step.

The above stated constraints can usually be satisfied by several solutions. Since we are looking for a solution which minimizes a given design criterion, a cost function should be defined to guide the selection of the best solution.

Example 3.7: Let us consider the problem of finding a binding for the example given in Figure 3.2 with the schedule presented in Figure 3.10. For this example, the binding of the two adders has to fulfill the following constraints:

$$binding_{i1} + binding_{i2} = 1, \text{ for } i = 1, 2, 4, 6, 7, 8;$$

$$\sum_{i \in \{1, 2, 4, 6, 7, 8\}} binding_{i1} \cdot schedule_{ik} \leq 1, \text{for } k = 1, 2, 3, 4;$$

$$\sum_{i \in \{1, 2, 4, 6, 7, 8\}} binding_{i2} \cdot schedule_{ik} \leq 1, \text{ for } k = 1, 2, 3, 4.$$

There are 16 different solutions which satisfy these constraints. One possible solution is:

$binding_{11} = 1$	$binding_{12} = 0$
$binding_{21} = 1$	$binding_{22} = 0$
$binding_{41} = 0$	$binding_{42} = 1$
$binding_{61} = 1$	$binding_{62} = 0$
$binding_{71} = 1$	$binding_{72} = 0$
$binding_{81} = 0$	$binding_{82} = 1$

The above binding solution is depicted in Figure 3.10.

□

The ILP-based definition of the binding problem can be extended in several ways to include more complex and realistic requirements. It can also be defined to include both scheduling and binding.

3.3.2 Clique Partitioning and Graph Colouring

Allocation and binding can also be defined as graph problems. They can be formulated either as the problem of finding cliques in a compatibility graph or that of coloring vertices in a conflict graph.

A *compatibility graph* is used to represent information on resource sharing.

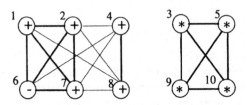

Figure 3.11: A resource-compatibility graph example.

Two operations are compatible, and can share the same hardware resource, if they are of the same type and are not executed at the same time. A compatibility graph $G_{comp}(V, E)$ is built of vertices V denoting operations and edges E denoting the compatibility relation between the operations. A vertex v_i is connected to a vertex v_j if the two operations they represent can be assigned to the same resource. An example of a compatibility graph is depicted in Figure 3.11.

To solve the binding problem using compatibility graphs, we have to find a maximal set of compatible operations. This can be formulated as a maximal *clique partitioning* problem. A clique is defined as a subgraph where all nodes are connected to each other. It is maximal if it is not contained in any other clique.

> **Example 3.8:**   Let us consider the example depicted in Figure 3.10. Based on the given schedule and the assumption that addition and substraction operations can share the same resource, a compatibility graph can be built as shown in Figure 3.11. Three maximal cliques can be identified as depicted by bold edges in Figure 3.11. Two cliques {1, 2, 6, 7} and {4, 8} represent the binding of the addition/substraction operations to two adders while the clique {3, 5, 9, 10} represents the binding of the multiplication operations to one multiplier. Note that the solution is not unique and other solutions are also possible. For example, we can also identify other two cliques, {2, 4, 7, 8} and {1, 6}, for the addition/ substraction operations.
>
> □

A conflict graph, on the other hand, captures the opposite information as the compatible groups. It denotes explicitly the operations that cannot share the same resource. A conflict graph $G_{conflict}(V, E)$ is built of vertices V denoting operations and edges E denoting the conflict relation between them. A vertex v_i is connected to a vertex v_j if the two operations they represent cannot be assigned to the same resource. An example of a conflict graph is depicted in Figure 3.12.

The binding problem in this case is solved using a *graph coloring* algorithm. The algorithm assigns different colors to vertices connected by edges while minimizing the number of colors.

Example 3.9: Let us consider the previous example. The resource-conflict graph, depicted in Figure 3.12, has only two edges representing two resource conflicts for addition/substraction operations. Addition operations 1 and 4 are executed in parallel during the first control step and addition/substraction operations 6 and 8 are executed in parallel in control step 2. In this case, the subgraph for addition/substraction needs be colored using two colors while the multiplication subgraph requires only one color. A coloring scheme is captured in Figure 3.12, which again is not unique.

□

Both the maximal clique-partitioning problem and the minimal graph-coloring problem are intractable for realistic-size examples. Thus heuristics, which generate solutions quickly without guaranteeing optimality, have widely been used. We can also make use of the fact that these two problems form each other's dual. The two graphs are complementary to each other and the complexity of one formulation could be much less than the other. This makes it possible to select the formulation with a less complex graph for a given binding problem in order to speed up the binding process.

3.3.3   Left-Edge Algorithm

To show other approaches to allocation and binding we discuss the left-edge algorithm which is used very often for register allocation and binding.

The left-edge algorithm was originally introduced to perform the channel routing task. It was used for assigning interconnections (trunks) into a number of tracks in a channel. In the original formulation of the algorithm, the channel was oriented horizontally and the algorithm sorted trunks in an increasing order of their left end-points. The algorithm assigned trunks into successive tracks starting from the first one. For every track it scanned the ordered list of unplaced trunks and placed them one by one into successive available parts of the track. We will use this algorithm for the assignment of variables into registers.

The left-edge algorithm requires information about the life-time of variables. The life-time of a variable is the time interval when the variable is used by the computation. It can be represented as a bar drawn in parallel to our scheduled

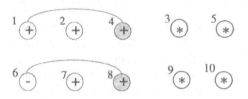

Figure 3.12: A conflict-graph example.

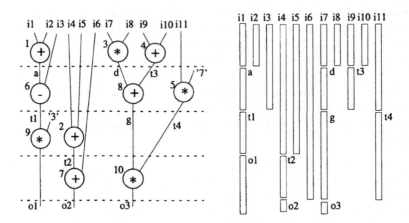

Figure 3.13: Variable life-times

design. The starting point of the bar represents the time when the variable is set and the end of the bar represents the time when the variable is released and its value is not used any longer. In other words, the bar represents define-use time for the variable. On the right hand side of Figure 3.13 there are bars representing the life-times of the 21 variables used in the example given on the left hand side.

A possible solution to the register-assignment problem can use the original formulation of the left-edge algorithm. In this solution all bars representing the life-times are sorted in increasing order of their starting points. The algorithm proceeds then with one register at a time. It assigns the first variable from the list, represented by a bar, into the first register. It then scans the sorted bars and the first encountered unbound bar which has start time higher than the end-time of the already assigned bar is placed into the same register. It continues this process by assigning the next variables into the same register. If the life-time of this register reaches the last scheduled step the algorithm starts to assign variables into the next register using the same method.

Example 3.10: Consider the assignment of the 21 variables used in the example depicted in Figure 3.13. The first step sorts all bars and the resulting sorted list of variables is depicted in Figure 3.14a. The final assignment of variables into registers is depicted in Figure 3.14b.

□

The left-edge algorithm will allocate the minimum number of registers, but has two disadvantages. First, not all life-time tables might be interpreted as intersecting intervals on a line. For example, the existence of conditional branches prohibits the interpretation of intersecting intervals on a line, since values occuring in mutural exclusiove branches may share a register although they seem to overlap in life-time [MLD92]. Second, the allocation produced is

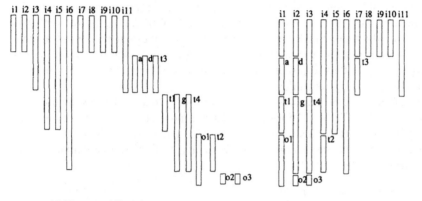

(a) The sorted list of variables   (b) Assignment of variables into registers

Figure 3.14: Applying the left-edge algorithm for register allocation

neither unique nor necessarily optimal in terms of, for example, the number of multiplexors required.

Example 3.11: Consider the sequencing graph depicted in Figure 3.15a. Figure 3.15b depicts its variable life-times while in Figure 3.15c the conflict graph for all variables is shown. This graph can be colored with three colors which gives an assignment of three registers.

□

3.4 Controller Synthesis

As stated earlier, system synthesis deals with systems usually specified as a set of communicating concurrent processes. In this case the controller synthesis is

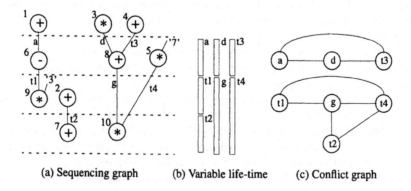

(a) Sequencing graph   (b) Variable life-time   (c) Conflict graph

Figure 3.15: Variable life-times and conflict graph.

highly dependent not only on the functionality of the controllers but also on the interaction and synchronization requirements. Traditional high-level synthesis methods, which are limited to synthesis of one concurrent process at a time, produce poor results or cannot deal with such designs. One important system synthesis task is thus to propose an efficient control engine which may include several cooperating controllers.

An efficient method to implement a controller is to use a finite state machine (FSM) notation. The FSM formalism makes it possible to specify a number of states together with transitions between them. An FSM can be defined using either Mealy or Moore machine style. Formally, an FSM can be defined by the following 5-tuple:

$<S, I, O, \delta, \lambda>$, where

S is a set of states;

I is a set of inputs (conditions);

O is a set of outputs (control signals);

δ is a next-state function, $\delta: S \times I \rightarrow S$; and

λ is an output function, $\lambda: S \times I \rightarrow O$ for Mealy machine or $\lambda: S \rightarrow O$ for Moore machine.

In this book, we assume that a controller is finally implemented as one or several FSMs.

The above formulation defines an FSM as a machine which has a number of states. The machine can go from one state to another by executing the next-state function. The next state is determined based on the current state and the input signals. The output signals are determined by the output function. For Mealy machines the output signals are generated based on the current sate and the input signals while for Moore machines they depend only on the current state.

A controller is usually graphically represented by a state diagram. The state diagram is a directed graph with nodes denoting states and arcs denoting transitions from one state to another. The control signals can be assigned either to states or arcs depending on the machine style while guarding conditions are assigned to arcs. A transition from a given state to the next state takes place if and only if the controller is in the given state and the guard assigned to the transition arc connecting both states is true.

In our formulation the controller, consisting of one or several FSMs, is used to control data path activities. The output signals are used to control data path operations while the input signals are conditions generated from the data path to influence the execution of the controller. Figure 3.16 represents schematically the general view of the controller and its relation to the data path.

After generating the FSM specification it is a logic synthesis task to perform further optimization steps, such as state minimization and state encoding, using

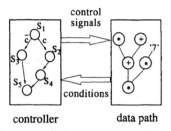

control
signals

conditions

controller   data path

Figure 3.16: A controller/data path architecture.

standard FSM synthesis methods [DeMi94a]. However, some decisions regarding the selection of the control structure have to be made during system synthesis. At this stage it should be decided if a single controller will be used to control the whole design or rather several cooperating controllers will be used. It can also be decided if a hierarchical controller can be used. The selection of controller style depends on the synthesis problem formulation and design criteria, such as performance, area and testability. These decisions deal mainly with the overall architecture of the controller rather than its detailed implementation.

3.4.1   Controller-Style Selection

In this section, we discuss several design styles which can be used to implement complex controllers. The main idea of the presented approaches is to implement a complex control structure by several smaller controllers in order to simplify the design of the generated FSMs.

Single Controller

In this style, the controller is modeled by a single FSM, which is the simplest solution to the controller-style selection problem. Since an FSM is a sequential machine parallelism is only allowed for operations assigned to the same state. Otherwise operations have to be statically scheduled and assigned to consecutive states. The method is very efficient for simple computations but in the case of several parallel execution threads it can lead to the state explosion problem, especially when parallel loops are involved.

Hierarchical Controller

A *hierarchical controller* consists of a number of simple controllers organized in a hierarchy. It is assumed that the hierarchy of controllers is ordered by their parent-child relation. The parent controller distributes an execution task among

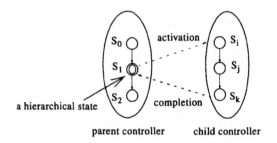

Figure 3.17: An example of a hierarchical controller.

a number of child controllers. Every child controller starts its execution upon receiving the activation signal and it sends a completion signal upon execution termination. The parent controller continues its execution when the child controller sends a completion signal. The child controller can be a parent controller for another lower-level child controller.

Hierarchical controllers usually make use of two special internal control signals, called *activation* and *completion*. The activation signal is used to start another controller and the completion signal provides information about termination of the execution of a controller. They are only used for controller synchronization.

> **Example 3.12:**   Figure 3.17 presents an example of a hierarchical controller. In this example, state S_1 is hierarchical and is implemented by a child controller consisting of three states S_i, S_j and S_k. It is activated by the parent controller, with an activation signal. The parant controller will continue its execution upon receiving the completion signal from the child controller.
>
> □

Hierarchical controllers can be used to implement in a natural way several control structures which are commonly used in algorithmic languages, such as procedure and loop constructs. Procedures are well suited for this because their purpose is to create a hierarchy of subprograms. A process calls a procedure and continues upon the procedure return. It matches the hierarchical controller principles very well. A large program can also be restructured by partitioning it into loop-free parts by structuring loops as subprograms. These structures can later use the hierarchical controller concept to implement efficient control engines [DeMi94a].

Parallel Controller

The function of a controller can also be distributed into several smaller controllers which are executed in parallel. These parallel controllers can

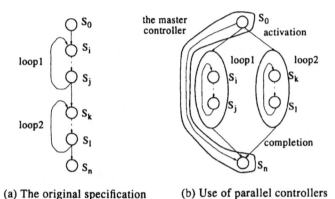

(a) The original specification   (b) Use of parallel controllers

Figure 3.18: Using parallel controllers for two loops.

communicate to exchange data or synchronize their execution. Generally speaking, decomposition of a controller into a number of parallel controllers reduces the overall controller complexity and solves some design problems. However, It requires new design methods.

In some cases, several parallel controllers do not need to synchronize. They are executed independently of each other. The activation/completion signals can then be used for correct sequencing of operations between a master controller and these parallel controllers, which is illustrated in the following example.

Example 3.13: A design specification includes two consecutive loops as depicted in Figure 3.18a. The computations of the loops are data independent and can be executed in parallel to speed up the computation. The generation of one sequential controller for the parallel computations is impossible in practice since the generated controller should include all combination of states in the two loops which results in a combinational explosion of states. It is possible, however, to generate one master controller which activates two independent parallel controllers, one for each loop, as illustrated in Figure 3.18b. This solves the performance problem while keeping the controller size small.

□

Conditions are used to specify synchronization between parallel controllers, when needed. The controller which is to be synchronized has a conditional state transition. This transition from one state to another will take place only in the case when the condition assigned to the edge connecting these two states is TRUE. The condition can be set by another controller. Using conditions different communication protocols between two or more parallel controllers can be implemented. Using this strategy we can specify hierarchical controllers

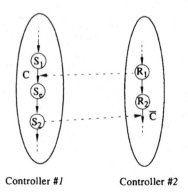

Controller #*1*   Controller #*2*

Figure 3.19: Two parallel controllers communicating using condition C.

with very complex synchronization schemes.

Communication between parallel controllers can be achieved using a special protocol implemented using the basic technique described above together with data exchange through shared data path objects.

Example 3.14: Let us consider two parallel controllers which communicate as depicted in Figure 3.19. They use a handshake protocol to synchronize their execution and exchange data. Controller #1 waits in state S_1 for condition C to become TRUE. The condition is set by controller #2 in state R_1. In state S_e controller #*1* can fetch data set by controller #2. Then controller #*1* resets condition C and controller #2 continues its execution.

□

Many current design methods offer efficient algorithms for the synthesis of synchronous FSMs. In many cases, however, this limits possible design space exploration and results in over-synchronized controllers. It should be noted that a proper combination of asynchronous and synchronous design styles can solve some of these problems. For example, basic controllers can be implemented as synchronous machines controlled by one clock or several independent clocks while their communication can follow an asynchronous protocol. This solution gives the freedom of using asynchronous design rules for controller synchronization and communication while using well-known synchronous design methods for single controllers.

It should be noted that this solution with ctivation and completion signals is the only possible one for the class of designs which have operations with unbounded-delays. We cannot schedule these operations to any particular clock cycle or a number of clocks cycles since the execution time can not be determined beforehand (see, for example, [DeMi94a]).

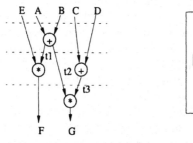

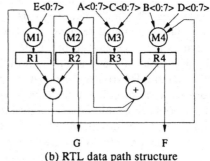

(a) Scheduled data-flow description   (b) RTL data path structure

S0: M3=0, Load R3, M4=0, Load R4 **next** S1;

S1: Add, M2=1, Load R2, M1=1, Load R1, M3=1, Load R3, M4=1, Load R4 **next** S2;

S2: Add, M1=0, Load R1, Mul, M4=2, Load R4 **next** S3;

S3: Mul, M2=0, Load R2 **next**...

(c) Control description (FSM)

Figure 3.20: A design example with the generated FSM.

3.4.2   Controller Generation

When the controller style has been selected and the number of FSMs is decided, we need to generate these FSMs. The FSMs can be represented, for example, as state diagrams or symbolic FSMs and later implemented using standard FSM synthesis methods.

The controller generation task has to decide whether the Moore or the Mealy machine should be used. This decision is mainly dependent on the design representation used and the available back-end logic synthesis tools. However, every Mealy machine can be converted into an equivalent Moore machine and vice versa. In this book, we will assume that the Moore machine is used.

Example 3.15: Consider the scheduled data-flow graph given in Figure 3.20a which assumes the use of one adder and one multiplier. The schedule for this graph consists of three control steps. A data path implementation after the binding of functional units, registers and multiplexors, is depicted in Figure 3.20b. Variables E and t3 are assigned to register R1, variables G, t1 and t2 to register R2, variables A and C to register R3, and variables B, D and F to register R4. Modules M1, M2, M3 and M4 represent multiplexors. The controller is given in Figure

3.20c in the form of a symbolic FSM instead of a state diagram. It has four states. The first state, S0, loads registers R3 and R4 with the values of variables A and B respectively. Control signals for multiplexors and registers are generated in this state. The signals M3=0 and M4=0 open the first input of the respective multiplexors while control signals Load R3 and Load R4 trigger the register loading. The next states perform the computations specified in the data-flow graph.

The VHDL description depicted on Figure 3.21 gives a possible synthe-sizable code for this FSM. The FSM has two input signals, reset and clock, and a number of control output signals. In this case, there are no input conditions which decide about the next-state selection. The description contains three processes: The process state_decode_logic implements the next-state function and the process output_decode_logic implements the output function. If the design would have had input condition signals the process state_decode_logic would have implemented the next-state selection as additional conditional statements, such as if-statement, in stead of the main case-statement. The process state_register imple-ments the change of the state every clock cycle as well as the resetting of the state register to state S0 on the reception of the reset signal.

□

3.4.3   Controller Implementation

Depending on the selected style of the controller a different controller structure has to be implemented. However, the basic structure is based on the implemen-tation of a single FSM. This basic implementation, together with additional hardware for complex controller synchronization and communication, is used for other types of controllers.

The single FSM controller can be implemented using random logic, microcode or PLAs. The general implementation structure is very similar, not matter which technique is used. A state register is used to store the current state while combinational logic is used to generate the next state based on the current state and the conditions coming from the data path. The current state is also used to generate control signals for the data path since we assume the use of Moore machines. In some cases, additional decoding and coding can be used for the control signals and the conditions respectively. Figure 3.22 illustrates the general implementation structure of the basic controller.

The basic controller can then be used as a building block to create complex controllers described previously. Both control signals and conditions generated by the basic controller are used for synchronization purpose. In addition, data path elements can be used to create complex communication facilities required by the given controller.

```
entity FSM is
  port (reset, clock : in BIT;
        M1, M2, M3, Add, Mul, Load_R1, Load_R2, Load_R3,
        Load_R4 : out Bit;
        M4 : out Bit_vector (0 to 1));
end FSM;

architecture controller of FSM is
  type state is (S0, S1, S2, S3);
  signal present_state, next_state : state;
begin -- controller
state_register: process (reset, clock)
        begin
          if (reset = '0') then
            present_state <= S0;
          elsif (clock = '1' and clock'EVENT) then
            present_state <= next_state;
          end if;
        end process state_register;

output_decode_logic : process (present_state)
        begin
          M1 <= '0'; M2 <= '0'; M3 <= '0'; M4 <= '00';
          Load_R1 <=  '0'; Load_R2 <=  '0'; Load_R3 <=  '0';
          Load_R4 <=  '0'; Add <= '0'; Mul <= '0';
          case present_state is
            when S0 => M3 <= '0'; Load_R3 <= '1'; M4 <= '00';
              Load_R4 <='1';
            when S1 => Add <= '1'; M2 <= '1'; Load_R2 <= '1';
              M1 <= '1'; Load_R1 <= '1'; M3 <= '1';
              Load_R3 <= '1'; M4 <= '01'; Load_R4 <= '1';
            when S2 => Add <= '1'; M1 <= '0'; Load_R1 <= '1';
              Mul <= '1'; M4 <= '10';
            when S3 => Mul <= '1'; M2 <= '0'; Load_R2 <= '1';
              :
          end case;       -- present_state
        end process output_decode_logic;

state_decode_logic : process (present_state)
        begin
          case present_state is
            when S0 => next_state <= S1;
            when S1 => next_state <= S2;
            when S2 => next_state <= S3;
            when S3 => next_state <= ...;
              :
          end case;       -- present_state
        end process state_decode_logic;
end controller;
```

Figure 3.21: VHDL specification of the FSM given in Figure 3.20c.

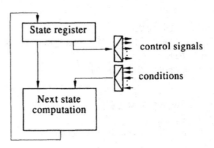

Figure 3.22: The general structure for the controller implementation.

Example 3.16: Figure 3.23 depicts a possible implementation of hierarchical controllers introduced earlier. The parent controller uses one of the control signals as an activation signal for a child controller. The child controller gets this signal as a condition signalling the start of the controller. After finishing its execution the child controller sends another control signal which is interpreted as a completion signal by the parent controller. Both the parent and the child controller have to use the same clock in this case.

□

Parallel controllers can synchronize using a method similar to the one sketched for hierarchical controllers. In more complex situations they require additional hardware for synchronization and data exchange. In general a protocol must be implemented between parallel controllers to provide correct means for communication. The most frequently used protocol is called the handshaking protocol. It uses request-acknowledge signals to establish communication. The controller which starts communication sends a request signal to the other controller and waits for the acknowledge signal. When the other controller receives the request signal it executes the needed operations and after that sends back the acknowledge signal. Both controllers continue their normal execution after the communication.

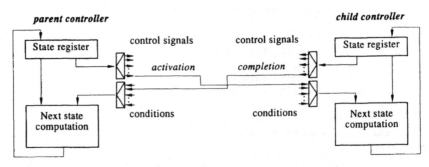

Figure 3.23: A hierarchical controller implementation.

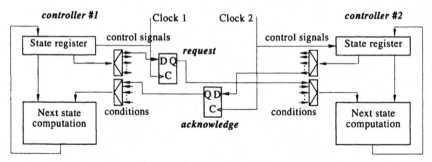

Figure 3.24: An implementation of a handshaking protocol between two parallel controllers.

Example 3.17: Figure 3.24 presents a possible implementation of the request-acknowledge protocol between two parallel controllers using two D flip-flops. Controller *#1* sends a request signal by setting a D flip-flop to 1. It stays in this state until controller *#2* sends an acknowledge signal by setting another D flip-flop to 1. This state is recognized by the first controller which continues its execution. It can be noted that this solution makes it possible to use independent clocks for controllers. For example, controller *#1* uses Clock 1 while controller *#2* uses Clock 2, as illustrated in the figure. This means that although both controllers are synchronous themselves the communication between them are performed asynchronously.

□

4

System-Level Synthesis

According to the design methodology outlined in the previous chapters, system-level synthesis is the design step dealing with a system represented at the highest abstraction level. At this level the specification is formulated as a set of interacting processes, and the basic system components considered during this design step are processors, ASICs, memories, buses, etc. Thus, system-level synthesis operates at the highest level of system design where fundamental decisions are taken which have great influence on the structure, cost and performance of the designed system.

We consider that the *input* for system-level synthesis is an executable specification of interacting processes and a set of design constraints. The constraints can be defined together with the functional specification or be given separately. They refer to characteristics of the system such as cost, speed, input/output rate, power consumption and testability. The fact that the input specification is executable is of great practical importance, as it allows system verification by simulation and debugging in an early phase, thus avoiding additional costs produced by several redesign steps.

The selection of a suitable language for system specification is a very important aspect of the design methodology and is currently the subject of intense debate and research. Very often several languages are used to specify different parts of the same system. In this chapter, we do not discuss the aspect of system specification. The discussion here is independent of any particular specification language. However, in the later chapters, the discussion will focus on several aspects specific to the synthesis of systems described as interacting VHDL processes.

The *output* produced by system-level synthesis is a set of behavioral specifications, each corresponding to a module which is assigned to a system component. Each of these modules can be directly synthesized to hardware using high-level synthesis tools, or, if it has been assigned to a programmable processor component, it can be compiled to software (see Figure 4.1). Thus, the issues discussed in this chapter also include the topics often referred to in the current literature as hardware/software co-synthesis. We will further elaborate on this aspect in Section 4.6.

System-level synthesis can decide that all processes should be implemented in hardware on an ASIC. The processes can be synthesized as a single controller with a data path. But it can also be decided that several controllers will implement the processes on the ASIC, communicating through shared data path elements or through a bus. Another possible choice is to assign some of the processes to hardware implementation, and the other processes to software implementation on a processor which can be an off-the-shelf microprocessor, a micro-controller, a DSP, or an ASIP. A more complex architecture can be selected for implementation, consisting of several processors, ASICs, memories, buses, etc.

System-level synthesis has to decide both on the kind and number of system components as well as on the distribution of the specified functionality over these components. The decision process is guided by certain optimization strategies which also have to take into consideration the restrictions imposed by the design constraints. Distribution of system functionality over different components generates the need for communication between those components. Both hardware and software required for communication have to be generated during system-level synthesis and be placed in the context of the overall system.

There is no general agreement yet on the specific tasks which define the system-level synthesis step, nor on the methods and algorithms which have to be employed. No matured commercial tool exists which addresses design automation at this level. However, some algorithms and methodologies have already been published and experimental tools designed in the research community have been described. These algorithms and methodologies address some particular issues of system-level synthesis. Surveys concerning some aspects of the design at system level can be found in [DeMi94a, DeMi94b, DeMi96, CaWi96, BuRo95, BIJe95, Wolf94, GaVa95, LSH96, AdTh96, KAJW96]. In [GVNG94] several issues concerning system design are extensively discussed and a systematic view of the topic is given.

In the following sections we will give a coherent view of system-level synthesis and elaborate on some of the most important aspects concerning this design step.

4.1   The System-Level Synthesis Tasks

In the previous section system-level synthesis has been defined as the design step which, starting from the initial specification captured as a set of interacting processes, generates the behavioral modules of the system and their assignment to system components. The following are the three main tasks which have to be performed during this step (see Figure 4.1):

1. *Allocation of system components*: this task defines a system level view of the architecture by selecting a set of processing, storage, and communication components on which one can implement system functionality. The components can be, for example, microprocessors, micro-controllers, DSPs, ASIPs, ASICs, FPGAs, memories, or buses.

2. *Partitioning*: system partitioning distributes the functionality captured by the specification among the allocated system components. Thus, the behavior captured by the processes is partitioned among microprocessors, micro-controllers, DSPs, ASIPs, or ASICs while variables are partitioned and mapped to registers and memories.

3. *Communication synthesis*: communication synthesis produces both the hardware and software needed for the interconnection of the system

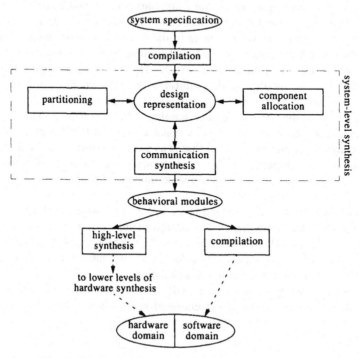

Figure 4.1: Design flow with system-level synthesis.

modules and their communication. As part of this task, communication channels have to be assigned to shared registers, buses, or ports.

The three tasks identified above are interrelated and in order to get an optimal implementation they should be, ideally, performed together and not in a certain predefined order. However, such an ideal approach leads to a problem of much greater complexity than the complexity of each task taken separately, one at a time.

First allocating the components and then partitioning functionality to the fixed architecture, reduces the complexity of the problem but limits the space of possible solutions to those which can be implemented on the selected components. Alternatively, different possible allocations could be considered during partitioning. Similarly, decisions taken at communication synthesis have a great influence on system performances, an influence which should not be ignored during the search for an optimal partition. The resulting partitioning, on the other side, which defines the placement of modules on system components, is the starting point for the generation of the communication interfaces.

In order to reduce the complexity of the design process, very often the design tasks are solved one at a time and, after estimation of the resulting performance and cost, a new design iteration is started with, for example, a modified component allocation. Very fast partitioning and estimation tools as well as high speed simulators are needed to run several such design cycles.

Another strategy is to start from an initial design and to optimize it step-wise by performing both allocation and partitioning transformations while taking into consideration communication issues. This is a typical transformational approach and the quality of the solutions depends on the efficiency of the heuristics used for exploration of a very large design space.

Regardless of the specific strategy employed for performing the system-level synthesis task, the overall goal is to reach a point in the design space which corresponds to an implementation that is as close as possible to the optimal one and fulfills the imposed design constraints. Constraints and optimization criteria can refer to cost, speed, chip area, number of pins, power consumption, testability, memory size, etc.

System-level synthesis can lead to an architecture which includes both hardware and software components. The design of such a mixed hardware/software system entails several specific difficulties. They are generated by the heterogeneous nature of the system which affects not only the synthesis phase but also system specification, verification, and simulation.

In the following sections, we will discuss some of the main aspects concerning the three tasks of system-level synthesis that have been identified above.

4.2   Allocation of System Components

Allocation has to decide on the kind and number of components used for implementation of the system. Basically three classes of components can be allocated [GVNG94]:

1. processing elements: microprocessors, micro-controllers, DSPs, ASIPs, ASICs, FPGAs;
2. storing elements: memories, register files, registers;
3. communication elements: buses.

Because of its extreme complexity, this task is performed, according to the current state of the art, manually, based on the skills of experienced designers. The number of possible alternatives is very large and they have to be carefully analyzed taking into consideration design constraints, available budget, accessible technologies, and the expected production volume. The designer can, for instance, first consider an implementation based on off-the-shelf processors or processor cores. Such a software-based solution is very cost effective, as the custom hardware, if needed at all, is practically reduced to some glue logic. In order to improve speed several processors can be connected to form a MIMD or SIMD machine. If performance constraints are not fulfilled by an all-software solution, part or all of the functionality can be implemented as hardware on ASIC. Another choice, in order to meet design requirements, can be an implementation based on ASIPs. A specific instruction-set architecture can often provide much better performance than off-the-shelf processors, and this can be a cost effective choice especially if production in large volumes is expected [CaWi96].

Several alternatives are also available for the interconnection of components with simple port-to-port connections or buses, and various communication protocols.

Automation of all these decisions is very difficult to imagine, with the exception of some particular situations. However, powerful tools can support the designer in taking the right decision in the shortest possible time. For example, fast and accurate estimators allow fast evaluation of certain design alternatives. Using retargetable compilers an application can be tested on different hardware structures and its performance can be evaluated. The hardware itself can be simulated using adequate platforms.

If the choice of possible architectural structures is limited, allocation can be performed, in principle, automatically, using an optimization strategy which leads to a certain architecture and a partitioning of functionality over the components. If, for example, we limit our choice to a generic architecture consisting of microprocessors of certain kinds and ASICs, a heuristic can be developed in order to optimize the number and kind of processors, ASIC chips, and buses for a given application. We will discuss such approaches in Section 4.7.

4.3   System Partitioning

The purpose of partitioning is to assign certain objects to clusters, so that a given objective function is optimized and design constraints are fulfilled. Partitioning is performed not only for cost/performance optimization but it also addresses the increasing complexity of system synthesis by dividing the design into smaller components which can be managed by available design tools.

Definition 4.1   Given a set of n objects $V=\{v_1, v_2, ..., v_n\}$, a k-way partitioning $P^k=\{C_1, C_2, ..., C_k\}$ consists of k clusters, $C_1, C_2, ..., C_k$ so that $C_1 \cup C_2 \cup ... \cup C_k=V$, and $C_i \cap C_j=\varnothing$ for all i,j, $i \neq j$.

Definition 4.2   The partitioning problem: find a partitioning P^k of a set V of n objects, so that the cost determined by an objective function $FOb(P^k)$ is minimal and a set of constraints $Cnstr(P^k)$, is satisfied.

Partitioning is performed at various abstraction levels during the design process. *Structural partitioning* is typical for the lower levels of abstraction. At this level a structural design is partitioned in order to determine how some hardware objects have to be grouped to satisfy, for instance, certain packaging constraints. At the layout level such a partitioning approach is used with the goal to optimize area and/or propagation delay.

However, essential decisions with a high impact on the final quality of the implemented system are taken before the functional specification has been synthesized to structure. At the system level we are interested in partitioning functionality in order to divide the behavior of the system between multiple components [VLH96]. There are no hardware objects which have to be clustered at this level, but behaviors, variables, and channels [GVNG94]. Behaviors have to be clustered and assigned to processing elements (microprocessors, ASIPs, ASICs, etc.) so that, for example, constraints on silicon area, memory size, or execution time are satisfied, and the number of interconnections between the clusters is minimized. Variables are clustered and assigned to memories considering various objectives like, for instance, minimization of the probability of simultaneous access from parallel behaviors to the same memory module. For communication synthesis, channels are grouped and assigned to buses in order to reduce the number of bus conflicts and the number of behaviors which are connected to a given bus.

Behavioral modules are generated as the result of *functional partitioning* (see Figure 4.1). During later design steps these modules will be synthesized to software or to hardware structures.

4.3.1   Granularity

Partitioning can be performed at different levels of granularity, depending on the size of the objects into which the specification is decomposed. Functional partitioning at a *coarse granularity* deals with processes, subprograms, or blocks of statements. Such an approach is typical for system-level synthesis. It has the advantage of dealing with a relatively small number of objects which reduces the complexity of the partitioning problem. Due to the high abstraction level the manipulated objects are familiar to the designer and, thus, user interaction and modification of the design are facilitated.

Functional partitioning with *fine granularity* is performed at the operation level. It is used very often during high-level synthesis, after decomposition of a process into basic operations. Partitioning at this granularity allows a fine tuning of the results and can, potentially, produce good quality results. However, the large number of objects considered increases the complexity of the partitioning problem and makes it very difficult to get a result which is close to the global optimum for large designs.

4.3.2   Abstract Representation

A common design representation for partitioning is based on graph notations. The graph model can be different, depending on the level of abstraction and the features of the design which have been captured. Common representations for functional partitioning are the data-flow graph (DFG), the control/data-flow graph (CDFG), and the process graph.

The DFG and CDFG will be the right representations if fine grain partitioning, at the operation level, is performed. The process graph, on the other hand, captures the design at an abstraction level which is appropriate for coarse grain partitioning.

Example 4.1: Let us consider the three VHDL processes shown in Figure 4.3. Figure 4.2 presents the corresponding process graph and the DFG for the sequence inside process P1. We also show possible parti-

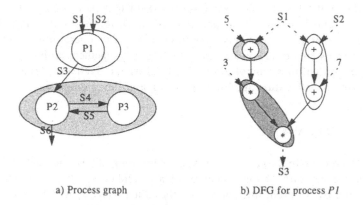

      a) Process graph   b) DFG for process *P1*

Figure 4.2: Graph representations for VHDL example in Figure 4.3.

```
. . .
signal S1, S2, S3, S4, S5, S6: INTEGER;
. . .
P1: process
    variable A, B: INTEGER;
begin
    . . .
    A:=(S1+5)*3;
    B:=S1+S2+7;
    S3<=A*B;
    . . .
end process;

P2: process
    variable X, Y: INTEGER;
begin
    . . .
    wait on S3;
    S4<=S3+X;
    . . .
    wait on S5;
    S6<=S5*Y;
end process;

P3: process
    variable Z: INTEGER;
begin
    wait on S4;
    . . .
    S5<=S4+Z;
end process;
```

Figure 4.3: A VHDL example.

tionings of the two graphs. We have assumed that partitioning of the
process graph is based on an objective which favors the clustering of
processes that are intensively communicating. The operations in the
DFG are clustered based on a metric which groups together similar operations, but only if they cannot be executed in parallel. Such an approach is
often used for allocation and binding during high-level synthesis.

□

4.3.3   Objective

In order to produce high-quality results, a partitioning algorithm has to rely on
a quantitative measure of the goodness of a candidate solution. This can be
characterized by different attributes, usually called *metrics*, which have to be
expressed quantitatively. In order to express the overall quality of a certain

partitioning, the metrics are combined into an *objective function* which determines the goodness of the solution. The simplest combination is a weighted sum of the relevant metrics:

$$FOb = \sum_i w_i \cdot M_i$$

where M_i are metric values weighted by constants w_i. The values of the constants are decided by the designer according to the desired trade-off among the different metrics.

Hardware/software partitioning in the Vulcan system [GuDM96], for example, is based on the following objective function which has to be minimized:

$$FOb = w_1 \cdot S_H - w_2 \cdot S_S + w_3 \cdot B - w_4 \cdot P + w_5 \cdot m$$

S_H expresses the implementation cost of the hardware partition in terms of number of cells; S_S expresses the cost for the software partition in terms of the total program and data size; B represents the bus utilization and is a measure of the total amount of communication between the hardware and software partitions that has to take place in a time unit; P indicates the processor utilization; m is the total size of variables transferred across the hardware-software boundary. By minimizing this objective function, partitioning will attempt to reduce the size of the hardware partition and the amount of inter-domain communication, and to move as much functionality into software as possible.

Design constraints, for instance certain deadlines on execution time, are not captured by the function above. This means that solutions which do not satisfy the design constraints have to be filtered out by the partitioning algorithm and should not be considered for evaluation of the objective function. An alternative solution is to include the constraints into the cost function so that solutions that violate constraints are penalized. In [VaGa92] the following cost function is used for partitioning a system specification in order to satisfy chip-capacity constraints (number of chips, chip area, number of pins) while considering also system-performance constraints:

$$FOb = w_1 \cdot \sum_i \left(100 \cdot \frac{violate\_area(Cl_i)}{max\_area(Cl_i)}\right)^2 + w_2 \cdot \sum_i \left(100 \cdot \frac{violate\_pins(Cl_i)}{max\_pins(Cl_i)}\right)^2$$

$$+ \ w_3 \cdot \left(100 \cdot \frac{violate\_nrchips}{max\_nrchips}\right)^2 + w_4 \cdot \sum_j \left(100 \cdot \frac{violate\_exectime(b_j)}{max\_exectime(b_j)}\right)^2$$

The *violate* value is equal to the actual value of the metric estimated for the respective cluster minus the imposed constraint, if the constraint has been not satisfied, and 0 otherwise. For example:

$$violate\_area(Cl_i) = max\ (0,\ area(Cl_i) - max\_area(Cl_i))$$

The partitioning goal is to achieve the value 0 for the objective function, which means that all constraints are satisfied. The sums related to area and pins are calculated for all clusters, Cl_i, of the actual partitioning. The sum for execution time is over all behaviors, b_j, of the specification. For normalization the metric values have been divided by the maximal accepted value. To favor balanced over unbalanced violations of the constraints, each term is multiplied by 100 and squared. This facilitates, during exploration of the design space, convergence towards a solution which satisfies all constraints.

The metrics used in objective functions like those given above are defined for a complete partitioning in which each object of the system has been assigned to a cluster. They are so called global metrics and the corresponding objective functions evaluate the quality of an implementation of the whole system. Such metrics are used by iterative-improvement partitioning strategies in which a given partitioning is iteratively modified and for each new alternative the global cost is evaluated. Other partitioning strategies are not based on stepwise improvement of a complete partitioning, but are successively grouping objects into clusters, one at a time. During such a constructive approach there is no global partitioning of the system and, thus, no global metrics are available.

Clustering relies on *closeness metrics* which are based on a local rather than on a global view of the system. Closeness metrics measure the benefit gained from grouping two objects into the same partition [VaGa95b]. Several such metrics can be combined into a *closeness function*. So, for example, we can use as a measure of closeness the number of bits transferred between two objects, the quantity of hardware that can be shared between the objects, or the degree to which operations in the two objects have to be executed sequentially. In [McF83] functions are partitioned using the following measure of closeness between two functions f_i and f_j:

$$Close(f_1, f_2) = w_1 \cdot \frac{cost(f_i) + cost(f_j) - cost(f_i \cup f_j)}{cost(f_i \cup f_j)} - (w_2 \cdot par(f_i, f_j))$$

Here *cost* returns the implementation cost for the operations in the respective functions. $cost(f_i \cup f_j)$ represents the cost if the two functions are grouped together and share resources. $par(f_i, f_j)$ is 1 if f_i and f_j can be executed in parallel and 0 otherwise. Sharing resources between potentially parallel functions reduces the expected performances and, thus, the benefit gained by such a clustering is smaller.

Determination of the metric values used by objective or closeness functions plays an essential role in any partitioning method. An accurate *estimation* of these metrics is essential in order to guide the design space exploration towards solutions which satisfy cost and timing constraints. The most accurate method would be to create an implementation and to determine the metrics. However, such a method is much too expensive and time consuming and only allows for the exploration of very few design alternatives. A practical method is to create a rough implementation which ignores all details that are not relevant from the estimation point of view but is sufficient to derive the required metrics with an appropriate precision. Such an estimation is based on design models for the hardware components which have to be synthesized. Estimations for the software components rely on models of the processors on which the software will be executed. Estimation based on such models can be sufficiently accurate and fast so that a high number of design alternatives can be explored and a good system implementation can finally be produced.

Several estimation techniques for different metrics are presented in [GVNG94]. Estimation of hardware size is discussed in [VaGa95a] and [OKDX95] while [HeEr95] presents a technique for hardware runtime estimation. Estimation of software cost and especially of execution time has recently been the subject of extensive research [GGN95, LiMa95, SuSa96, Mal96]. Analysis of constraint satisfiability and performance estimation is of crucial importance for the synthesis of real-time embedded systems [NaGa96]. Constraint satisfiability in the context of hardware/software co-synthesis is discussed in [GuDM96, Gup96a]. Performance estimation of embedded real-time systems implemented on distributed architectures has been addressed in [Wolf94, YeWo95a, YeWo97].

4.3.4   Algorithm

For small systems with very well understood structures and functionalities, the number of realistic alternatives for partitioning at the system level is reduced. Thus, assignment of functionality to system components can be performed using an *ad hoc* approach based on the designers experience and intuition. Design environments like those presented in [COB95a, BIJe95], for instance, do not support automatic partitioning but mainly concentrate on the communication synthesis task.

If the specified system is large, with a complex functionality resulting from a high number of interacting components, the number of partitioning alternatives is extremely large and their exploration has to be based on high-performance algorithms. Starting from an abstract graph representation of the system, and based on objective and/or closeness functions as well as on adequate estimations of certain metrics, the partitioning algorithm maps functional objects to system components with the overall goal of meeting design requirements.

Partitioning is known to be an NP-complete problem. Thus, it is practically impossible to find an optimal solution by performing an exhaustive search through the design space. Partitioning algorithms of practical applicability have to explore the design space according to a certain strategy which converges towards a solution close to one which yields the minimal cost. In the next section we discuss some basic partitioning heuristics.

4.4   Partitioning Algorithms

There are two basic classes of partitioning approaches: *constructive* (or clustering) and *iterative* (or transformation-based) [AlKa95, Joh96]. Constructive algorithms utilize a bottom-up approach: each object initially belongs to its own cluster, and clusters are then gradually merged or grown until the desired partitioning is found. Decisions on clustering of objects are based on closeness metrics which do not require a global view of the system but only rely on local relations between objects. Choosing the right strategy and adequate metrics for object grouping can lead to a final partitioning of the required quality. In the next subsection we discuss *hierarchical clustering* as an example of the constructive approach.

Iterative strategies are based on a design space exploration which is directly guided by an objective function that reflects the global quality of the partitioning. A starting solution is modified iteratively, by passing from one candidate solution to another based on evaluations of an objective function. The final goal is to reach a good, i.e., near-optimal solution. The design space exploration is based on a so called *neighborhood structure* that is defined over the set of feasible solutions. The *neighborhood* $N(x)$ of a solution x is a set of solutions that can be reached from x by a simple operation. Such an operation, often called *move*, is typical for a given application. It can be, for example, the transfer of an object from one cluster to another or the interchange of objects between two clusters.

The algorithm in Figure 4.4 shows the typical steps of an iterative strategy. Essential features of such an algorithm are the way a new solution x' is selected from the neighborhood of the current solution x^{now} and the criterion for accepting such a solution. *Greedy algorithms* always move from the current solution to the best neighboring solution. Thus $x' \in N(x^{now})$ is an acceptable solution if $FOb(x') \leq FOb(x)$, $\forall x \in N(x^{now})$. The process terminates when the algorithm reaches a solution for which all neighbors have a greater cost. This solution is a *local minimum* which can be a poor solution, far from the optimum which is the *global minimum*.

More sophisticated iterative heuristics have been elaborated. They allow so-called *hill-climbing* moves, in order to escape local minima. This means that,

Construct initial configuration $x^{now} := x_0$
repeat
　　Select new, acceptable solution $x' \in N(x^{now})$
　　$x^{now} := x'$
until stopping criterium met
return solution corresponding to the minimum cost function

Figure 4.4: The iterative partitioning approach.

under certain conditions, moves which increase the cost are accepted, because they can lead to subsequent solutions that are better than any solution reached so far. In the following subsections, we will discuss the *Kernighan-Lin algorithm* as a typical iterative approach, as well as hill-climbing methods based on *simulated annealing* and *tabu search*.

Iterative strategies, which are guided by a global view of the system as reflected by the objective function, can produce better results than constructive algorithms. However, estimations for evaluation of the cost function can be very complex and exploration of the solution space is sometimes highly time consuming. Reducing the size of the solution space by decreasing the number of objects to be partitioned, can be of great advantage especially if this does not affect very much the quality of the final solution. A very efficient technique is to merge some very close objects by using a constructive algorithm before running an iterative heuristic on the resulting clusters [AlKa95, VaGa95b, VaGa95c].

Being a typical optimization task, partitioning can also be formulated as a mathematical programming problem [AlKa95]. Integer linear programming (ILP) formulations are used in several approaches today [NiMa96, NiMa97, Ben96]. We will discuss ILP based solutions in Subsection 4.7.1.

In the next subsections we present some basic partitioning algorithms. Some of the approaches are further discussed in Chapter 5 which is dedicated to general optimization problems.

4.4.1   Hierarchical Clustering

Hierarchical clustering is introduced in [Joh67] as a general clustering algorithm. An application of this method for high-level synthesis in the BUD system is described in [McF86] and [McKo90]. The clustering process is performed in several iterations with the final goal to group a set of objects into partitions according to some measure of closeness. The two closest objects are clustered first and are considered for further clustering as a single object. At each of the following iterations, the two closest objects (which can be individual objects or clusters resulting from previous iterations) are grouped together. This process is iterated until a single cluster is produced. As result of hierarchical clustering a so called "hierarchical cluster tree" is formed. The leaf nodes of this tree are the original objects. Internal nodes represent clustered objects. Each non-terminal node of the tree has an associated height (relatively to the leaves, which are considered to be at height zero). This height reflects the distance between the objects that have been merged into the corresponding cluster. A small distance means high closeness. Thus, nodes closer to the leaves represent clusters in which the objects are strongly connected.

To select among different possible partitionings the cluster tree has to be cut by a "cut line". Each subtree below the cut line becomes one resulting partition.

Thus, the height at which the tree is cut, determines the number and size of the partitions. A line at a low level will result in a high number of relatively small partitions, while a cut close to the root of the tree, results in a small number of large clusters. Usually several possible partitionings are evaluated by cutting the tree at different levels. The alternatives are evaluated according to some design criteria such as area or number of interconnections, and the most convenient one is selected.

The closeness function used for hierarchical clustering is defined between the initial objects to be partitioned. However, at successive iterations, closeness between different groups of objects, or between an individual object and a group have to be computed. This is solved by estimating the closeness between two clusters or between a cluster and an individual object, based on the closeness between individual objects. The resulting closeness can be, for instance, the minimum, maximum, or the average of the closeness between the individual objects in the given groups.

> **Example 4.2:** In Figure 4.5 we illustrate the hierarchical clustering algorithm using a set of five objects $\{v_1, v_2, v_3, v_4, v_5\}$. The closeness values between pairs of objects are depicted on the edges connecting the respective objects. The succession of graphs, as indicated in the figure, shows the sequence of clusters generated by the algorithm. For this example the closeness between two clusters (or between a cluster and an elementary object) has been estimated as the maximum closeness between the component objects.
>
> Figure 4.6 illustrates the cluster tree produced by the algorithm. For each

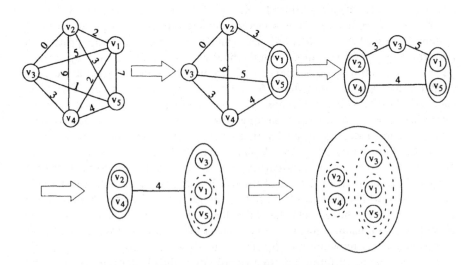

Figure 4.5: Successive steps in Hierarchical clustering.

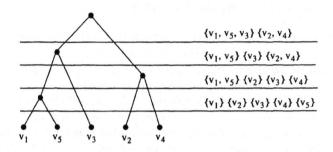

$\{v_1, v_5, v_3\} \{v_2, v_4\}$

$\{v_1, v_5\} \{v_3\} \{v_2, v_4\}$

$\{v_1, v_5\} \{v_2\} \{v_3\} \{v_4\}$

$\{v_1\} \{v_2\} \{v_3\} \{v_4\} \{v_5\}$

$v_1 \quad v_5 \quad v_3 \quad v_2 \quad v_4$

Figure 4.6: A cluster tree.

cut line we also show the corresponding partitioning alternative. The lowest line, for example, corresponds to a partitioning into five clusters, each one containing one single object. The line on the top indicates partitioning into two clusters of three and two objects respectively.

□

One of the problems with hierarchical clustering is that the same closeness function is used throughout the whole partitioning process. Thus, all closeness criteria which have to be considered, such as area, amount of communication, or parallelism, have to be combined into this function (see also Section 4.3). The only way to control trade-offs between these criteria is by assigning proper weights to each closeness metric. Another problem is that closeness computation between groups of objects is based on approximations like minimum, maximum or average of closeness between individual objects. As the partitioning process proceeds and clusters are growing, the errors produced by such approximations increase.

Some of the problems highlighted above are solved by multi-stage clustering introduced in [LaTh91], as part of the APARTY partitioning tool. Multi-stage clustering performs several successive hierarchical clustering processes. The clusters resulting from one partitioning are considered as initial objects for the next clustering stage. Each clustering is based on a particular metric. The order of applying the different closeness metrics reflects their priority, and has a decisive influence on the resulting partitioning.

Another advantage of multi-stage clustering is that closeness between clusters is recalculated at the start of each new stage, according to the current metric. Thus, the error produced by the approximation of closeness during the great number of successive iterations performed by classical hierarchical clustering, is limited.

4.4.2   Kernighan-Lin (KL) Algorithm

The Kernighan-Lin algorithm [KeLi70] partitions a given graph into two clusters of equal size by trying to minimize the total cost of the edges between the two clusters. Such a formulation, known also as min-cut partitioning, is typical for structural partitioning at layout optimization.

Let us consider a graph $G(V, E)$ with the set of nodes $V=\{v_1, v_2, ..., v_{2n}\}$, and the set of edges E. Given an edge e_{ij} connecting node v_i and v_j, c_{ij} is the cost associated to the edge. As result, the KL algorithm generates two clusters C_1 and C_2, each with n nodes, so that the following objective function is minimized:

$$FOb(C_1, C_2) = \sum c_{ij}, \quad \text{for } i, j \text{ so that } v_i \in C_1, v_j \in C_2, e_{ij} \in E$$

Given a certain configuration (C_1, C_2), we define the external cost Ext_i of a node v_i as the total cost of the edges which connect v_i to nodes in the other cluster; the internal cost Int_i, is the total cost of the edges to nodes in the same cluster:

$$Ext_i = \sum_{v_j} c_{ij}, \text{ where } v_j \text{ belongs to a different cluster than } v_i, \text{ and } e_{ij} \in E$$

$$Int_i = \sum_{v_j} c_{ij}, \text{ where } v_j \text{ belongs to the same cluster as } v_i, \text{ and } e_{ij} \in E$$

By moving a node v_i from one cluster to the other, the new clusters C_1' and C_2' are created. We have the following difference between the cost function for the old and the new configuration:

$$FOb(C_1, C_2) - FOb(C_1', C_2') = D_i = Ext_i - Int_i$$

We call D_i the gain obtained by moving v_i.

The gain of interchanging nodes v_i and v_j between the two clusters will be:

$$G_{ij} = D_i + D_j, \qquad \qquad \text{if there is no edge connecting } v_i \text{ with } v_j, \text{ and}$$
$$G_{ij} = D_i + D_j - 2 \cdot c_{ij}, \qquad \text{if an edge } e_{ij} \text{ is connecting } v_i \text{ with } v_j.$$

The KL algorithm starts with two arbitrary clusters C_1 and C_2 of equal size n. It then swaps pairs of nodes between the two clusters, with the aim to reduce the cost of the final solution. The algorithm performs successive passes. During each pass, every node moves exactly once, either from C_1 to C_2 or from C_2 to C_1. At the beginning of a pass each node is unlocked, which means that it is free to move. The node is locked after it has been moved. By locking a node, the algorithm avoids cycling that could occur if the same nodes are moved back and forth. Inside each pass, the algorithm iteratively swaps the pair of unlocked nodes with the highest gain G. The interchanges are performed until all nodes are locked. The lowest-cost solution encountered during the pass is preserved,

Construct initial configuration $x^{now}:=(C_1, C_2)$, with $|C_1|=|C_2|=n$
repeat
    $S_0:=0$
    Unlock all nodes
    for $k:=1$ **to** n **do**
        Find the pair $(v_i \in C_1, v_j \in C_2)$ so that v_i and v_j are unlocked and G_{ij} is maximal
        $S_k:=S_{k-1}+G_{ij}$
        $tentative_k:=(v_i, v_j)$
        Lock v_i and v_j
        Update D values for each node considering as if v_i and v_j are swapped
    end for
    Find p so that S_p is maximum of all partial sums S
    if $S_p>0$ **then**
        Generate new solution x^{now} starting from current solution x^{now}, by performing all
        the interchanges $tentative_l$, $1 \le l \le p$
    end if
until maximal gain $S_p \le 0$
return x^{now}

Figure 4.7: Kernighan-Lin algorithm.

and considered as starting point for the next pass.

The algorithm is shown in Figure 4.7. The *for* loop executes the successive iterations inside a pass. The selected interchanges are not directly executed but are successively stored in the sequence *tentative*. The total gain obtained by executing a succession of interchanges is cumulated in the sequence of sums S_i. After all nodes are locked (which means exactly n iterations) the sequence of moves which produces the maximal gain is executed. The algorithm terminates when the solution found by a pass is not better than the previous one (the maximal gain S_p is negative or zero).

Considering the succession of passes, the KL algorithm is greedy, as it iterates from one pass to the other only if the cost improves. However, the algorithm can climb out of local minima, since inside each pass two nodes can be swapped even if the resulting gain is negative. This will happen if for all pairs of unlocked nodes $G_{ij}<0$.

Example 4.3: We illustrate the hill climbing strategy of the KL algorithm using the example in Figure 4.8. It illustrates two partitioning steps starting from the initial configuration shown in Figure 4.8a. The internal and external costs, and the gains for each possible interchange, in the initial configuration, are:

$Ext_1 =0$	$Int_1 =200$	$G_{15}=-290$	$G_{35}=-1200$
$Ext_2 =10$	$Int_2 =210$	$G_{16}=-410$	$G_{36}=-1120$
$Ext_3 =100$	$Int_3 =1010$	$G_{17}=-1210$	$G_{37}=-1920$
$Ext_4 =0$	$Int_4 =1000$	$G_{18}=-1200$	$G_{38}=-1910$
$Ext_5 =110$	$Int_5 =200$	$G_{25}=-310$	$G_{45}=-1090$

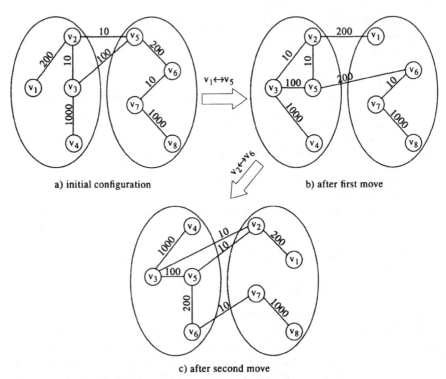

a) initial configuration   b) after first move

c) after second move

Figure 4.8: Node interchange with Kernighan-Lin strategy.

$Ext_6 = 0$	$Int_6 = 210$	$G_{26} = -410$	$G_{46} = -1210$
$Ext_7 = 0$	$Int_7 = 1010$	$G_{27} = -1210$	$G_{47} = -2010$
$Ext_8 = 0$	$Int_8 = 1000$	$G_{28} = -1200$	$G_{48} = -2000$

As we can see, all possible interchanges produce negative gains, which means that no cost improving step is possible. Interchange of nodes v_1 and v_5 produces the smallest cost increase (from 110 to 400), so it will be selected as the next tentative step. The new configuration is shown in Figure 4.8b. For the next iteration nodes v_1 and v_5 are locked. After updating Ext, Int, and D values the following gains are computed:

$G_{26} = 370$	$G_{36} = -920$	$G_{46} = -810$
$G_{27} = -830$	$G_{37} = -2120$	$G_{47} = -2010$
$G_{28} = -820$	$G_{38} = -2110$	$G_{48} = -2000$

The next tentative swap will be v_2 with v_6 which improves the cost function to 30. All subsequent steps selected throughout this pass produce negative gains. Thus, steps $v_1 \leftrightarrow v_5$ and $v_2 \leftrightarrow v_6$ will be performed, and the configuration in Figure 4.8c is the final one for this pass. The next pass can produce no improvement of the cost function so the algorithm terminates with this partitioning (which actually is the global optimum).

A simple greedy algorithm would not be able to improve the initial configuration, which represents a local optimum. However, by accepting hill-climbing moves, the KL algorithm was able to reach the global optimum by performing an intermediate step with a negative gain.

□

The complexity of the KL algorithm is $O(n^2 \log n)$, where n is the number of nodes [KeLi70]. Several improvements of the algorithm have been proposed. They are based on the same overall strategy but use improved data structures for representation of the graph and efficient scanning methods to locate the node pair with maximum gain. The algorithm proposed in [Dutt93] has a worst-case complexity of $O(\max(e*d, e*\log n))$, where e is the number of edges, and d is the average degree of the nodes.

Classically, the KL algorithm is formulated for the partitioning of a given

```
Construct initial configuration x^now:=(C_1, C_2), with |C_1|+|C_2|=m
actual_cost:=FOb(x^now)
actual_solution:=x^now
loop
    best_cost:=∞
    Unlock all nodes
    for k:=1 to m do
        Find node v_i so that v_i is unlocked and FOb(x^i) is minimal, where x^i is the solu-
        tion obtained from x^now by moving v_i to the other cluster
        x^now=x^i
        new_cost:=FOb(x^i)
        Lock v_i
        if new_cost < best_cost then
            best_cost:=new_cost
            best_solution:=x^now
        end if
    end for
    if best_cost < actual_cost then
        actual_cost:=best_cost
        actual_solution:=best_solution
        x^now:=best_solution
    else
        return actual_solution
    end if
end loop
```

Figure 4.9: Kernighan-Lin based functional partitioning.

graph into two clusters (two-way partitioning). It can be extended for multi-way partitioning, for example, by repeatedly applying the algorithm, first to the initial graph, and then to the resulting subgraphs

There are some features in the original formulation of the KL algorithm, as presented in Figure 4.7, which make it mainly applicable for structural layout partitioning. For example, it is assumed that clusters of equal size have to be generated, and that optimization should be in terms of interconnection cost. Nevertheless, the general strategy proposed by Kernighan and Lin can be successfully applied to a much broader area of partitioning problems [VaLe97].

In Figure 4.9 we present a Kernighan-Lin based algorithm which partitions a graph into two clusters of arbitrary size, by minimizing a given objective function *FOb* [GVNG94]. The algorithm starts with two arbitrary clusters C_1 and C_2. It then moves one node at each step, from one cluster to the other. During each pass, every node is moved exactly one time. Inside each pass the algorithm iteratively moves the unlocked node which produces the minimal value of the objective function. The lowest-cost solution produced inside the pass (*best_solution*) is preserved as a starting point for the next pass. As typical for the KL approach, the algorithm terminates when the cost produced by a pass is greater or equal to the previous one (*best_cost* $\geq$ *actual_cost*). Thus, we have the same greedy approach to considering the successive passes as in the version in Figure 4.7. The hill-climbing potential of the algorithm is given by

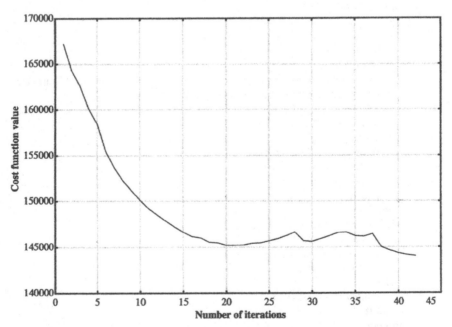

Figure 4.10: Variation of the cost function during partitioning with KL algorithm.

the possibility inside a pass to move one node between the two partitions even if the cost function is not improved. In Figure 4.10 we illustrate how the value of a cost function is changing during partitioning of a 100 nodes graph, using the algorithm in Figure 4.9. It shows how the succession of improving and worsening moves are driving the search towards a solution which, hopefully, is close to the optimum.

4.4.3   Other Iterative Partitioning Approaches

As shown in the previous subsection, the KL algorithm has a certain potential to escape from local minima. This ability is restricted to moves inside a partitioning pass, but the succession of passes themselves is greedy. The algorithm converges rapidly, however, the possibility to climb out from local minima is limited. *Simulated annealing* (SA) [KGV83] does not select the neighbor solution which produces the best cost, but selects randomly one of the solutions in the neighborhood of the current state. If this randomly selected solution is an improved one, it will always be accepted. But worse solutions can also be accepted, and this provides the ability of the algorithm to escape local minima. Uphill moves are accepted with a certain probability that depends on how large the cost deterioration is, and on a control parameter T called temperature.

In contrast to simulated annealing, *tabu search* (TS) [GTW93] controls uphill moves not purely randomly but in a systematic way. The TS approach accepts uphill moves and stimulates convergence toward a global optimum by creating and exploiting data structures to take advantage of the search history at selection of the next move. Similar to the KL algorithm, TS also performs the move which, at a certain moment, producers the best change of the cost function. In order to avoid cycling, KL locks *all* the nodes that have been moved during a pass. The strategy used by TS for avoiding cycles is more flexible. It maintains the history of *a certain number* of recent moves and considers their reverse moves as forbidden (tabu). If, at a given step, none of the allowed moves is an improving one, TS performs an uphill move and is, in this way, able to escape from local minima.

Details on the SA and TS, as well as on the genetic approach, are given in Chapter 5. In Subsection 8.4.2 hardware/software partitioning algorithms based on SA and TS will be presented.

4.5 Communication Synthesis

As result of system partitioning, processes specifying the behavior of the system have been assigned to components. These components can be both processors executing software processes and application-specific hardware components (Figure 4.11a). According to their system level specification, processes are communicating through abstract channels. They also have to interact with peripheral devices and application-specific interfaces. At the system level communication is specified using mechanisms like synchronous or asynchronous message passing, rendezvous, remote procedure call, and monitors. All implementation details are hidden at this level.

Communication synthesis generates the hardware and software which interconnects the system components, and enables processes to communicate with each other as well as with peripheral devices and other external interfaces.

Communication synthesis is a complex and time-consuming task. High diversity of devices and processors which have to co-operate, complex performance issues, and the lack of appropriate support for software development and debugging are some of the difficulties which have to be faced during this design step. The support of design tools for communication synthesis is essential in order to improve design productivity. Some approaches addressing aspects of communication synthesis are presented in [COB95a, BAJ94, SrBr95, NaGa94a, NaGa94b, NaGa95].

If we try to approach communication synthesis as a top-down design task, we can identify the following three main steps:

1. Channel binding,
2. Communication refinement,
3. Interface generation.

During these three steps a high-level specification without any details concerning complex communication protocols and low-level issues is successively refined, finally resulting in a specification which can be synthesized directly into the hardware and software components of the final implementation. The resulting system not only has to produce the initially specified behavior, but it also has to satisfy performance and cost constraints as imposed by the design specification.

4.5.1   Channel Binding

The abstract channels defined by the system specification have to be implemented using physical communication components. Such components can be, for instance, a point-to-point communication line or a shared bus. Thus, the first task is to allocate the resources which support communication throughout the system. Then, abstract channels have to be partitioned and the resulting groups bound to the allocated resources. Messages corresponding to channels in one group are multiplexed on a shared communication component. This task is very similar to the allocation/binding task in high-level synthesis, where operations are bound to allocated functional units (see Section 3.3).

Similar to other system components, allocation of communication components is done, according to current state-of-the-art, mostly manually, taking in consideration cost/performance trade-offs. There are also very few attempts to automate the partitioning task for communication channels. The main goals of channel grouping in [NaGa94b] are to avoid bus conflicts and to reduce the total number of connecting wires. The first goal is achieved by grouping of channels which do not access the bus concurrently. The grouping of channels which are accessed by the same behaviors reduces the total number of wires.

The approach presented in [DBJ95] selects from a library of communication units the components which are able to provide the requested communication services and binds a certain number of abstract channels to each of the selected units. There are several communication units which can provide a certain service and they differ in cost and performance. Depending on its features, a communication unit can support a certain number of channels to be multiplexed on it, without reducing the communication rate below a required minimum. The algorithm in [DBJ95] is based on the elaboration of a decision tree. It tries to minimize cost also taking into consideration performance constraints. The constraints are given in terms of the average and the peak rate for each channel. The approach is based on an internal representation for system-level design, called SOLAR, which is discussed in Subsection 4.7.6.

In [FKCD93] a synthesis technique is presented which aims at the sharing of the same physical component by several channels so that any access conflict is excluded by proper scheduling of communication operations. There are two approaches supported. The first one considers the physical lines as functional units which are shared by the communicating processes, and reduces the problem to classical scheduling with resource sharing (see Section 3.2). Thus, after scheduling, operations on the channels assigned to a given physical component are automatically serialized. A second approach does not enforce any serialization of communication at scheduling. In this case the timing of operations as resulting from scheduling is investigated in order to determine which channels can share the same communication component.

> **Example 4.4:** Figure 4.11 illustrates the channel-binding task for an example consisting of eight processes which have been partitioned over two processors executing the software part and two ASICs. One of the processors is also connected to a peripheral device. The processes communicating through abstract channels are depicted in Figure 4.11a. Figure 4.11b shows how the channels have been grouped and bound to a shared bus and to two point-to-point connections.
>
> □

4.5.2   Communication Refinement

After allocation and binding in the channel-binding step, communication is still described at a high abstraction level (Figure 4.11b). There are still several alternatives open for implementation of communication, and the specification has to

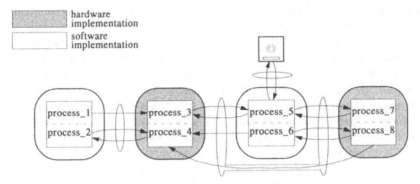

(a) Partitioned behaviors communicating through abstract channels

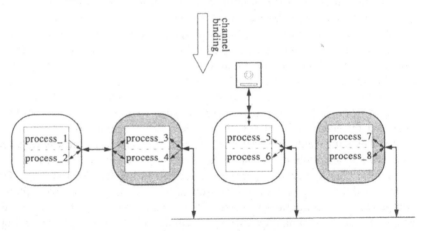

(b) Communication after channel binding

Figure 4.11: Channel binding.

be refined with several details, going down towards the final implementation.

After the previous step, we basically know the interconnect topology of the system, based on point-to-point communication, on buses, or a on a hybrid scheme. It has been also determined which channels are bound to a given communication component. Several decisions have to be taken now, in order to determine the particular features of the communication support:

• The *widths of the communication lines* have to be determined, depending on constraints concerning data transfer rates, number of available pins, and cost. A bus has to support a minimum average transfer rate for each of the channels implemented on it. It also has to provide a certain minimal peak rate so that transfer of a single data item is provided in an acceptable time. Given these performance constraints, a minimum width can be determined in order to reduce the number of requested wires and pins [NaGa94b, GaGl96].

- If communication buses are shared, an adequate *control strategy* has to be decided. An implementation with a single master (only one module, the master, can initiate transmissions on the bus) is simple and cheap as it does not require any bus arbitration. However, such an implementation requires every communication between slave components to be performed through the bus master, which can result in serious limitations on the performance. A multi-master implementation, on the other side, requires extensive *arbitration* in order to resolve conflicts at bus access.

- In order to provide communication between system components in a consistent manner, the *communication protocol* has to be defined for each communication link. The protocol defines the exact mechanism of data transfer over the communication support and can be, for example, full-handshake, half-handshake, fixed-delay, hardwired ports, or a complex layered protocol [NaGa94a]. It is important that the appropriate high-level behavior of the logic channels can be implemented on top of the protocol selected for the respective communication support.

4.5.3   Interface Generation

During the previous two steps, the system specification has been progressively refined adding details concerning the communication support. With all this information provided, the interfaces needed for a correct functionality of the system can be now generated, which implies the synthesis of the following items:

- *access routines* inside the communicating processes;
- *controllers* consisting of buffers, FIFO queues, and arbitration logic, which implement the correct access to the communication support;
- *adapters* needed to interface components which use incompatible communication protocols; this can be the case if pre-designed or custom components are used [NaGa95];
- *device drivers* which support the access to peripheral devices and application specific interfaces;
- *low level support for communication related tasks*, like interrupt control, DMA, memory mapped I/O;

The generation of interfaces is based on parameterized libraries of software code and hardware specifications, which can be customized or described by the user. The access routines to communication services are expanded as executable code into the software components and as hardware specification (in VHDL, for example) into the hardware components. Software routines, on their side, can rely on operating-system (OS) services, bus-access routines, or specialized drivers (Figure 4.12). Most of the interface components can be

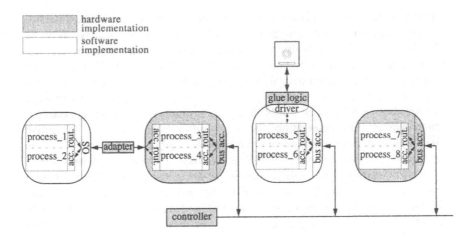

Figure 4.12: Implementation of the example in Figure 4.10 after interface generation.

implemented both in software and in hardware. The decision on which part of the interface, if any, should be implemented in hardware depends on the imposed time constraints. Two kinds of constraints have to be considered [COB95a]: low-level constraints, concerning the signalling conventions of connected devices, and performance requirements expressed usually as rate constraints or response-time requirements.

> **Example 4.5:** Figure 4.12 shows a possible implementation of the system presented in Figure 4.11. Communication on the shared bus is managed by a controller implemented in hardware. An adapter, also implemented in hardware, interfaces one processor and the ASIC which are communicating by two different protocols through a point-to-point connection. Communication with the peripheral device is implemented by a mixed hardware/software interface which consists of a software driver and additional glue logic.
>
>   □

4.6   Hardware/Software Co-Design

The term "hardware/software co-design" is used today to denote the co-operative development process of both the hardware and the software components of a system. According to traditional methodologies the hardware and software parts of an application are separately developed and their final integration is based on *ad hoc* methods. This increases time-to-market, cost, and, very often, decreases quality and reliability of the product.

Research in hardware/software co-design aims to develop methodologies and tools for a systematic and concurrent development of the hardware and software components, providing interaction and exploration of trade-offs along the whole design process [BuRo95, CaWi96, DeMi94b, DeMi96, GaVa95, LSH96, AdTh96, KAJW96, Wolf94].

System-level synthesis, as discussed in this chapter, covers the situation, when a hybrid system, consisting of hardware and software components has to be synthesized. Thus, it includes all synthesis-related aspects of co-design, often referred to as "hardware-software co-synthesis". After the system-level synthesis steps, as presented in Figure 4.1, it can turn out that part of the system functionality has been assigned (as software) to programmable processors, and the rest is implemented (as hardware) on one or several ASIC or FPGA chips. This will be the case, for instance, if an all-software implementation on custom processors does not meet some performance constraints, but, on the other hand, an all-hardware implementation is not required.

There are several aspects which are specific to the design and implementation of hybrid systems, compared to that of homogeneous hardware or software systems:

- The heterogeneous nature of the system implies different specification styles, execution models, and different design techniques to be considered simultaneously and in interaction. These aspects have to be carefully considered not only at synthesis but also for system specification, verification, and simulation [CKL96, LSH96, KAJW96];
- Hardware/software partitioning is very often treated as a particular two-way partitioning in which performance has to be maximized and hardware size has to be minimized [GVNG94].
- The interaction of hardware and software components creates additional interfacing problems which have to be considered during communication synthesis [BCO96, OrBo97, BIJe95, VRBM96];
- Specific aspects have to be considered if an ASIP or a commercial DSP is used as part of the system architecture. Such aspects include the selection of the proper ASIP architecture, the development of retargetable compilers and of architecture-specific compiler-optimization techniques [Goo96, Pau96, CaWi96].

In the next section we briefly survey some published approaches to system-level synthesis of hardware/software systems.

4.7 Synthesis Approaches at the System Level

In this section we briefly review some representative approaches to synthesis at the system level.

4.7.1   Integer Programming Based Approaches

Both allocation of processing and communication components and partitioning are optimization problems which can be solved using mathematical programming. The most ambitious approaches [PrPa92, Ben96] try to formulate the allocation, the partitioning, and the scheduling of processes as well as data communication as a mixed integer linear programming (MILP) problem. In order to master the high complexity of the problem, the approach in [NiMa97] does not include process scheduling in the MILP formulation. Scheduling is solved with some heuristics and the resulting time values are used for the mathematical formulation of the partitioning problem.

Regardless of some differences, all these mathematical-programming-based approaches formulate a set of constraints that have to be satisfied to ensure proper functioning of the system, and an objective function that has to be optimized. The constraints, basically, have to express restrictions imposed by data dependencies, by design constraints, and have to avoid resource conflicts.

For instance, the following inequality has to be written for each pair of processors P_k and P_m ($k \neq m$) which potentially can be allocated for system implementation, and each pair of processes (tasks) T_i, T_j, so that T_j is data dependent on T_i [Ben96]:

$$s_i + t_{ki} + c_{ij} \leq s_j + (2 - d_{ki} - d_{mj}) \cdot C$$

The above inequality expresses the data-dependence constraint between processes T_i and T_j by imposing that T_j can start only after it has received the data produced by T_i. s_i and s_j are real variables representing the start time of the two processes; d_{ki} and d_{mj} are integer 0/1 variables which have the value 1 if process T_i (T_j) is executed on processor P_k (P_m) and value 0 otherwise; t_{ki} and c_{ij} are constants given by the designer, expressing the execution time of task T_i when executed on processor P_k and, respectively, the communication time for sending the data between processes T_i and T_j when they are executed on different processors. C is a sufficiently large number and has the effect that the inequality is practically observed only if T_i is executed on processor P_k and T_j on P_m ($d_{ki}=d_{mj}=1$).

If both tasks are executed on the same processor, no communication time has to be considered. Thus, for each processor P_m and each pair of data dependent tasks, T_i, T_j, the following inequality has to be added [Ben96]:

$$s_i + t_{mi} \leq s_j + (2 - d_{mi} - d_{mj}) \cdot C$$

Similar inequalities have to be added to express dependency between data communication and the producer and receiver processes respectively.

Resource conflicts have to be avoided by expressing that two processes must not be executed at the same time on a programmable processor (this restriction does not hold for the hardware). Similar restrictions have to be imposed for communication on a bus. For data dependent processes resource conflicts are implicitly excluded by inequalities like those presented above. For the other processes, special inequalities have to be introduced in order to express that their execution times are not allowed to overlap if they are bound to the same programmable processor. Thus, the following inequalities have to be written for each pair of tasks T_i, T_j between which there is no (direct or indirect) data dependency and each programmable processor P_m [Ben96]:

$$s_i + t_{mi} \leq s_j + (3 - d_{mi} - d_{mj} - y_{ij}) \cdot C$$
$$s_j + t_{mj} \leq s_i + (3 - d_{mi} - d_{mj} - y_{ji}) \cdot C$$

y_{ij} and y_{ji} are integer 0/1 variables; y_{ij} (y_{ji}) has the value 1 if process T_i (T_j) is executed before T_j (T_i) and 0 otherwise. Thus, the two inequalities impose that if both tasks are assigned to processor P_m, T_j has to start after T_i finishes (the first inequality) or T_i has to start after T_j finishes (the second inequality). Similar inequalities are introduced in order to constrain the sequencing of communications on buses.

Time constraints imposed by the designer are easily expressed by additional inequalities related to the start time of tasks. The cost function to be minimized can be, for instance, the total cost of the system, which can be estimated as the sum of the costs of allocated programmable processors, hardware implemented processes, and buses.

After solving the MILP model we get the values for the time and integer 0/1 variables, which produce a minimum of the cost function. The time variables (like s_i above) correspond to an optimal schedule of processes and communications while integer 0/1 variables (like, d_{mi} above) define their optimal partitioning on the allocated programmable processors, ASICs, and buses.

Formulation of an MILP model is an elegant and relatively easy solution to the allocation and partitioning problem. Using existing MILP solvers it can produce an exact, optimal solution. However, the big disadvantage of this approach is the complexity of solving the MILP model. The size of such a model grows very quickly with the number of processes and the number of the resources which can be allocated. Thus, an MILP based approach is realistic only for small problems.

4.7.2   Distributed Real-Time Systems

The approach presented in [YeWo95a, YeWo95b, YeWo95c, YeWo97, Mal96] addresses the synthesis of distributed embedded systems for real-time applications. It considers both the allocation of processing and communication elements, the partitioning of processes and communication channels, as well as scheduling of processes and communication.

The abstract model on which the approach is based is typical for distributed real-time systems. Each process is characterized by its computation time (which can be given also as an interval between a minimum and maximum execution time). A set of possibly data-dependent processes, represented as an acyclic data-flow graph, is considered as a task. Each task has a given *rate constraint* (period) which is the time between two consecutive initiations and a *deadline* which is the maximum time allowed from initiation to termination of the task. A problem specification may contain several such tasks, each with its own rate constraint and deadline. As a final output, the co-synthesis algorithm produces three main results: (1) the architecture consisting of processing elements and interconnection buses, (2) the partitioning of processes onto the processors and of communication channels onto the buses, and (3) the scheduling of processes and message-transfer operations. The co-synthesis process guarantees the fulfillment of deadlines and rate constraints and also considers imposed limits on hardware and software cost.

One of the specific aspects which make this approach suitable for some real-time applications is its ability to handle periodic tasks and a pre-emptive scheduling policy. Thus, applications can be considered in which several tasks are running with different periods and processes can interrupt each other depending on their priorities.

The synthesis algorithm is based on an iterative-improvement strategy [YeWo95c]. At each step there are several alternatives the algorithm can select from: moving a process from one processing element to another, moving a message transfer from one bus to another, or allocating a new processing element or a new bus. The effect of each possible alternative is estimated and the one which produces the best improvement in terms of the design goal is selected. This is basically a greedy approach which could end up in a local minimum. However, the search strategy has some potential of moving out from local minima by computing a particular cost function, during the early stages of synthesis. During this stage the elimination of lowly utilized processing elements or buses is encouraged. This is solved by allowing processes to be moved from the least-utilized processor or bus even if the performance is worsened and the overall cost does not change for the moment. During later co-synthesis stages the strategy concentrates on balancing the utilization of the remaining processing elements and buses and, thus, optimizing the performance.

An essential aspect of this approach is the performance estimation which underlies the decisions taken during exploration of the design space. As mentioned, the scheduling strategy considered by this approach is a pre-emptive one, which means that it is always the ready process with the highest priority which will be executed on a processor. Performance estimation is based on rate-monotonic analysis (RMA) [LiLa73] in order to derive worst-case delays for each task. Priorities are assigned to processes in inverse order of their deadline [LeWh82]. A heuristic has been developed which determines deadlines for each process starting from the user-imposed deadlines on the tasks [YeWo95a, YeWo95b, YeWo95c].

Given the specified tasks, the priorities assigned to each process, a binding of the processes and message transfers to processing elements and buses, the heuristics presented in [YeWo95a, Mal96] derive bounds on the delays of each tasks. The estimation algorithms are an extension of RMA to support multiprocessor systems and data dependency between processes.

4.7.3   SpecSyn, a System Design Environment

SpecSyn is an environment which assists the designer in performing several system-design tasks [VaGa92, GVNG94, Gaj96]. SpecSyn supports a system-design methodology consisting of the following steps [GVN94, GaVa95]:

1. *Specification capture*: the result of this step is a functional specification which does not prescribe any implementation detail. This specification has to be in a machine-readable and simulatable form which also may serve as input to automatic synthesis tools. System specifications accepted by SpecSyn can be formulated in VHDL and in an extension of VHDL called SpecCharts [NVG92, VNG95]. SpecCharts is based on a new conceptual model, called Program State Machines which combines the hierarchical/concurrent Finite-State Machine model with the programming-language paradigm. Thus, it permits each state of a hierarchical/concurrent FSM to be described by means of instructions in a programming language, specifically in VHDL. SpecSyn provides tools for translation of specifications formulated in SpecCharts to VHDL.

2. *Exploration*: during this step, several design alternatives are explored before one, suitable for implementation, is selected. During the exploration, system components have to be allocated and the input specification has to be partitioned over these components.

3. *Specification refinement*: this step transforms the initial, purely functional, specification into a new one that reflects the results of the exploration phase [GGB96]. The resulting model, which functionally is equivalent to the original specification, consists not only of the partitioned functionality but also of the allocated system components. Several decisions have to be made during refinement, such as mapping of variables to local or global memory, insertion of interface protocols between components, or addition of bus arbiters. The resulting specification is complete in the sense that not only functionality and system components are included but also their communication interfaces are defined. Thus, the refined specification can be simulated for verification and can serve as input for lower-level synthesis tools which produce the software and hardware components of the implementation.

The SpecSyn system supports extensive designer interaction during all design steps. Several decisions are taken interactively by the designer while other design tasks are performed automatically. SpecSyn also considers all three kinds of functional objects contained in an input specification: behaviors, variables, and communication channels. During the partitioning phase behaviors are partitioned over processing elements, variables over memories and communication channels over buses. The partitioning algorithm incorporates a hill-climbing heuristic, such as simulated annealing, inside a binary search procedure [VGG94].

An essential part of the SpecSyn system is the estimation tools which provide the information necessary to properly guide the exploration of the design space. Estimators are implemented for hardware parameters like clock cycle, execution time, and number of pins [VaGa95a]. Execution time, program, and data size can be estimated for the software components [GGN95].

4.7.4   The Vulcan Co-Synthesis System

The Vulcan system supports the hardware/software co-design of embedded systems consisting of parts that operate at different speeds, communicate with each other, and are constrained by requirements on their timing [GuDM93, GCD94, Gup95, GuDM96]. The input specifications are formulated in HardwareC, a C-like language with an associated hardware semantics [KuDM92]. As part of such a specification the designer can formulate two kinds of timing constraints: *delay constraints* which are bounds on the time interval between initiation of the execution of two operations, and *execution-rate constraints* which bound the interval between successive initiations of the same operation.

The synthesized system will be implemented on an architecture consisting of one processor, one hardware component (ASIC), one system bus, and one global memory through which all hardware/software communication takes place. Timing analysis, hardware/software partitioning, and software implementation are the main tasks performed by the Vulcan system.

The internal model on which the system operates is a set of hierarchically related data-flow graphs. The graph nodes are capturing operations while arcs are representing the flow of data. Timing analysis is based on path analysis performed on the graph model. For this purpose the graph nodes are weighted with the execution delay for each operation, depending on the actual implementation (in hardware or software) of the respective node. Execution times for the software operations are derived using a processor cost model of the target processor [GuDM94]. Operations with a fixed, statically known delay are considered deterministic delay operations. Operations with a delay which depends on the value or timing of input data (data-dependent loops or wait operations) are called non-deterministic delay (ND) operations.

Given the weighted graph corresponding to an implementation alternative of an application, timing analysis in the Vulcan system verifies if the imposed constraints can be satisfied [Gup96a]. A timing constraint is considered to be satisfiable if it is satisfied for all possible (maybe infinite) delay values of the ND operations. If a constraint is not satisfiable, Vulcan can possibly determine a marginal satisfiability of the constraint; this means that the constraint can be satisfied for all possible values within a specified bound of the delay of the ND operations.

Exploring different implementation alternatives, hardware/software partitioning in the Vulcan system attempts to find the lowest cost implementation which still satisfies the design constraints. Partitioning is performed on the internal graph representation at the granularity level of operations. The partitioning strategy is based on a greedy iterative improvement algorithm [GuDM96]. It starts from an initial configuration in which the ND operations depending on input data are placed into the software, and all other operations into the hardware. During the next steps deterministic delay operations are successively moved into software, in order to reduce implementation cost, as long as timing constraints are satisfied (constraint satisfiability for each partitioning alternative is checked by the timing analysis tools).

Implementation of the software component on the target processor requires serialization and scheduling of the operations represented as graph nodes. A complete, static scheduling in the presence of ND operations is not acceptable. Thus, the software component is implemented in Vulcan as a set of concurrent threads. A thread is a linearized set of operations that may begin with an ND operation but does not contain any ND operation inside. Threads are implemented with coroutines and are executed by a non-pre-emptive run-time scheduler [GCD92, GuDM94]. Generation of the threads, starting from the graph structure, is based on a heuristic which tries to find a linearization which not only corresponds to the partial order imposed by data dependencies but also satisfies operation deadlines [GuDM94, Gup96b].

The hardware components of the resulting system are synthesized by the *Olympus* synthesis system [DKMT90]. Vulcan also automatically generates the interface logic between processor and application-specific hardware [GuDM93, GCD94].

4.7.5   The Cosyma Co-Synthesis System

Cosyma is an experimental framework for the co-design of small embedded real-time systems [EHB93]. Input specifications are formulated in C^x, which is an extension to the ANSI C standard to accept time constraints and the task concept. The target architecture is similar to that used in Vulcan, with the restriction that hardware and software components execute in an interleaved manner (processor and application specific hardware execute in mutual exclusion).

The approach taken in Cosyma is often called a software-oriented approach (as opposed to that in Vulcan, for example) because of two main reasons: first, the initial specification is formulated as a software program; second, starting from this all software alternative, a minimal amount of functionality is extracted for migration into hardware in order to meet performance constraints.

The input specification is translated into an extended syntax-graph representation which also captures data and control-flow information. Profiling is used to identify computation-time intensive parts of the system. Some of this parts will be considered, during the partitioning step, for hardware extraction.

Hardware/software partitioning in Cosyma is performed at the granularity level of basic blocks. The partitioning algorithm is based on simulated annealing and is guided by a cost function which captures estimations of the potential speed-up obtained by moving a certain basic block into hardware, and of the resulting communication overhead [HeEr95]. This algorithm attempts to achieve performance constraints at a minimum hardware cost.

After a partitioning has been produced the software component is compiled and the application specific hardware is synthesized to a RT-level structure [HEHB94]. Timing analysis is then performed on the resulting implementation in order to check if time constraints are satisfied. This analysis is based on a hybrid approach, combining RT-level simulation and formal analysis [YEBH93]. Violation of time constraints can be the result of inaccurate speedup or communication overhead estimations which are used to guide the hardware/software partitioning. In this case estimation values are updated by using the feedback received after analysis of the synthesized hardware/software implementation. Starting with these new values another partitioning is generated. This iterative process is continued until it converges to a satisfying implementation.

4.7.6   The COSMOS Co-Design Environment

COSMOS is a design environment for the modeling and synthesis of complex hardware/software systems [BAJ94, BIJe95]. One of the key issues in COSMOS is the communication synthesis aspect, providing facilities for the abstraction of complex communication protocols and for the re-use of communication modules.

System specifications accepted by COSMOS are formulated as communicating processes. The actual implementation of the environment accepts specifications in SDL [BHS91]. However, front-ends for other specification languages can easily be added, as all the design tools in the system are operating on an internal representation called SOLAR, to which input models are translated.

SOLAR is an intermediate representation designed to support the main concepts handled at the level of system specification [JeOB95]. In SOLAR a system is modeled as communicating design units (DU). A DU can contain either other units (DUs or channel units) or a set of transition tables which are specifying hierarchical and communicating FSMs. Communication between DUs is performed through channel units (CU). The communication services provided by a CU are accessible via procedures (methods) which can be called from the DUs. These methods are the only part of the CU visible from outside. Thus, all details concerning the communication process are abstracted away during the specification and partitioning steps. One of several communication units implementing the methods of a given CU, can be selected from an external library during later design steps.

After the translation of a system specification into SOLAR, the design flow in COSMOS assumes three main design steps: partitioning, communication synthesis, and architecture generation.

Partitioning distributes the SOLAR model into communicating modules which, in the final implementation, will be mapped onto separate processors or onto application specific hardware. Partitioning in COSMOS is performed interactively through the manual application of several transformation primitives which allow processes not only to be moved, but also to be merged, or split [BOJ94].

Communication synthesis in COSMOS comprises two steps. During the first step, the designer manually selects, from a library of communication models, the appropriate units to carry out the desired communication. The selected CUs are assigned to each of the abstract CUs in order to implement the effective communication protocols that have been selected. Selection of a CU from the library depends on the particularities of the communication to be performed, on the required performance, reliability, costs, or implementation technology. During the second step, each CU is removed and the information encapsulated in it is partially distributed between the communicating processes, and partially embodied in a communication controller. Such a communication controller can correspond to an existing interface circuit, can result in an ASIC to be synthesized, or can be implemented as software on a (possibly dedicated) microprocessor.

At architecture generation a virtual prototype is produced first, by translating SOLAR into VHDL and C code. Hardware components are modeled in VHDL while software is mapped into C code. The resulting virtual architecture can be simulated. Finally, an implementation can be produced, by compiling the software components and synthesizing the hardware part.

4.7.7   The Chinook Co-synthesis System

Chinook is a tool for the co-synthesis of real-time embedded systems [COB95a, CWB94]. The support provided by Chinook concentrates mainly on the interfacing, communication, and scheduling aspects, which are considered to be the most error-prone and time-consuming design tasks. The main goal is to provide the designer with the ability to easily map a specification on a variety of target architectures and, thus, to get a fast feedback which allows for a better exploration of the design space and for better informed trade-offs [BCO96].

Chinook accepts input specifications formulated in Verilog [ThMo91]. Such a specification has to describe not only the intended behavior of the system and the imposed design constraints, but also consists of a structural part. This part lists all the processors, peripherals, devices, and communication interfaces to be used for implementation. In the behavioral specification the designer also has to tag tasks and modules with the system component that will be used for their implementation. As an output, Chinook produces all the components of the desired system including software code for each processor and netlists for the glue logic needed to construct the system.

There are two levels of timing constraints considered by the Chinook system. At the low level there are the signaling conventions that must be used to connect the system to its peripheral devices and communication interfaces. At the higher level performance requirements, such as rate and response time constraints, are considered.

As discussed before, the initial specification delivered to Chinook already contains the allocation of system components and partitioning of the system modules over these components. Thus, system integration and software scheduling are the main tasks which still have to be performed with the support of Chinook.

System integration produces all the hardware and software required to interconnect the system components to enable them to communicate. In Chinook, this is performed in two steps called *interface synthesis* and *communication synthesis*, respectively.

Interface synthesis is aimed at generating hardware and software elements needed for communication between the system components. At the lowest level this implies synthesis of device drivers and generation of customized code for the processor to be connected to the respective device. Considering the specific signaling requirements of the device, the speed of the processor, and the degree to which processor time is available for input/output handling, Chinook decides which portions of the driver, if any, that have to be implemented as external hardware. The software part is then statically scheduled, taking into consideration low level timing constraints. Finally, access routines to the connected device are updated, taking into account the particularities of the generated driver.

When generating the connection between a processor and a device, the interface synthesis algorithm tries to minimize the amount of glue logic while respecting timing constraints [COB92, COB95b]. It first attempts to use I/O ports if the processor has them. If there are any unconnected device ports remaining but no more I/O ports available, the algorithm connects them using memory-mapped input/output.

Communication synthesis provides hardware and software support for data communication between modules allocated on different processors. Considering timing constraints, Chinook supports choices between different communication protocols, different interconnection topologies and automatically synthesizes buffers, FIFO queues and arbiters.

An essential component of the Chinook system is the processor/device library which captures necessary knowledge for system integration and performance estimation. It contains detailed generic specifications of device interfaces (in the form of timing diagrams and Verilog code), specifications of processor interfaces, and schemas for software run-time estimation.

The generation of the software component for the resulting system is performed by the scheduler, following a static, non-pre-emptive scheduling policy. The algorithm presented in [ChBo94, ChBo95] attempts to find a static schedule which meets timing constraints. It always succeeds in finding a valid schedule if one exists. Heuristics have been added to the algorithm for finding schedules which are not only valid, but are also close to the shortest possible schedule.

5

Optimization Heuristics

5.1 Introduction

As have been shown in the discussions up till now, many system synthesis tasks deal with the problem of finding the best possible solution from a set of solutions which satisfy some design constraints. The problems of system partitioning, scheduling and allocation discussed earlier in this book are good examples of such tasks.

These tasks can usually be formulated as the following optimization problem:

Minimize $f(\mathbf{x})$
Subject to $g_i(\mathbf{x}) \geq b_i;$ $i = 1, 2, ..., m;$

where $\mathbf{x}$ is a vector of *decision variables*; f is the *cost (objective) function*; and the inequalities are a set of *constraints*.

If both the cost function f and the g_i's are linear functions, and the decision variables are of real values, we have a *linear programming* (LP) problem. LP problems can be solved by the simplex algorithm [Fou81], which is an exact method, i.e., it will always identify the optimal solution if it exists.

5.1.1   Combinatorial Optimization

In system synthesis, the decision variables, $\mathbf{x}$, of the formulated optimization problem are usually discrete, i.e., the solution is a set, or a sequence, of integers or other discrete objects. For example, the problem of resource allocation might deal with decision on whether an ALU should be used or an adder and a comparator should be implemented. When formulating system partitioning as a k-way partitioning problem, a decision variable, x_i, that decides which cluster the object v_i will be assigned to, can only have an integer value between 1 and k in order to produce a feasible solution. This type of problems is called *combinatorial optimization* problems.

Among all combinatorial optimization problems the *travelling salesman problem* (TSP) is probably the best known. In this problem a salesman, starting from his home city, is to visit each city on a given list exactly once and then return home, in such a way that the length of his tour is minimal. The importance of TSP lies in the fact that it combines the typical features of a large class of combinatorial optimization problems and contains both the elements that attract mathematicians to a particular problem, namely simplicity of statement and difficulty of solution [Gar85]. The difficulty of TSP is due to the fact that the number of feasible solutions, called the solution space, is huge. A feasible solution of TSP can be represented as a permutation of the numbers from 1 to n, where n is the total number of cities. The size of the solution space is therefore $(n - 1)!$, which grows exponentially with the number of cities. If a computer is used to enumerate all of the possible solutions, and it is capable of doing that for a 20-city problem in an hour, it will take that computer 20 hours to enumerate the solutions for a 21-city problem, 17.5 days for a 22-city problem, and 6 centuries (!) for a 25-city problem.

It is therefore unpractical to use exhaustive search to solve a combinatorial optimization problem unless the problem size is very small. In most of the system synthesis problems, we have to deal with many decision variables at the same time and therefore exhaustive search cannot be used.

5.1.2   Branch-and-Bound Technique

One way to remove the need to search the whole solution space, while guaranteeing that the optimal solution will be found, is to use the *branch-and-bound* technique [AHU83]. The basic idea of the branch-and-bound technique is to work with partial solutions at hand, and then step by step add more information to these solutions, until they become complete. Here a partial solution represents the set of those complete solutions that can be derived from it. For example, in TSP, a complete solution can be built step by step by adding the next city to visit at each step. A partial solution, in this case, consists of a sequence of m ($m < n$) cities which have been visited up till now. The set of complete solutions which have the same sequence of m visited cities in the first part of the salesman's tour will be represented by this partial solution, since they can all be built by extending the partial solution with a particular permutation of the rest of the cities.

Starting with a partial solution, the branch-and-bound technique forms a tree where the branches from each node lead to new partial solutions which have been taken one step further, and the leafs of the tree are the complete solutions. A branch-and-bound technique for TSP can, for example, form a tree where the root represents the partial solution that the salesman's tour starts with the home city. A branch from the root leads to a new partial solution which represents the situation that a particular city is visited next. Since there are $n-1$ cities to be selected, the root has $n-1$ possible branches. From each of these $n-1$ partial solutions, new partial solutions can be reached by adding one more city to them. The next layer of the nodes will have $n-2$ branches each. Finally the complete solutions will be represented by the leaf nodes. If the complete tree is built, the number of leaf nodes will be the same as the total number of feasible solutions.

Example 5.1: Let us consider a simple 4-city TSP example. Suppose the travelling salesman starts from city 1. The distance from city i to city j is the same as that of travelling in the opposite direction and is denoted as $d_{i,j}$. And $d_{1,2} = 3$, $d_{1,3} = 6$, $d_{1,4} = 41$, $d_{2,3} = 40$, $d_{2,4} = 5$, $d_{3,4} = 4$. A branch-and-bound tree for this example is illustrated in Figure 5.1. The root of the tree, denoted by {1}, represents the partial solution that the salesman's tour starts from city 1. Since there are three possible cities to visit, the root has three branches, denoted by {1,2}, {1,3} and {1,4}, respectively, representing three partial solutions. From the partial solution {1,2} which means visiting city 2 after leaving city 1, two new partial solutions can be derived: one to visit city 3 while the other to visit city 4, leading to the branches denoted by {1,2,3} and {1,2,4}. From the partial solution {1,2,3}, the only way to complete the tour is to visit city 4 and return to the home city. The length of the complete tour {1,2,3,4} equals $d_{1,2} + d_{2,3} + d_{3,4} + d_{1,4}$, which is 88. In Figure 5.1, the leaf nodes,

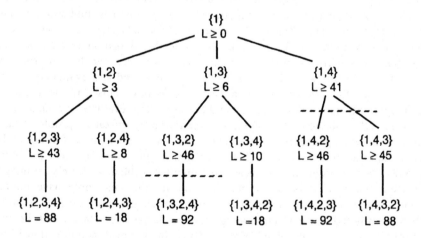

Figure 5.1: A branch-and-bound tree for a 4-city TSP example.

representing all feasible solutions, are labelled with the length of the tour respectively. In this example, since the distance function is symmetric, the feasible solutions can be grouped in pairs. For example, the solutions {1,2,3,4} and {1,4,3,2} are the same, both having a tour length of 88. The best pair of solutions in this example has a tour length of 18.

□

A branch-and-bound technique traverses the tree in some order to find the best leaf in it, without building explicitly the complete tree. Vital to the efficiency of the technique is the possibility of calculating, for a given partial solution, a lower bound on the cost function for all complete solutions that can be derived from it. By keeping track of the cost of the best complete solution found so far, a partial solution together with all its derivatives can be simply discarded if the lower bound is higher than the cost of the best solution found so far.

In the branch-and-bound technique for TSP, a simple lower bound for a partial solution is the current length of the salesman's tour of the m cities visited up to now, since the final length will be the sum of the current length and the length of the rest of the tour (we assume that there is no negative distance between any two cities). If this length is longer already than the best length found of a complete tour, there is no point to search further for solutions extending from the current tour, and the sub-tree with this partial solution as the root can be cut away. In Figure 5.1, each partial solution of the 4-city TSP example is attached with such a lower bound. Two possible cuts for the sub-trees which will not lead to better solutions are depicted by the two dashed lines in Figure 5.1.

It is easy to see that the above algorithm guarantees that the best solution will always be found, since the only unvisited solutions are those which are worse, with respect to the cost function, than some other visited solution. The efficiency of the algorithm, on the other hand, depends very much on the way the tree is searched. It is important to guide the algorithm towards a good solution as quickly as possible so that more sub-trees can be cut from the search tree. In this way, the number of nodes generated and evaluated will be smaller and the algorithm has a better efficiency. This can be done by employing heuristic rules in selecting the next node to evaluate in each step.

Many other methods for solving combinatorial optimization problems have also been developed during the recent years. One method is to formulate the combinatorial optimization problems as *integer programming* (IP) problems [Fou81]. An IP problem is a linear programming problem with only an integer solution space. That is, all of the decision variables can only have integer values. The integer programming problem is inherently more difficult to solve than the original linear programming problem. It has been shown that the size of problems that can be solved successfully by integer programming algorithms is an order of magnitude smaller than the size of LPs that can be easily solved.

In order to solve IP problems of large size, different techniques have been developed and among them heuristics methods are widely used in system synthesis.

5.2 Heuristics

Heuristics are techniques which seek good (i.e., near-optimal) solutions at a reasonable computational cost without being able to guarantee either feasibility or optimality. Due to the complexity of the many combinatorial optimization problems in system synthesis, heuristic algorithms are widely used.

For many NP-complete problems in general, heuristic techniques are the only feasible solutions, since many exact algorithms involve a huge amount of computation effort. Further, the decision variables in system synthesis frequently have complicated interdependency. For example, to improve a design in respect to the total cost of its implementation by sharing one physical module by two functional units, one will have to perform several modifications of the current design which involve the changes of several decision variables simultaneously. This includes removing one unit from the data path, introducing additional multiplexers, and adding control signals. It is very difficult to use an IP algorithm to solve such problems.

Another important argument for heuristics is that we have often nonlinear cost functions and constraints in system synthesis. Sometimes we don't even have mathematical functions. The cost function can, for example, be defined by a computer program.

Finally, it is important to note that what we are optimizing is actually a model of a real-world problem. There is no guarantee that the best solution to the model is also the best solution to the underlying real-world problem. Therefore, a near-optimal solution could be as good or even better than the optimum when the system is finally implemented. The importance here is to quickly produce a good enough solution. This is also extremely important in the case when cutting down the time-to-market is much more important than producing the best product. Therefore heuristics have been widely used in different steps of the system synthesis process.

5.2.1 Classification

Based on the basic characteristics of a heuristic approach, it can be classified into different groups.

The first group consists of heuristics which are based on the *constructive* approach. In such an approach a solution is generated by adding individual components one at a time until a feasible solution is obtained. For example, the ASAP scheduling algorithm discussed in Section 3.2 is a constructive algorithm which constructs a schedule step by step until all operations are scheduled.

Most of the constructive algorithms are of *greedy* nature; i.e., the algorithm tries to maximize improvement at each step. For example, the list scheduling algorithm sorts the operations that are available to be scheduled in each constructive step by a priority function. This function is defined to reflect the advantage of scheduling each operation, and the one with the maximal advantage will be selected first. Since the algorithm analyses in each step only locally the implication of the step, the best possible improvement in each step does not guarantee the optimality of the final solution.

Another group of heuristics is based on stepwise improvement of feasible solutions. Such an algorithm begins with a feasible solution and successively improves it by a sequence of moves (transformations). This type of optimization algorithms is especially interesting since it can be used directly in a transformational approach to system synthesis, which is the main theme of this book.

If the moves of a stepwise improvement algorithm are restricted to small changes of the current solution, the algorithm searches only a small area around the current solution in each step. This type of algorithms is called neighborhood-search (or local-search) algorithms.

> **Example 5.2:** In the travelling salesman problem, a neighborhood structure N_k, called the k-change, defines for each solution i a neighborhood S_i consisting of the set of solutions that can be obtained from the given solution i by removing k edges from the tour corresponding to solution i and replacing them with k other edges such that again a complete tour is obtained [Lin65].
>
> $\square$

A heuristic algorithm can also use the principle of *decomposition*. In this case, the algorithm is decomposed into a number of discrete steps, the output of one being the input to the next. In this way, the complexity of the algorithm is reduced.

Another principle often used by a heuristic algorithm is partitioning; the problem is broken into smaller sub-problems, each of which is solved independently. This is referred to as *divide-and-conquer*.

A heuristic algorithm can also be either problem-specific or more generally applicable. A problem-specific algorithm makes use of domain specific information of a design. It works for one problem but cannot be used to solve a different one. Recently there is increasing interest in techniques which can be applicable far more generally to a large number of problems. This chapter will present mainly heuristics which are more generally applicable; while the rest of the book includes discussion of many problem-specific heuristics, such as the left-edge algorithm used for register allocation and binding presented in Chapter 3, and the hierarchical clustering and Kernighan-Lin algorithm for system partitioning presented in Chapter 4.

5.2.2   Evaluation of Heuristics

When a heuristic algorithm is used, it is important to know how good it will perform. Evaluation of heuristic algorithms includes two basic aspects. The first aspect is how good the generated solution really is, and the second aspect is what is the typical time taken by the algorithm to find the solution.

It is usually very difficult to state how close a heuristic solution is to the optimum one, since the optimal solution is unknown. One often used technique is to find a lower bound which is as close as possible to the optimal solution for minimization problems. The distance between the heuristic solution and the lower bound can then be used to estimate the closeness of the generated solution and the optimal one. Such a lower bound may usually be found by using some form of relaxation of the problem to one which can be solved optimally much more easily.

As an example, there is the well-known assignment relaxation of the travelling salesman problem, which has been exploited in a number of ways. The assignment problem is given as follow:

Example 5.3: A set of n people is available to carry out n tasks. If person i does task j, it costs c_{ij} units. Find an assignment $\{\pi_i, \pi_2, ..., \pi_n\}$ which minimizes

$$\sum_{i=1}^{n} c_{i\pi_i}$$

The solution to the assignment problem (AP) can be represented as a permutation $\{\pi_i, \pi_2, ..., \pi_n\}$ of the numbers $\{1, 2, ..., n\}$, and such a solution can be found by a polynomial algorithm. If an instance of a TSP is solved as if it were an AP, there will typically be sub-tours in this solution. Clearly any valid tour must be generated by breaking these sub-tours and re-connecting them by links that cannot be shorter. Therefore the total distance represented by the AP solution must be a lower bound on the distance of a TSP solution.

$\square$

Another general approach to finding bounds is to put conditions on some partial aspect of a solution. For example, a simple bound for the TSP can be based on the necessary condition that every city must be connected to exactly two other cities. Therefore a valid lower bound on the optimal tour length is given by

$$\frac{\sum_{i=1}^{n} (d_{i,[1]} + d_{i,[2]})}{2}$$

where $d_{i,[k]}$ is the distance of the kth nearest city to city i.

It is also a common practice to compare the performance of a heuristic with that of existing techniques on a set of benchmarks [Ree93]. Benchmarks can be collected from real applications. However, real examples are difficult to obtain, and may represent only a small fraction of the possible instances of the optimization problem. Therefore benchmarks are often generated randomly.

Using benchmark examples from reality or generated randomly, it is also possible to study the typical execution time of heuristic algorithms. For example, random graphs are used to evaluate different heuristic algorithms for hardware/software partitioning [EPKD97] in terms of their execution time, which is explained in Chapter 8.

By testing a heuristic algorithm across a wide range of problem instances, it is hoped that we can obtain information about how well it performs in general in terms of both the generated solution's quality and the execution time. To do this properly, we must use a rigorous experimental procedure. The analysis of the results can then be carried out using standard statistical methods such as analysis of variance [Ree93].

5.2.3   Neighborhood Search

Suppose we have a minimization problem over a set of feasible solutions X and a cost function f, which can be calculated for all $x \in X$. In theory, the optimal solution could be obtained by an exhaustive search. However, as discussed before, in most system synthesis problems the set X will be far too large for this to be practical. Neighborhood search overcomes this problem by defining a neighborhood structure on it and searching only the neighborhood of the current solution for an improvement in each iteration.

The basic algorithm of neighborhood search is illustrated in Figure 5.2, where $N(x)$ denotes the neighborhood of x, which is a set of solutions reachable from x by a simple transformation. The algorithm starts with a starting solution which is assigned as the current solution. The neighborhood of the current solution is then examined to find a new solution according to some choice criteria. The new solution is then assigned as the current solution and the above procedure will be repeated until either no new solution can be found based on the choice criteria or the terminating criteria apply. During the above procedure, the best solution found so far will be remembered. When the algorithm terminates, the best solution found so far will become the final solution to the problem.

Depending on the selection of the choice criteria used to find the next solution, we have different neighborhood-search algorithms. The *descent* method exams a set of neighborhood solutions and selects the best among them. If the selected solution is better than the best solution found so far, it will be accepted as the next solution and the process is repeated with the updated best

Step 1   (Initialization)
(A)   Select a starting solution $x^{now} \in \mathbf{X}$;
(B) $x^{best} = x^{now}$, $best\_cost = c(x^{best})$;

Step 2   (Choice and termination)
          Choose a solution $x^{next} \in N(x^{now})$;
          If the choice criteria cannot be satisfied by any member of $N(x^{now})$,
              or the terminating criteria apply, **then** stop.

Step 3   (Update)
          Re-set $x^{now} = x^{next}$;
          If $c(x^{now}) < best\_cost$ **then** $x^{best} = x^{now}$, $best\_cost = c(x^{best})$;
          Goto Step 2.

Figure 5.2: The basic neighborhood-search algorithm.

solution found so far. Otherwise, the algorithm terminates and reports the best solution found so far as the final solution. The descent method can be further divided into the steepest descent method and the random descent method.

The steepest descent method searches the whole neighborhood and selects the one which results in the greatest improvement to the cost function. The random descent method selects neighboring solutions randomly and accepts the first solution which improves the cost function.

Neighborhood-search algorithms are very attractive for many combinational optimization problems since these problems have a natural neighborhood structure. If the number of elements in the optimal solution is fixed and known, a neighborhood can be defined as the set of solutions obtained by swapping a fixed number of elements in the current solution for the same number of non-solution elements.

As stated in Example 5.2, a neighborhood of the current solution for a travelling salesman problem can be obtained by removing two links and replacing them by two new links. If k items are swapped, these neighborhoods are called k-neighborhoods, and the solutions obtained by neighborhood search using the k-neighborhoods are called k-optimal solutions. There is a trade-off between solution quality and computation time represented by the different values of k. For small values of k the neighborhoods are small and can be searched efficiently, but the process is likely to converge to a local optimum. For large values of k the neighborhood size will increase dramatically, but there will be a smaller number of local optima and therefore the chances of obtaining the global optimum will be increased.

By its nature, the descent method often terminates with a local optimal solution, especially when the neighborhood size is small. To avoid stuck at local optima, we can enlarge the neighborhood size, as discussed before. We can also apply the same algorithm with several different starting solutions. In this way, the different runs of the algorithm will hopefully exploit the different parts of the solution space and the chances of finding the global optimum is again increased.

Another method to escape local optima is to allow some uphill moves. An uphill move or a sequence of them can often guide an algorithm from a local optimum towards a new area of the solution space. However, uphill moves must be controlled carefully to avoid the algorithm going into a cycle. Many schemes of allowing uphill moves have been developed during the recent years and the most popular methods are simulated annealing and tabu search which will be presented in the following two sections.

5.3   Simulated Annealing

A *Simulated annealing* (SA) algorithm is similar to the random descent method in that the neighborhood is sampled at random. It differs in that a neighbor giving rise to an increase in the cost function may be accepted. This acceptance will depend on a control parameter (called temperature) and the magnitude of the increase [KGP83]. By allowing uphill moves in a controlled manner, SA provides a mechanism to allow the algorithm to escape from local optima.

5.3.1   Introduction and the Algorithm

The origin of SA goes back to 1953 when it was used to simulate, on a computer, the annealing process of materials. Annealing is the slow cooling of a material after first heating it pass its melting point. Cooling is then slowly done and when cooling stops, the material settles into a low energy state. Intuitively at high temperature, the atoms are randomly oriented due to their high energy status caused by heat, as illustrated in Figure 5.3a. When the temperature reduces, the atoms tend to line up with their neighbors, but different regions may still have different orientation directions, as illustrated in Figure 5.3b. According to thermodynamic theory, if cooling is done slowly, the

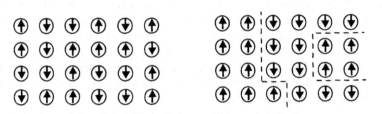

(a) At melting temperature   (b) When temperature reduced

Figure 5.3: Orientation of atoms at different temperatures.

final frozen state will have a near-minimal energy state.

An annealing process can be simulated using thermodynamic simulation, which was done by Metropolis in 1953. Such a simulation generates a *random* perturbation of atoms and calculates the resulting energy change. If the energy has decreased, the system moves to this new state. If energy has increased, on the other hand, the new state is accepted according to the laws of thermodynamics which states that at temperature t, the *probability* of an increase in energy of magnitude ΔE is given by $p(\Delta E) = \exp(-\Delta E/kt)$, where k is called the Boltzmann's constant.

In 1983, Kirpatrick showed that the SA algorithm could be applied to combinatorial optimization by making use of the similarity between thermodynamic simulation and combinatorial optimization. This similarity is summarized in Table 5.1 and the basic SA algorithm for the combinatorial optimization problems is given in Figure 5.4. The algorithm starts with an initial solution. A neighbor of this solution is then randomly selected. If the selected solution is better than the current solution, it will always be accepted and become the next solution, as in the case of the random descent method. If the selected solution is worse, it will be accepted with the probability $\exp(-\delta/t)$. Therefore, at the beginning when the temperature t is high, the probability of accepting a worse neighbor and make an uphill move is high. With the reduction of temperature, this probability decreases. Further, the probability of accepting a very much worse solution is smaller than accepting a solution which is only slightly worse than the current solution. Note here that t is only a control parameter and has no physical analogy. Therefore Boltzmann's constant is not needed here.

Table 5.1: Relation between thermodynamic simulation and combinatorial optimization

Thermodynamic simulation	Combinatorial optimization
system states	feasible solutions
energy	cost
change of state	moving to a neighboring solution
temperature	"control parameter"
frozen state	heuristic solution

The above process of neighborhood search will be repeated for *nrep* times and then the control parameter (temperature) will be reduced. The same procedure is iterated again and this will continue until the stopping condition is true. The final solution will then become the solution to the optimization problem.

Many research results have been reported in the literature on the use of SA for different system synthesis and other combinatorial optimization problems.

Select an initial solution $x^{now} \in X$, an initial temperature $t_0 > 0$, and a temperature
reduction function α;
Repeat
  Repeat
        Randomly select $x^{next} \in N(x^{now})$;
        $\delta = f(x^{next}) - f(x^{now})$;
        If $\delta < 0$ then $x^{now} = x^{next}$ else
        Begin
            generate a random number p uniformly in the range (0, 1);
            If $p < \exp(-\delta/t)$ then $x^{now} = x^{next}$;
        End
    Until *iteration_count = nrep*;
    Set $t = \alpha(t)$;
  Until stopping condition = true;
  Return x^{now} as the solution.

Figure 5.4: The basic simulated annealing algorithm.

Some theoretical results regarding the general properties of SA have also been
reported [Ree93]. It has been shown that for a given temperature, as long as it is
possible to move from any solution into any other with non-zero probability,
SA converges towards a stationary distribution which is independent of the
starting solution. As the temperature tends to zero, the form of the stationary
distribution tends to a uniform distribution over the set of optimal solutions. To
obtain solutions arbitrarily close to the optima, the number of iterations at a
single temperature will be at least quadratic in the size of the solution space. As
the solution space is usually exponential in problem size, the running time for
SA which guarantees optimality will be exponential. Therefore for practical use
of SA in system synthesis, we have to find an efficient implementation of SA
algorithms which can provide good enough solutions in a short amount of time.
This can be done by careful selection of the generic parameters used in the
algorithm and making the right problem-specific decisions, as to be discussed
in the next two sub-sections respectively.

5.3.2   Generic Decisions

The generic decisions of SA involve first the *cooling schedule*, including the
starting temperature, the number of iterations on a given temperature, the rule
for lowering of the temperature, and the stopping criterion. All together, these
parameters influence the total number of moves performed, and implicitly the
execution time and the quality of the final result.

The *initial temperature* (t_0) must be "hot" enough to allow an almost free
exchange of neighborhood solutions, if the final solution is to be independent
of the starting one. In some cases there is enough information in the problem to

estimate t_0. For example, if the maximum difference, *maxd*, in cost between neighboring solutions is known and we want the accepting probability for any solution to be at least p, we can let $t_0 = - maxd/(ln\ p)$.

If it is impossible to make any estimation, a system can be heated rapidly until the proportion of accepted moves to rejected moves reaches a given value. Cooling will then start. This method corresponds to the physical analogy in which a solid material is heated quickly to its melting point before being cooled slowly according to the annealing process.

The rate at which the temperature parameter is reduced is vital to the success of an SA algorithm. This is determined by the number of repetitions *nrep* at each temperature and the rate at which the temperature is reduced. There are two basic temperature reduction schemes: a large number of iterations at few temperatures or a small number of iterations at many temperatures.

One commonly used temperature reduction scheme is the geometric reduction function $\alpha(t) = at$, where $a < 1$. Experience has shown that relatively high values of a perform best and most reported successes in the literature use values between 0.8 and 0.99 [Ree93], which corresponds to fairly slow cooling. Further the reduction rate can also be changed during the annealing process; for better results, it should be slower in middle temperature ranges.

The number of repetition (*nrep*) at each temperature should also be carefully selected. In theory, a number of iterations exponential in problem size is needed. In practice, *nrep* is a polynomial of the problem size. It may vary from temperature to temperature. In this case, it should be larger at lower temperatures to ensure that a local optimum has been fully explored. The repetition number *nrep* can also be determined by feedback from the annealing process. Typically the algorithm will accept a given number of moves and then change the temperature. A maximal iteration bound should also be given to force the termination at the current temperature if not enough accepting moves will be made.

Other cooling schemes have also been reported and one of them is to have only one iteration at each temperature with $\alpha(t) = t/(1 + bt)$ where b is a suitably small value [Ree93].

The last generic decision to make involves the stopping condition. In theory, the temperature should be allowed to decrease to zero before the stopping condition is satisfied. However, there is no need to decrease the temperature this far. The criterion for stopping can be expressed either in terms of a minimum value of the temperature or the freezing of the system at the current solution.

If the algorithm wants to produce a solution which is within ε of the optimum with probability θ, it can stop when

$$t \le \frac{\varepsilon}{\ln[(|S| - 1)/\theta]}$$

where S is the solution space. Other rules can also be used as stopping conditions. For example, we can use the number of iterations or temperatures which has passed without an acceptance. The algorithm can also stop when the proportion of accepted moves drops below a given value. Finally, it can stop when a given total number of iterations have been completed. In all these cases, the best parameters of SA have usually been determined after much experimentation.

5.3.3   Problem-Specific Decisions

The problem-specific decisions of an SA algorithm deal with the solution space, neighborhood structure, and the cost function. The main objectives of these decisions are to make sure that the validity of the algorithm is maintained, the computation time should be used effectively for as many iterations as possible, and the solution should be close to the optimum.

Different from the generic decisions, there are no general rules for the problem-specific decisions since they depend usually on the special characteristics of the problem to be solved. Further, many of them must be fine-tuned based on experimentation with typical input data. Finally, it may be beneficial to use different parameters for problems of different sizes.

When making decisions about the neighborhood structure, the first thing to consider is that every solution should be reachable from every other, which is required for convergence of the solutions. For the algorithm to be efficient, it must be easy to generate randomly a neighboring feasible solution. If the solution space is constrained by stringent feasibility conditions, the restrictions may be overcome by relaxing the feasibility conditions and including a penalty term in the cost function. This means that the algorithm will accept some solutions which are infeasible by the original definition. However, these solution are associated with a penalty in the cost function to discourage their acceptance and avoid the final solution to be an infeasible one. The relaxation of feasibility conditions, however, increases the size of the solution space, which consequently increases the number of iterations needed, since many studies show that the number of iterations depends on the size of the solution space. Therefore, the relaxation must be done with care so that the solution space will not be expanded dramatically.

Generally speaking, the size of the neighborhood should also be kept reasonably small in order to allow adequate search in fewer iterations. The cost difference between a neighbor and the current solution should also be able to be efficiently calculated.

Let us use the node coloring problem as an example to illustrate some of the problem-specific decisions for a SA algorithm. The node coloring problem is a well-known graph problem with many applications in system synthesis, such as the register assignment problem discussed in Chapter 3.

Example 5.4: A graph consists of a set of nodes, some of which are

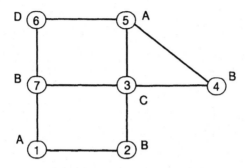

Figure 5.5: A graph with 4-coloring.

connected to each other by edges. Two nodes connected by an edge are said to be adjacent. The node coloring problem deals with allocating a color to each node such that adjacent nodes are not given the same color. Figure 5.5 depicts a 4-coloring with colors A, B, C, and D. The objective is to find an allocation with the smallest possible number of colors [Ree93].

□

The natural definition of the solution space of the node coloring problem is the set of feasible colorings. These correspond to the partitions of the nodes into k subsets such that there are no edges between any two nodes in the same subset. The nodes within a subset can then be colored in a single color. And the objective is to minimize the number of subsets used. For the example given in Figure 5.5, the partition is given by the subsets: {1, 5}, {2, 4, 7}, {3}, and {6}. A straightforward cost function of the node coloring problem is given by the number of subsets used.

The neighborhood of a given solution can be defined by swapping some of the nodes between two subsets. Since arbitrary swaps will often lead to solutions which fall outside the solution space, random selection of two nodes to swap does not work well. For example, if nodes 1 and 4 are selected in Example 5.3, swapping them will result in that nodes 4 and 5, which are connected by an edge, to be colored by the same color, violating the feasibility requirement. This problem can be overcome by using Kempe chains [Ree93]. If we consider only two colors, we have a number of disconnected components, which are called the Kempe chains. Figure 5.6 illustrates the Kempe chains for the colors A and B for Example 5.4.

If any node from a given Kempe chain is moved from its current subset into the other, all the other nodes in the same chain will have to swap subsets in order to maintain feasibility. For example, to move node 1 from color A to B, nodes 2 and 7 should be moved from color B to A. In this way, feasibility can be maintained by defining the neighboring solutions as those obtained by

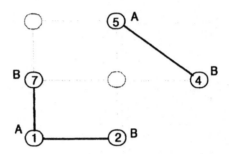

Figure 5.6: AB-Kempe chains.

swapping the colors of nodes in a single Kempe chain [Ree93]. This is achieved
at the expense of calculating or storing/updating the Kempe chains in each
iteration, which could be quite time consuming with large subsets of nodes.

An alternative method is to relax the definition of feasibility, and add a
penalty in the cost function to ensure that infeasible solutions are expensive.
The feasible solution can now be defined as any partition of the nodes into
subsets. A term involving the product of a penalty factor and the total number
of edges between nodes in the same subset can be added to the cost function. In
this way, a neighborhood solution can be generated simply by moving a node
from one subset to another.

5.3.4   Discussion

Generally speaking, SA is most useful in complex situations for which it is
difficult to design robust heuristics which take into account the problem
structure. Implementing an SA algorithm is also relatively straightforward. The
main weakness of SA is the long running time of the algorithm. Further there
are no clear rules for parameter selections and neighborhood structure
decisions. Fine-tuning of the cooling schedule and other parameters is also very
time consuming. Such a tuning for an SA-based algorithm for hardware/
software partitioning will be described in Chapter 8.

5.4   Tabu Search

Tabu search (TS) is a neighborhood-search method which employs intelligent
search and flexible memory technique to avoid being trapped at local optima.
As in the case of simulated annealing, TS is a "higher level" heuristic
procedure used to guide other methods towards an optimal solution. TS is based
on the assumption that intelligent search should be based on more systematic
forms of guidance rather than random selection. It also exploit flexible *memory*

to control the search process. The chief mechanism for exploiting memory is to classify a subset of the neighborhood moves as forbidden (called tabu). TS maintains a selective history of the states encountered and/or the moves executed during the search. TS has succeeded in obtaining optimal and near optimal solutions to a wide variety of classical and system synthesis problems. It outperforms, in many cases, classical heuristics and other neighborhood-search methods, such as simulated annealing.

5.4.1   The Basic Scheme

Tabu search is a neighborhood-search method, which allows both downhill and uphill moves. Moves are selected intelligently (best *admissible* moves are selected). *Tabus* are used to restrict the search space during the neighborhood-search process and avoid cyclic behavior (deadlock). The classification of tabus is based on the history of the search.

In the travelling salesman problem, given a solution represented by a permutation of the numbers from 1 to n, where n is the total number of the cities, a neighborhood structure can be defined, as discussed before, by the set of solutions obtained by removing two edges from the current tour and replacing them with two other edges (this corresponds to the swap of two numbers in the current permutation and reversing the order of all cities in the tour between these two cities). Using TS, the best of all the 2-change neighboring solutions will then be selected as the next solution. To avoid repeating or reversing the swaps done recently, the most recent swaps are classified as tabus. Typically a tabu will only be valid for a number of iterations (called tabu tenure). When a tabu move would result in a solution better than any visited so far, its tabu classification may be overridden. A condition that allows such override to occur is called an *aspiration criterion*.

The tabus are stored in a recency-based memory (short-term memory), which is complemented by a frequency-based memory (long-term memory) to diversify the search into new regions. For example, the frequency of the swaps of the different pairs of cities in the TSP can be stored. When no admissible moves exist, a diversification can be performed to guide the algorithm to the part of solution space which has not be visited frequently. This can be done by selecting the swap which has been least performed up till now. Finally, a so-called intermediate-term memory can also be used for intensification, i.e., to reinforce move combinations and solution features historically found to be good. This memory is also frequency-based.

Figure 5.7 illustrates the basic algorithm for the TS method, without diversification or intensification, which follows the basic strategy of a steepest descent algorithm. The main differences here are the use of a history record H; the selection of a set of candidate solutions $Candidate\_N(x^{now})$ from the set of neighboring solutions, taking into account H, $N(H, x^{now})$; and the evaluation of

Step 1	(Initialization)
(A)	Select a starting solution $x^{now} \in \mathbf{X}$;
(B)	$x^{best} = x^{now}$, $best\_cost = c(x^{best})$;
(C)	Set the history record H empty;
Step 2	(Choice and termination)
	Determine $Candidate\_N(x^{now})$ as a subset of $N(H, x^{now})$;
	Select x^{next} from $Candidate\_N(x^{now})$ to minimize $c(H, x)$ over this set;
	Terminate by a chosen stopping condition and return x^{best} as result;
Step 3	(Update)
	Re-set $x^{now} = x^{next}$;
	If $c(x^{now}) < best\_cost$ **then** perform Step 1(B);
	Update the history record H;
	Goto Step 2.

Figure 5.7: The basic tabu search algorithm.

the cost function, taking into account H, $c(H, x)$. These issues will be discussed in the following sections.

5.4.2   Tabus and Tabu Tenure

A tabu is usually specified by some attributes of the moves. Typically when a move is performed that contains an attribute e, a record is maintained for the reverse attribute $\bar{e}$, in order to restrict a move that contains some subset of such reverse attributes. Tabus are therefore used to preventing reversals of recent moves as well as repetitions.

> **Example 5.5:** In the travelling salesman problem, if a 2-change neighborhood structure is used, some examples of tabus can be: (a) the cities involved in the swap, which is used to prevent the cities from moving; and (b) the positions the cities occupy before the swap, which is used to prevent the cities from returning to the original positions.
>
> □

A tabu restriction is typically activated only under certain conditions. For recency-based restriction, its attributes should occur within a limited number of iterations prior to the present iteration (the number of these iterations equals the tabu tenure). For frequency-based restriction, they should occur with a certain frequency over a longer span of iterations. The condition of being tabu-active or tabu-inactive is called the *tabu status* of an attribute.

The tabu restrictions and tenure should be selected to achieve cycle prevention and induce vigor into the search.

It is generally preferable to select an attribute whose tabu status less rigidly restricts the choice of available moves. For example, in Example 5.5, Tabu(a) is much more restrictive than Tabu(b): if a city is moved in an iteration, it cannot subsequently be moved at all during the tabu tenure of Tabu(a), while in the case of Tabu(b), it can still be moved except moving back to the original position.

Implementation of recency-based memory can be done in two ways. In the first approach, we use an array $tabu\_end(e)$, where e ranges over the tabu attributes, to represent the ending iterations. On iteration i, to activate a tabu attribute e, let $tabu\_end(e) = i + t$, where t is the tabu tenure. To test if e is tabu-active, check if $tabu\_end(e) \geq current\_iteration$.

The second approach to implement the recency-based memory is to allocate a given number of memory slots for keeping the tabu list. Old tabus will be dropped when the slots are filled out.

Tabu Tenure

The tabu tenure, t, must be carefully selected. Generally speaking, for highly restrictive tabus, t should be smaller than that for lesser restrictive tabus. It should be long enough to prevent cycling, but short enough to avoid driving the search away from the global optimum.

Tabu tenure can be determined using static or dynamic rules. The static rule chooses a value for t that remains fixed. Two approaches can be used for the static rule. One is to let t be a constant, typically between 7 and 20 [Ree93]. The other is to let $t = f(n)$, where n is the problem size (typically between $0.5n^{1/2}$ and $2n^{1/2}$). Experimentation must be carried out to choose the best tenure.

Dynamic rules allow the value of t to vary during the search process. In a simple dynamic scheme, t varies (randomly or systematically) between two fixed bounds. In an attribute-dependent dynamic scheme, the bounds are determined based on the properties of attributes. For example, they will be larger for more "attractive" attributes. That is, good quality moves will imply a larger t. A weaker restriction should usually have a larger t. For example, in TSP, the tabu list may consist of the edges recently added and the edges recently dropped. In this case, the tabus preventing edges from being dropped should have a shorter tenure, since they are restrictive due to the small number of edges in the tour (which equals n, the number of cities). On the other hand, the tabus preventing edges from being added can have much longer tenure, since there are many other edges that can be considered to be added to the tour ($n^2 - n$ edges can be considered).

5.4.3   Aspiration Criteria

Aspiration criteria (AC) are used to determine when the tabu restrictions can be overridden, in order to encourage selection of high-quality solutions. They contribute significantly to the quality of the TS technique.

ACs are often selected based on the concept of *influence*, which measures the degree of change in solution structure of a particular move or feasibility of the generated solution. There are two kinds of aspirations: *move aspirations* which revoke a move's tabu classification and *attribute aspirations* which revoke an attribute's tabu-active status. Some examples of AC are presented as follows.

Aspiration by Default

If all available moves are classified as tabu, and are not rendered admissible by some other AC, then a "least tabu" move will be selected.

This AC should always be implemented, e.g., by selecting the tabu with the shortest time to become inactive.

Aspiration by Objective

In this case, the AC is given as: $c(x^{trial}) < best\_cost$ so far.

Aspiration by Search Direction

Let *direction(e)* be 'improving' if the most recent move containing attribute e was an improving move, and *direction(e)* be 'non-improving', otherwise.

An attribute aspiration for e is satisfied, if *direction(e)* is 'improving' and $c(x^{trial}) < c(x^{now})$.

5.4.4   Diversification and Intensification

Frequency-based memory is used to complement recency-based memory, in order to broaden the foundation for selecting preferred moves. A frequency measure is given by the ratio of the number of occurrences of a particular event to a given denominator. An example of denominators are the total number of occurrences of all moves. Frequency-based memory is used for diversification and intensification.

> **Example 5.6:** For the TSP, a TS algorithm can keep track of the edges visited so far by maintaining a two dimensional array of occurrence of each edge. After each move, the entries corresponding to all the edges in the new tour are incremented by one. After a specified number of iterations, a new starting tour is generated based on the edges that have been visited the least. This is an example of diversification, since the algorithm is forced to the area which has not been visited frequently before.
>
> □

Diversification encourages compositions of attributes significantly different from those encountered previously during the search. It is used mainly for the algorithm to jump out from local optima.

Intensification, on the other hand, encourages the incorporation of "good attributes" in the solutions. It is used to converge to local optima or the global optimum.

Intensification and diversification counterbalance and reinforce each other.

Let S denote a subsequence of the solution sequence generated to the current point. For diversification, S is usually chosen to be a significant subset of the full solution sequence. For intensification, S is chosen to be a small subset of the elite solutions (high quality local optima) that share a large number of common attributes, and whose members can reach each other by relatively small numbers of moves [Ree93].

5.4.5   Neighborhood Selection Techniques

A simple TS algorithm can evaluate all moves from the entire neighborhood and select the best move in each iteration. This is however very time consuming. To avoid evaluating moves from the entire neighborhood, different methods can be used. The simplest method is to sample the neighborhood space at random by using the Monte Carlo principle. Another method is to use systematic selection. The neighborhood is decomposed into critical subsets. One subset is selected for evaluation at each iteration. The algorithm ensures that the subsets not examined on one iteration will be evaluated in the subsequent iterations. We can also use a master list to keep the best moves. The master list is used for move selections for several iterations. A threshold of acceptability triggers the creation of a new master list.

5.4.6   Stopping Conditions

Different conditions can be used to terminate a TS algorithm. One simple condition is that a fixed number of iterations has elapsed in total. We can also use the condition that a fixed number of iterations has elapsed since the last best solution was found, or a given amount of CPU time has been used.

Generally speaking, a TS algorithm does not converge naturally, as in the case of simulated annealing, and the stopping condition must be fine-tuned with experimentation.

5.4.7   Discussion

TS is extremely useful when the feasibility condition is very strong and the randomly generated neighborhood solutions are usually unfeasible ones. This is partly due to that TS emphasizes complete neighborhood evaluation to identify moves of high quality, while SA samples the neighborhood solutions randomly.

Another main difference between TS and SA is that TS makes heavily and intelligently use of both short-term and long-term memory, while SA is memoryless.

The main weakness of TS is that no theory has yet been formulated to support TS and its convergence behavior. Good understanding of the problem structure is also required. Domain specific knowledge is needed for selection of tabus and aspiration criteria. Further, experiments might have to be done using different tabu classification schemes and aspiration criteria.

An application of TS to the hardware/software partitioning problem will be presented in Chapter 8.

5.5   Genetic Algorithms

A *genetic algorithm* (GA) [Hol75] performs a multi-directional search by maintaining a population of potential solutions. The population undergoes a simulated evolution from one generation to another: at each generation the relatively good solutions reproduce, while the relatively bad solutions die. The goodness or badness of the solutions are defined by a cost function.

5.5.1   Natural Selection and the Basic Algorithm

GA uses the population genetics metaphor. An optimization problem is mapped into the problem of finding the most fit *individual* within a *population* during an evolution process. Fitness is measured by a fitness function, which is related to the objective function of the optimization problem. Individuals are the equivalent of solutions and a population is a set of N individuals.

In nature, individuals most fit survive. Adapting to a changing environment is essential for the survival of individuals of each species. While the various features that uniquely characterize an individual determine its survival capacity, the features in turn are determined by the individual's genetic content. Specifically, each feature is controlled by a basic unit called a gene. The sets of genes controlling features form the *chromosomes*. Although evolution manifests itself as a secession of changes in species' features, it is the changes in the species' genetic material that form the essence of evolution. Specifically, evolution's driving force is the joint action of natural selection and the recombination of genetic material that occurs during reproduction [SrPa94].

Evolution is initiated when the chromosomes from two parents recombine during reproduction. New combinations of genes can be generated by the exchange of genetic material among chromosomes, called *crossover*.

Holland developed genetic algorithms in the early 1970s as algorithms that mimic the evolutionary process in nature [Hol75]. A GA starts with a set of initial solutions. Each solution is encoded as a *chromosome* which is represented as a string of bits from a binary alphabet. To generate new solutions there are two typical operations which are performed on the solutions of the present generation: *crossover* and *mutation*. For the crossover, two solutions S_1 and S_2 of the current generation are selected and the chromosome corresponding to a new solution is produced. The new chromosome is the result of mixing a part of the chromosome of S_1 with a part of that corresponding to S_2. This means that the new solution inherits certain features of its two parent solutions. The mutation operator, on the other side, produces a small, random perturbation to a given solution (chromosome).

An essential feature of a GA is the so called *selection scheme*. It determines the solutions which are considered as parents for crossover operations or as candidates for mutation. An essential criterion at selection for crossover is the *fitness* of the solution, defined by the cost function. Thus, selection of fit solutions for crossover ensures the propagation of high quality features into the next generation. Usually selection is based on a probabilistic scheme which favors candidates having a high fitness. Mutations are very often performed on randomly selected chromosomes in the current generation.

Producing several successive generations, the average fitness of the solutions is increasing. The algorithm is usually stopped after a certain number of iterations or when no further improvements are produced. The best solution that has been produced is one which hopefully is close to the optimum. Figure 5.8 illustrates the simple genetic algorithm, which is a "high level" heuristic. To apply a GA to a problem, it is necessary to identify: (1) a meaningful representation for the candidate solutions; (2) a fitness function to assess different solutions; and (3) a set of useful genetic operators, that can efficiently recombine and mutate candidate solutions.

5.5.2   Encoding

An essential issue in GAs is the encoding mechanism for the representation of the optimization problem's variables. The encoding mechanism depends on the

```
begin
    Initialize population;
    Evaluate population;
    while termination criterion not reached do
    begin
        Select solutions for the next population;
        Perform crossover and mutation;
        Evaluate population;
    end
end
```

Figure 5.8: The simple genetic algorithm.

nature of the problem variables, and it should map each solution to a unique binary string.

Since most system synthesis tasks are combinatorial optimization problems and the problem variables are integer, they can be directly coded using a fixed number of binary bits. The binary codes of all the variables can then be concatenated to obtain a binary string.

A drawback of encoding variables as binary strings is the presence of Hamming cliffs: large Hamming distances between the binary codes of adjacent integers [SrPa94]. For example, 01111 and 10000 are the integer representations of 15 and 16, respectively, and have a Hamming distance of 5. For GA to improve the code of 15 to that of 16, it must alter all bits simultaneously. Such Hamming cliffs present a problem for GA, since both mutation and crossover cannot overcome them easily. A technique called *gray codes* ensures that the codes for adjacent integers always have a Hamming distance of 1. However, the Hamming distance does not monotonously increase with the difference in integer values in gray codes, which introduces Hamming cliffs at other levels [Ree93]. The selection of an appropriate coding scheme depends on the characteristics of the application.

It is also very useful to have an encoding scheme which has such property that when a mutation or crossover operation is performed, the resulting chromosome always corresponds to a meaningful solution. For example, hardware/software partitioning can be formulated as a two-way partitioning problem, where decision variable, x_i, indicates whether the function v_i should be implemented in hardware ($x_i = 0$) or software ($x_i = 1$). We can use a binary string of length n to represent a solution, where n is the total number of functions in the input specification. Each bit position in this string represents a decision variable. If the ith bit is 0, it indicates that v_i is assigned to the hardware implementation; otherwise, the ith bit is 1 and v_i is assigned to the software implementation. With this encoding scheme, no matter what mutation or crossover operation is perform, the resulting binary string always corresponds uniquely to a solution for this hardware/software partitioning problem.

It is, however, not always possible to have an encoding scheme which has the above property. In this case, a mutation or crossover operation could generate a string which dose not correspond to a solution. For example, if we have a three-way partitioning problem where a decision variable, x_i, can assume the value of 0, 1, or 2. Two binary bits must be used to represent such a variable, and the pattern '11' should not be used. A mutation operation can nevertheless generate a string where some of the bit pairs have the '11' pattern, by simply flipping a bit from 0 to 1. To solve this problem, *evolution programs* can be used [Mic94]. In an evolution program, the chromosomes are not restricted to consisting of binary strings. They can be built with arbitrary data structures, such as strings of integers. For example, in the case of the three-way partitioning problem, a string of integers in the range of 0 to 2 can be used to uniquely represent a solution. The operators for mutation and crossover in an evolution program can also be tailored to the data structure to make sure that all generated strings represent feasible as well as meaningful solutions. In the following subsections, we will, however, only discuss the classical GAs.

5.5.3   Fitness Function and Selection

The objective function of the original optimization problem (maximization problem) provides the mechanism for evaluating each string. However, its range of values varies from problem to problem. To maintain uniformity over various problems, we can use the fitness function to normalize the objective function to a convenient range between 0 and 1. The normalized value of the objective function becomes the fitness of the string, which will be used by the selection mechanism to evaluate the strings of the population.

The selection mechanism models the survival-of-the-fittest principle of nature. Fitter solutions survive while weaker ones perish. In the simple GA, a fitter string produces a higher number of offsprings and thus its features have a higher chance of surviving in the subsequent generation. In the *proportionate selection scheme*, a string with fitness value f_i is allocated f_i/f offsprings on average, where f is the average fitness value of the population [SrPt94].

To implement the proportionate selection principle, a *roulette wheel selection scheme* can be used. Each string is allocated a sector of a roulette wheel with the angle subtended by the sector at the center of the wheel equaling $2\pi f_i/f$. A string is allocated an offspring if a randomly generated number in the range 0 to 2π falls in the sector corresponding to the string. The algorithm selects strings in this way until it has generated the entire population of the next generation. Usually, the algorithm maintains the same number of individuals in the different generations.

5.5.4   Genetic Operators

During the reproduction phase of the genetic algorithm, the two classical genetic operators are crossover and mutation. In many applications, the simple crossover operator has been shown to be very effective; the addition of mutation merely helps to preserve a reasonable level of population diversity. There are also applications where more advanced operators have been found useful.

<u>Crossover</u>

A crossover is an exchange of genetic material between two strings (solutions). The simplest approach to crossover is a single-point crossover. In this case, each of the two strings is cut into two parts, with the cutting position being selected randomly, but identical in the two strings. The second parts of the two strings are exchanged, and the new strings are put together again.

Crossover is not always effected. After choosing a pair of strings, the algorithm invokes crossover only if a randomly generated number in the range 0 to 1 is greater than p_c, the *crossover rate*. Otherwise, the strings remain unaltered. The value of p_c lies in the range from 0 to 1.

The difficulty in defining a crossover operator is to find one which generates strings corresponding to meaningful solutions.

Mutation

After crossover, strings are subjected to mutation. Mutation of a bit involves flipping it: changing a 0 to 1 or vice versa. Just as p_c controls the probability of a crossover, another parameter, p_m (the *mutation rate*), defines the probability that a bit of the selected string will be flipped. The bits of a string are independently mutated.

Usually mutation is treated only as a secondary operator with the role of restoring lost genetic material. For example, suppose all the strings in a population have converged to a 0 at a given position and the optimal solution has a 1 at that position. Then crossover cannot regenerate a 1 at that position, while a mutation could. Generally speaking, mutation decreases the risk that the entire population will gather around a single point in the search space.

Other Operators

Another genetic operator is called survival. This means that the best individuals from a generation may survive to the next by simply coping itself. The number of surviving individuals are determined by a constant factor, p_s. Therefore there is no randomness in the surviving mechanism.

5.5.5   Selection of Control Parameters

The efficient utilization of GAs depends very much on the right choice of the different control parameters. This includes the crossover and mutation rates and the population size.

Generally speaking, increasing the crossover probability increases recombination of building blocks, but is also increases the disruption of good strings. Increasing the mutation probability tends to transform the genetic search into a random search, but it also helps to reintroduce lost genetic material. Increasing the population size increases its diversity and reduces the probability that the GA will prematurely converge to a local optimum, but it also increases the time required for the population to converge to the optimal regions in the search space [SrPt94].

The difficulty of control parameters selection is due to the interaction of the different genetic operators, and the dependence on the nature of the objective function. The general selection of control parameters remains an open question. Several researchers have nevertheless proposed control parameter sets that guarantee good performance on carefully chosen types of objective functions [SrPt94]. Two distinct parameter sets have been proposed: One has a small population size and relatively large mutation and crossover probabilities, while the other has a larger population size, but much smaller crossover and mutation probabilities. Typical parameters used for the first group are: Crossover rate of 0.9, mutation rate of 0.01, and population size of 30 [Gre86]. For the second group, we have crossover rate of 0.6, mutation rate of 0.001, and population size of 100 [DeSp90].

5.5.6   Discussion

GAs are very efficient in searching a wide area of a complex solution space for regions with a high probability to contain good solutions. However, they are known for their weakness in searching a certain region of the solution space for local optima or the global optimum. In order to avoid excessively long run times and to improve the quality of resulted solutions, GAs can be combined with other optimization heuristics into so called hybrid GAs. After a certain number of iterations using the genetic approach, the most promising solutions are used as starting points for SA or TS algorithms [Alp95].

PART II   TRANSFORMATIONAL APPROACH

6

Transformational
Design Basics

The second part of this book, Chapters 6 to 8, will present several transformational techniques to system synthesis. Typically, such techniques take a behavioral VHDL specification as input and map it first into a naive structural implementation. The implementation is then modified by iterative application of semantic-preserving transformations. These transformations do not change the level of abstraction, rather they explore the design space, looking for near-optimal solutions. When the transformation procedure terminates and a satisfactory solution is found, the design will be mapped into a lower-level description and the synthesis process continues to refine the design into a final implementation.

This chapter will address several basic issues of the transformational techniques. In particular, it will describe a design representation model which is used during the transformation process, the mapping of VHDL to this design representation, some basic transformations, and the selection of transformations during high-level synthesis.

6.1 Design Representation

Central to a transformational approach to synthesis is to have a design representation model that can be used to capture the intermediate results accurately to facilitate analysis of the current solution [Pen87]. This model must be able to represent designs with different degrees of completion. That is, it should be able to capture a fairly abstract design with a lot of unspecified information, for example, a pure behavioral specification. At the same time, it should be able to describe a very detailed implementation with physical parameters which is produced by the synthesis process. Thus, it is not necessary or always possible to have just one design representation model; a lot of synthesis systems actually use several different representation models during different stages of the synthesis process.

In most design environments, however, it is very useful to have a unified design representation which can be used to represent the design at different degree of completeness. The Extended Timed Petri Net (ETPN) representation model is such a unified design representation [Pen87, PeKu93]. ETPN consists of two separate but related parts: the control part and the data path. The data path is represented as a directed graph with nodes and arcs. The nodes are used to capture data manipulation and storage units. The arcs represent the connections of the nodes. The control part, on the other hand, is represented as a timed Petri net with restricted transition firing rules [PeKu93]. A very simple example of ETPN is shown in Figure 6.1. Figure 6.1b shows the data path where each data path node is depicted as a rectangle with a label indicating the basic operation of the node. The arcs of the data path represent the data flow between the nodes. Flow of data from one node to another is controlled by the control signals coming from the control part. The control relation is indicated

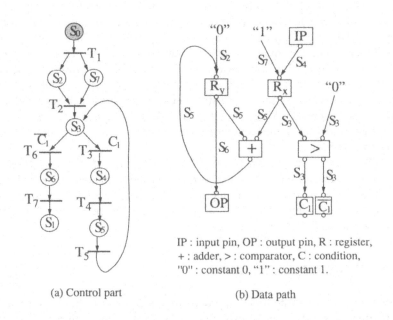

IP : input pin, OP : output pin, R : register,
+ : adder, > : comparator, C : condition,
"0" : constant 0, "1" : constant 1.

(a) Control part   (b) Data path

Figure 6.1: An example of ETPN.

by using control place names to guard the arcs. Figure 6.1a depicts the Petri net which represents the control flow of the example. A control state is represented as a marking of the Petri net, i.e., the possession of tokens in a subset of the places of the Petri net which are depicted as circles. The transitions of control states are represented as firings of one or several transitions of the Petri net which are depicted as bars.

A Petri net is a formal modeling technique which represents the states of a system by the markings of tokens in a graph. A Petri net consists of two different elements, the S-elements (places) and the T-elements (transitions), which are related to each other by a flow relation. Besides representing states by marking nodes, Petri nets also encode the permissible state transitions in the graph structure. They could be used, for example, to represent an event/ condition system where a partial ordering of the occurrence of events is specified but the contents of the events are ignored [Pet81]. Therefore they are weak devices for expressing computation. We have used Petri nets to represent the control structures of digital systems. They are extended with a data path notation to express the computation part.

The main application of ETPN is to explicitly capture the intermediate result of a design so as to allow the design algorithm to make accurate design decisions.

Example 6.1: Let us consider the problem of a floating point multiplication. It can be decomposed into three separate operations, the addition of the exponents (S_1), the multiplication of the mantissa (S_2), and the normalization of the above results (S_3). It is required that S_1 and S_2 be finished before S_3 can be started. Such design requirements can be expressed in the Petri net notation of ETPN as illustrated in Figure 6.2a where S_1 and S_2 are supposed to operate in parallel. However, the parallel execution of S_1 and S_2 is not necessary as Figure 6.2b and Figure 6.2c also present solutions to the multiplication problem which satisfy the original specification. Moreover, the solutions represented in Figure 6.2b and Figure 6.2c are less expensive but take longer time to execute the given function than the one given in Figure 6.2a. The ETPN design representation model can be used to capture all of these three designs explicitly with sufficient details (including also the data path, which is not depicted in the figure) so that an algorithm can be used to analyze them and decide whether any of them fits into the overall design of the system or one is better than the others in respect to certain design trade-offs.

□

With the use of the Petri net notation, the ETPN design representation model can be used to represent digital hardware with a high degree of concurrency and asynchrony, thus supporting the design of locally synchronous, globally asynchronous architectures.

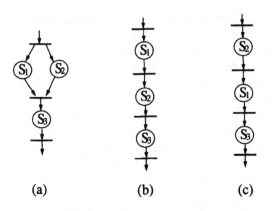

(a)   (b)   (c)

Figure 6.2: The Floating point multiplication example.

6.1.1   Partial Ordering

The most important concept of the ETPN design representation is the partial
ordering of computations, which is used to express concurrent and
asynchronous aspects of a hardware system. Traditionally most computation
models are based on the concepts of a linear order of atomic operations or
locksteps of atomic operations in some parallel models. That is, there exists a
total ordering relation between any two computation events. Many natural
phenomena and computation systems, however, can be simply and clearly
expressed as partial ordering of operations.

The use of a partial ordering as a basic notation in the ETPN model is due
not only to the simplicity of specification but also to its necessity. Some
systems can only be defined in terms of a partial ordering of events. For
example, a distributed system can be specified only as a partial ordering event
system.

We have chosen the partial ordering notation to represent concurrency and
asynchrony which are the two most important aspects of the control structure of
hardware systems. Other models of concurrent systems usually use the notation
of interleaving, i.e., if two events A and B are in a parallel relation, both first A
then B and first B then A are considered as permissible behaviors of the system.
The problem with the interleaving model of concurrency is that it depends upon
the atomic property of the system actions and contributes to the combinatorial
explosion of allowed behaviors.

The partial ordering of computations in the ETPN design representation
model is expressed in terms of the Petri net notation. A basic notion within
Petri nets is the partial ordering relation over the S-elements (places) and T-
element (transitions) of a system; if two events (represented, for example, by

the holding of tokens in a place, or firing of transitions) are not in the partial ordering relation, it indicates that they are causally independent and may occur in either order or simultaneously. Such a situation is illustrated in Figure 6.2 where the two events corresponding to S_1 and S_2 are independent of each other.

6.1.2   Basic Definitions

The ETPN representation is formally defined as a parallel computation model with data/control-flow notation, augmented with timing constraints. The data flow part of the model is captured as a data path. The control flow dictates the partial ordering of data path operations and is modeled by a Petri net notation.

The formal semantics of the proposed model is defined in terms of its inter-actions with the environment. That is, two pieces of hardware are considered to be semantically equivalent if they interact with the environment in the same way. This allows manipulation of the internal structure of the given design to improve performance as well as to reduce cost during the synthesis process.

Definition 6.1   A *data path*, **D**, over an algebraic structure is a five-tuple, **D** = (**V**, **I**, **O**, **A**, **B**), where

V = $\{V_1, V_2,...,V_n\}$ is a finite set of *vertices* each of which represents a data manip-ulation or storage unit;

I = $\mathbf{I}(V_1) \cup \mathbf{I}(V_2) \cup ... \cup \mathbf{I}(V_n)$ with $\mathbf{I}(V_j)$ = the set of *input ports* associated with ver-tex V_j;

O = $\mathbf{O}(V_1) \cup \mathbf{O}(V_2) \cup ... \cup \mathbf{O}(V_n)$ with $\mathbf{O}(V_j)$ = the set of *output ports* associated with vertex V_j;

A $\subseteq$ **O** $\times$ **I** = $\{<O, I> \mid O \in \mathbf{O}, I \in \mathbf{I}\}$ is a finite set of *arcs* each of which represents a connection from an output port of a vertex to an input port of another vertex or the same vertex;

B : **O** $\rightarrow 2^{\mathbf{OP}}$ is a mapping from output ports to sets of operations; **OP** = $\{OP_1, OP_2, ..., OP_m\}$ is a finite set of *operations* of the underlying algebraic structure; **OP** is divided into the *sequential* subset **SEQ** and the *combinatorial* subset **COM**.

Intuitively, a data path is a directed graph with each vertex (node) having possibly multiple input ports and output ports (an example of ETPN data paths is illustrated in Figure 6.1b). The vertices are used to model data manipulation and storage units, while the arcs are used to model their connections.

The above definition is concerned mainly with the structure rather than the function of the data path. How the data path is used to perform computation is not explicitly defined. We assume that there exists an implicit interpretation of the underlying algebraic structure which supports the computation rules. Such an algebraic structure should consist of a domain of values for constants and variables, an assignment of values to the constants and a function definition for each operator. This algebraic structure is not considered here as it does not directly affect the basic formulation of the model. Further, to define the semantics of the system independent of any particular interpretation makes it possible to cope with different implementation environments. However, we assume that some modules exist in a module library which can perform the defined operations of the data path. Some of these modules have memory capability, such as registers and RAMs, and their corresponding operations will be in the sequential subset. The other operations which will be mapped to combinatorial modules belong to the combinatorial subset.

The notion of ports is used here as a basic abstraction of the input/output behavior of a data manipulation unit. The operation of the vertices are defined only by the relation between the output ports and the input ports. It is assumed that the output port will present a value which has the given relationship with the values present in the input ports. Note also that the operations are associated with the output ports, not the vertices as a whole. Therefore, a vertex can simultaneously perform several different operations on its multiple output ports.

Definition 6.2   A *data/control flow system*, Γ, is an eight-tuple, $\Gamma = (\mathbf{D}, \mathbf{S}, \mathbf{T}, \mathbf{F}, \mathbf{C}, \mathbf{G}, \mathbf{R}, M_o)$ where

$\mathbf{D} = (\mathbf{V}, \mathbf{I}, \mathbf{O}, \mathbf{A}, \mathbf{B})$ is a data path;

$\mathbf{S} = \{S_1, S_2, ..., S_n\}$ is a finite set of *S-elements* or *control places*;

$\mathbf{T} = \{T_1, T_2, ..., T_m\}$ is a finite set of *T-elements* or *transitions*;

$\mathbf{F} \subseteq (\mathbf{S} \times \mathbf{T}) \cup (\mathbf{T} \times \mathbf{S})$ is a binary relation, called the *control flow relation*;

$\mathbf{C} : \mathbf{S} \to 2^{\mathbf{A}}$ is a mapping from control places to sets of arcs of the given data path; an arc A_i is *controlled* by a control place S_j if $A_i \in \mathbf{C}(S_j)$;

$\mathbf{G} : \mathbf{O} \to 2^{\mathbf{T}}$ is a mapping from output ports of data path vertices to sets of transitions; a transition T_i is *guarded* by an output port O_j if $T_i \in \mathbf{G}(O_j)$;

$\mathbf{R} : \mathbf{S} \to 2^{(\mathbf{O} \times \mathbf{OP})}$ is a mapping from control places to sets of pairs consisting of an output port and an operation; an output port/operation pair $<O_i, OP_j>$ is *selected* by S_k if $<O_i, OP_j> \in \mathbf{R}(S_k)$;

$M_o : \mathbf{S} \to \{0, 1\}$ is an *initial marking* function.

The definition of the data/control flow model is based on the marked Petri net notation. The Petri net S-elements (places) are used to capture the *control signal* concept. When a control place holds a token, a control signal will be sent to control the corresponding arcs in the data path specified by the control mapping function **C**. As there could be more than one control place which holds tokens, there exist multiple control signals that control parallel operations in the data path. The temporal relation between the control signals is defined by a partial ordering structure, which is captured by the control flow relation **F**. To represent the situation where the control flow is influenced by the results of some data path operations (corresponding to a conditional branch, for example), we must be able to use condition signals (as results of some data path operations) to direct the control flow. For this purpose, the guarding condition concept is introduced; a transition may be guarded by conditions produced from the data path represented at output ports of some vertices, which is defined by G. The mapping R is used to select the operation to be performed with an output port when it is assigned multiple operations (as in the case of an ALU).

The control mapping C specifies which arcs of the data path are controlled by, or said to be *associated* with, each control place of the control Petri net. We can also associate the data path vertices with the control signals by the following definition.

Definition 6.3 A data path vertex V_i is said to be *associated* with a control place S_j, if $\exists <O, I> \in A, (I \in I(V_i)) \wedge (<O, I> \in C(S_j))$, i. e., there exists an arc which is connected to one of the vertex's *input* ports and controlled by the place.

Definition 6.4 The arcs and vertices associated with a control place S_i, denoted as $\mathbf{ASS}(S_i)$, are said to be *active* under S_i.

Intuitively, the arcs will be open, i.e., allow data to pass, and the data manipulation units will perform the selected operations when their associated control signals are on (i.e., the control place holds a token). Before the behavior of a data/control flow system is formally specified, let us look at the simple example illustrated in Figure 6.1.

Example 6.2: The example depicted in Figure 6.1 corresponds to the following definition of a data/control flow system. $V = \{IP, OP, \text{"0"}, \text{"1"}, R_x, R_y, +, >, C_1, \overline{C}_1, \text{"0"}\}$. $I(IP) = \emptyset$; $I(OP) = \{OP_1\}$; $I(\text{"0"}) = \emptyset$; $I(\text{"1"}) = \emptyset$; $I(R_x) = \{R_{x\_1}\}$; $I(R_y) = \{R_{y\_1}\}$; $I(+) = \{+_{11}, +_{12}\}$; $I(>) = \{>_{11}, >_{12}\}$; $I(\text{"0"}) = \emptyset$; $I(C_1) = \{C_{1\_1}\}$; $I(\overline{C}_1) = \{\overline{C}_{1\_1}\}$. $O(IP) = \{IP_0\}$; $O(OP) = \emptyset$; $O(\text{"0"}) = \{0_0\}$; $O(\text{"1"}) = \{1_0\}$; $O(R_x) = \{R_{x\_0}\}$; $O(R_y) = \{R_{y\_0}\}$; $O(+) = \{+_0\}$; $O(>) = \{>_{01}, >_{02}\}$; $O(\text{"0"}) = \{0_0\}$; $O(C_1) = \{C_{1\_0}\}$; $O(\overline{C}_1) = \{\overline{C}_{1\_0}\}$. $A = \{<0_0, R_{y\_1}>, <1_0, R_{x\_1}>, <IP_{00}, R_{x\_1}>, <R_{y\_0}, +_{11}>, <R_{x\_0}, +_{12}>, <R_{x\_0}, >_{11}>, <0_0, >_{12}>, <>_{01}, C_{1\_1}>, <>_{02}, \overline{C}_{1\_1}>, <+_0, R_{y\_1}>, <R_{y\_0}, OP_1>\}$. $OP = \{\text{input\_p, output\_p, addition, comparison, register, condition, 0, 1}\}$. $COM = \{\text{input\_p, output\_p, addition, comparison, 0, 1}\}$. $SEQ = \{\text{register, condition}\}$. $B(IP_0) = \{\text{input\_p}\}$; $B(0_0) = \{0\}$; $B(1_0) = \{1\}$; $B(R_{x\_0}) = \{\text{register}\}$; $B(R_{y\_0}) = \{\text{register}\}$; $B(+_0) = \{\text{addition}\}$; $B(>_0) = \{\text{comparison}\}$; $B(0_0) = \{0\}$, $B(C_{1\_0}) = \{\text{condition}\}$; $B(\overline{C}_{1\_0}) = \{\text{condition}\}$. $S = \{S_0, S_1, S_2, S_3, S_4, S_5, S_6, S_7\}$. $T = \{T_1, T_2, T_3, T_4, T_5, T_6, T_7\}$. $F = \{(S_0, T_1), (T_1, S_2), (T_1, S_7), (S_2, T_2), (S_7, T_2), (T_2, S_3), (S_3, T_3), (T_3, S_4), (S_4, T_4), (T_4, S_5), (S_5, T_5), (T_5, S_3), (S_3, T_6), (T_6, S_6), (S_6, T_7), (T_7, S_1)\}$. $C(S_0) = \emptyset$; $C(S_1) = \emptyset$, $C(S_2) = \{<0_0, R_{y\_1}>\}$; $C(S_3) = \{<R_{x\_0}, >_{11}>, <0_0, >_{12}>, <>_{01}, C_{1\_1}>, <>_{02}, \overline{C}_{1\_1}>\}$;

$C(S_4) = \{<IP_0, R_{x\_1}>\}$; $C(S_5) = \{<R_{y\_0}, +_{11}>, <R_{x\_0}, +_{12}>, <+_0, R_{y\_1}>\}$; $C(S_6) = \{<R_{y\_0}^-, OP_1>\}$; $C(S_7) = \{<1_0, R_{x\_1}>\}$. $\bar{G}(C_{1\_0}) = \{T_3\}$; $\bar{G}(\bar{C}_{1\_0}) = \{T_6\}$. $M_0(\bar{S}_0) = 1$; $M_0(S_1) = 0$; $M_0(S_1) = \bar{0}$; $M_0(S_2) = 0$; $M_0(S_3) = 0$; $\bar{M}_0(S_4) = 0$; $M_0(S_5) = 0$; $M_0(S_6) = 0$; $M_0(S_7) = 0$.

$\square$

We use a *graphical* representation which follows the convention of Petri net graphs [Pet81] and data path notation [Pen87]. Figure 6.1b shows the data path of the example where each vertex is depicted as a rectangle with labels indicating its primary operation. The labels are drawn from a set of predefined symbols which represents all of the operations in **OP**. Further, subscripts are used to differentiate, when needed, vertices with the same operation. The small circles attached to a vertex rectangle represent the input/output ports of the vertex. The arcs of the data path represent the data flow between the vertices. Flow of data from one vertex to another is controlled by the control signals coming from the control part. The control relation is indicated by using control place names to guard arcs.

Figure 6.1a depicts the Petri net which represents the control flow of the example. The Petri net places (S-elements) are depicted as circles while the transitions (T-elements) as bars. The initial marking function of this example maps place S_0 to 1 and all other places to 0, i.e., initially there is only one token in place S_0, which is indicated by shading the circle representing S_0. The guarding relation between output ports of condition vertices and transitions is indicated by using port names (or vertex names when the vertex has only one output port, as in this example) to guard transitions.

Before formally giving the definition of semantics of the computation model, we will define its behavior model which is in turn based on the execution rules of the control Petri net and its interaction with the data path.

Definition 6.5   Given a data/control flow system $\Gamma = (D, S, T, F, C, G, R, M_o)$, its *behavior* is defined as follows:

1) A function $M : S \rightarrow N$ is called a *marking* of Γ ($N = \{0, 1, 2, ...\}$). A marking is an assignment of *tokens* to the places. Initially there is a token in each of the initial places, i.e., place S_i such that $M_o(S_i) = 1$.

2) A transition T is *enabled* at a marking M if and only if for every S such that $(S, T) \in F$, we have $M(S) \geq 1$; i.e., each of the transition's input places has at least one token.

3) A transition T may be *fired* when it is enabled and the guarding condition (i.e., the data value present at the guarding port[1]) is true. If a transition has more than one guarding condition, a logic *or* operation is applied to them.

4) Firing an enabled transition T removes a token from each of its input places and

---

1. It is assumed that the data values present at a port must be of Boolean type, if it guards some transitions.

deposits a token in each of its output places, resulting in a new marking which is said to be *immediately reachable* from the old one. All markings which can be reached from M_o by firing a sequence of transitions constitute the set of *reachable* markings denoted as **RM**.

5) $v(P)$ denotes the data value present at input or output port P. When a control place holds a token, its associated arcs in the data path will open for data to flow; i.e., the data value present at the input port is equal to that of the corresponding output port of the same arc.

6) For each output port O of a vertex V, $v(O) = OP(v(\mathbf{I}(V)))^1$, where $OP \in \mathbf{B}(O)$, $<O, OP>$ is selected by a control place (as defined by **R**) which holds a token if $|\mathbf{B}(O)| > 1^2$, and $OP \in \mathbf{COM}$. If $OP \in \mathbf{SEQ}$, it is assumed that O has memory capability and $v(O)$ remains the same until it is changed, which occurs when one of the vertex's associated place has held a token and the token is being removed[3].

7) If none of the connected arcs of an input port I is active, $v(I)$ is *undefined*. If the operation associated with an output port is not a sequential one and the value of the output port depends on an undefined input value, its value is also *undefined*.

The possible existence of undefined values and the intrinsic non-deterministic properties of the Petri net firing sequence could result in ambiguous behavior for some systems. We therefore introduce some restrictions with the following definition.

Definition 6.6   A data/control flow system $\Gamma = (\mathbf{D}, \mathbf{S}, \mathbf{T}, \mathbf{F}, \mathbf{C}, \mathbf{G}, \mathbf{R}, M_o)$ is properly designed if:

1) $\mathbf{ASS}(S_i) \cap \mathbf{ASS}(S_j) = \emptyset$, if $\exists M \in \mathbf{RM}$, $(M(S_i) > 0) \wedge (M(S_j) > 0)$. That is, if two subgroups of the data path can be activated at the same time, they must be disjoint.

2) $\forall S_i \in \mathbf{S}$, $M(S_i) \le 1$, for each $M \in \mathbf{RM}$. That is, the Petri net must be *safe*, or, in other words, there will not be any attempt to execute two sets of operations on the same piece of hardware at the same time.

3) If $(S, T_1) \in \mathbf{F}$ and $(S, T_2) \in \mathbf{F}$, then both T_1 and T_2 must be guarded and their guarding conditions must not be true at the same time. That is, the Petri net must be *conflict-free*[4], or, in other words, non-deterministic choice is not allowed.

---

1. Apply the function defined by OP to the values present at the input ports of vertex V, i.e., $v(\mathbf{I}(V))$.

2. $|A|$ denotes the cardinality, i.e., the number of elements of a set A.

3. This assumption is analogous to the falling-edge trigger operation of flip-flops at gate level. Note that a RAM is modelled as a register file together with a reading control mechanism and a writing control mechanism.

4) $\forall S_i \in \mathbf{S}$, $\mathbf{ASS}(S_i)$ must include at least one vertex that has an output port mapped to a sequential operation or used to guard a transition.

5) $\forall M \in \mathbf{RM}$, $M(S_i) + M(S_j) \leq 1$, if $<O, OP> \in \mathbf{R}(S_i)$ and $<O, OP'> \in \mathbf{R}(S_j)$. That is, if two places are used to select operation at the same output port, they are not allowed to have tokens simultaneously.

From now on, we consider only systems which are properly designed.

Example 6.3: Let us use the behavior model to check how the design given in Figure 6.1 works. Initially, there is a token in the initial place S_0 (Definition 6.5(1)). Since $\mathbf{C}(S_0) = \emptyset$, i.e., S_0 does not control any arc in the data path, no computation will occur. Since transition T_1 has only one input place, S_0, and S_0 has a token, T_1 is enabled and can be fired (Definition 6.5(2, 3)). Firing T_1 removes the token in S_0 and deposits a token in S_2 and S_7, since they are both output places of T_1 (Definition 6.5(4)). Since S_2 controls the arc $<0_o, R_{y\_i}>$, it will be open for data to flow, which means the input port ar register R_y will get the value of "0"; i.e., $\mathbf{v}(R_{y\_i}) = \mathbf{v}(0_o)$ (Definition 6.5(5). Since R_y is associated with S_2 and is a register, $\mathbf{v}(R_{y\_o}) = \mathbf{v}(R_{y\_i})$ (i.e., "0") at the time when the token in S_2 is removed, which means the constant "0" will be stored in register R_y (Definition 6.5(6). Similarly S_7 will control the storage of constant "1" into register R_x. Now since both S_2 and S_7 hold a token, transition T_2 can be fired. The firing of T_2 leads to a token being deposited in S_3, which will trigger a new set of operations in the data path. This process will continue until a token is deposited in S_1 when no more transition can be fired.

□

The main feature of the data/control flow model is that parallel operations are *explicitly* represented in the form of tokens existing in several places at the same time. As illustrated in Figure 6.1, for example, places S_2 and S_7 will both hold a token after the firing of their common input transition T_1. This represents the parallel execution of the two sets of operations controlled by S_2 and S_7, as has been discussed in Example 6.3. Further, the completion of the parallel sets of operations could independently initiate other operations (this does not apply to the example given in Figure 6.1, though). The flow of control along different sequences of operations can be totally independent of each other, thus representing truly concurrent systems. There exists also a second level of parallelism of the operations controlled by the *same* control signal. When the control signal is on, all of its associated arcs and vertices are active and if the operations of the vertices are not data-dependent, they are carried out concurrently.

---

4. If a place, S, has two output transitions, T_1 and T_2, whose guarding conditions are true at the same time, a token in S can be used to fire either T_1 or T_2, thus resulting in a conflict situation.

6.1.3   Semantics Definition

The minimal requirement for semantics equivalence is that the input to output functional relationship should be preserved. However, this requirement is usually too weak for most applications, because the temporal relationship among the set of input and output operations is also important. The temporal relationship can be expressed either as a partial ordering relation of input/output operations or as absolute timing constraints [Pen88]. In our model, we use a partial ordering relation. Two systems are considered to be equivalent if and only if the specified partial ordering relation of the input/output operations is the same in addition to the functional equivalence requirement.

Definition 6.7   For a data path $D = (V, I, O, A, R)$, there is a set of *external vertices*. The set of ports of the external vertices are called *external ports*. The set of arcs that are connected to the external ports are called *external arcs*.

> **Example 6.4:** As an example, in Figure 6.1b there are two external vertices $\{IP, OP\}$. One, IP, is used for input, and the other, OP, for output. An external vertex can also have both an input port and an output port. In that case, it is used both for input and output of data. We have also two external arcs: $<IP_o, R_{x\_i}>$ and $<R_{y\_o}, OP_i, >$.
>   □

Definition 6.8   An *external event* is a pair $<A_i, v_i>$, with A_i being an external arc and v_i a value passed over the arc. An external event is controlled by the Petri net control place that is associated with the arc. That is, the external event occurs at the time when the associated control place holds a token.

> **Example 6.5:** In Figure 6.1, an example of external events is $<A_1, 14>$ where $A_1 = <IP_o, R_{x\_i}>$. This external event denotes that when S_4 hold a token, the data value of 14 has been sent over the external arc $<IP_o, R_{x\_i}>$.
>   □

Definition 6.9   For a data/control flow system $\Gamma = (D, S, T, F, C, G, R, M_o)$, let **RM** denotes the set of reachable markings by firing a sequence of transitions from M_o, as defined in Definition 6.5

1) S_i $(S_i \in S)$ and S_j $(S_j \in S)$ are said to be in *parallel order* if there exists a $M \in$ **RM** such that $M(S_i) = 1$ and $M(S_j) = 1$, and $S_i \neq S_j$. That is, two places are in parallel order if they can both hold a token at the same time.

2) S_i and S_j are said to be in *sequential order* if they are not in parallel order.

3) If S_i and S_j are in sequential order and there is a token path from S_i to S_j, we say S_i is predecessor of S_j, which is denoted as $S_i \Rightarrow S_j$. If S_i and S_j are in sequential order, at least one of the relations, $S_i \Rightarrow S_j$ and $S_j \Rightarrow S_i$ must hold.

> **Example 6.6:** In the design depicted in Figure 6.1, S_2 and S_7 are in

parallel order since by firing T_1 one token will be placed into S_2 and another into S_7. On the other hand S_3 and S_1 are in sequential order, since they will never hold tokens simultaneously. We have also $S_3 \Rightarrow S_1$. It can be seen that S_3 and S_5 are also in sequential order; and both $S_3 \Rightarrow S_5$ and $S_5 \Rightarrow S_3$ hold. Further S_4 and S_6 are also in sequential order; and, in this case, only $S_4 \Rightarrow S_6$ holds, since a token can flow from S_4 to S_6 via the path $\{S_4, T_4, S_5, T_5, S_3, T_6, S_6\}$.

□

Definition 6.10 Given a data/control flow system $\Gamma = (\mathbf{D}, \mathbf{S}, \mathbf{T}, \mathbf{F}, \mathbf{C}, \mathbf{G}, \mathbf{R}, M_o)$, its *external event structure* is defined as a three-tuple $(\mathbf{E}, \angle, \cong)$ where

$\mathbf{E} = \{E_1, E_2, ..., E_n\}$ is a bag of external events[1];

$\angle \subseteq (\mathbf{E} \times \mathbf{E})$ is a binary relation, which denotes the *precedent* relation between external events. $E_i \angle E_j$ with $E_i = <A_i, v_i>$ and $E_j = <A_j, v_j>$, if and only if E_i occurs before E_j and $S_i \Rightarrow S_j$, where $A_i \in \mathbf{C}(S_i)$ and $A_j \in \mathbf{C}(S_j)$;

$\cong \subseteq (\mathbf{E} \times \mathbf{E})$ is a binary relation, which denotes the *concurrent* relation between external events. $E_i \cong E_j$ with $E_i = <A_i, v_i>$ and $E_j = <A_j, v_j>$, if and only if E_i and E_j occurs at the same time (marking) and $A_i \in \mathbf{C}(S)$ and $A_j \in \mathbf{C}(S)$, i.e., both arcs A_i and A_j are controlled by the same place S.

An external event structure specifies all the possible external events of a system as well as the temporal relationship between them. If two external events are in the precedent or concurrent relation, they must always occur in the specified order or simultaneously, respectively. On the other hand, if two external events are not in either of the two relations, they can occur in any order.

In the above discussion, we assume that when an external event occurs whose operation is to obtain a value from the outside world, the environment will supply a value of the appropriate type to the system. We also assume that a sequence of such values is implicitly predefined for each input vertex, when an external event structure is specified.

Definition 6.11 The semantics of a data/control flow system Γ, denoted as $\mathcal{S}(\Gamma)$, is defined by its external event structure. That is, $\mathcal{S}(\Gamma) = (\mathbf{E}, \angle, \cong)$.

Definition 6.12 Two data/control flow systems Γ and Γ' are semantically equivalent, denoted as $\Gamma \equiv \Gamma'$, if $\mathcal{S}(\Gamma) = \mathcal{S}(\Gamma')$.

Two systems are semantically equivalent if they exhibit the same external behavior via their external vertices. This means that two systems can have extremely different internal structure and behavior but are still semantically

---

1. A bag is a collection of elements over some domain which allows multiple occurrences of elements. If the same value v_i is passed over a given external arc A_i at n different times, $<A_i, v_i>$ will occur n times in the bag of external events, which represents n different external events.

equivalent, and a high-level synthesis system can be used to transform and optimize a design while maintaining its semantics.

Definition 6.13 The *domain* of a control place S, denoted as $\textbf{dom}(S)$, is defined as the set of *sequential* vertices that have some output port connected to an arc controlled by S. The *codomain* of S, denoted as $\textbf{cod}(S)$, is defined as the set of *sequential* vertices which have some input port connected to an arc controlled by S. A vertex is defined to be *sequential* if any of its output ports is mapped to a sequential operation (Definition 6.1); otherwise, it is called a *combinatorial* vertex.

Definition 6.14 S_i and S_j are *data dependent*, denoted as $S_i \leftrightarrow S_j$, if any of the following is true:

1) $\textbf{cod}(S_i) \cap \textbf{dom}(S_j) \neq \varnothing$.

2) $\textbf{cod}(S_j) \cap \textbf{dom}(S_i) \neq \varnothing$.

3) $\textbf{cod}(S_i) \cap \textbf{cod}(S_j) \neq \varnothing$.

4) S_i and S_j are in a *control dependent* relation; i.e., whether S_i will hold a token or not depends on the values on ports of a subset of $\textbf{cod}(S_j)$, or vice versa.

5) $C(S_i)$ and $C(S_j)$ both contain some external arcs.

Example 6.7: In the design depicted in Figure 6.1, S_4 and S_5 are data dependent, since $\textbf{dom}(S_5) = \{R_x, R_y\}$, $\textbf{cod}(S_4) = \{R_x\}$, and $\textbf{dom}(S_5) \cap \textbf{cod}(S_4) = \{R_x\}$. S_4 and S_3 are in a control dependent relation, since whether S_4 will hold a token or not depends on the value on the port C_{1_o} and $\textbf{cod}(S_3) = \{C_1, \overline{C}_1\}$. Therefore S_4 and S_3 are also data dependent. Since S_4 and S_6 both control some external arcs, they are also data dependent.

$\square$

Intuitively, the data-dependent relation is defined as a relation between places whose associated operations will contribute data to each other, or involve some interaction with the environment. If all operations that are data dependent are executed in the predefined order, the semantics of a given digital system will not be changed. Operations associated with control places that are not in a data-dependent relation can, however, be performed in any order without changing the semantics. The following two definitions will give some examples of design modification which does not change the semantics.

Definition 6.15 Given $\Gamma = (\textbf{D}, \textbf{S}, \textbf{T}, \textbf{F}, \textbf{C}, \textbf{G}, \textbf{R}, M_o)$ and $\Gamma' = (\textbf{D}, \textbf{S}, \textbf{T}', \textbf{F}', \textbf{C}, \textbf{G}, \textbf{R}, M_o)$, Γ and Γ' are data-invariantly equivalent to each other, if and only if for every $S_i \Rightarrow S_j$ and $S_i \leftrightarrow S_j$ in Γ ($S_i \in \textbf{S}$, $S_j \in \textbf{S}$), we have $S_i \Rightarrow' S_j$ and $S_i \leftrightarrow' S_j$ in Γ' ($\Rightarrow'$ and $\leftrightarrow'$ denote the sequential order relation and the data dependent relation defined in Γ' respectively); and vice versa.

The above definition ensures that all data dependent operations in the two systems are performed in exactly the same order, and two operations associated with different places are performed in parallel only if they are data independent. Therefore, the data-invariant equivalence relation implies the semantic equivalence relation. This means that we can reconstruct the control structure (without changing the data path) of a digital system to improve system performance, for example, by carrying out as many operations in parallel as possible.

Definition 6.16   Given $\Gamma = (D, S, T, F, C, G, R, M_o)$ with $D = (V, I, O, A, B)$ and $\Gamma' = (D', S, T, F, C, G', R', M_o)$ with $D' = (V', I', O', A', B')$, Γ and Γ' are control-invariantly equivalent to each other, if and only if Γ' is the result of merging vertex V_i into vertex V_j of Γ, both V_i and V_j are combinatorial and have the same port structure, and their associated control places will not hold tokens simultaneously in any of the reachable markings. The result of a *vertex merger* is defined as:

$V' = V - \{V_i\}$.

$I' = I - \{I(V_i)\}$.

$O' = O - \{O(V_i)\}$.

B' is the same as B except $B'(O_{j,k}) = B(O_{j,k}) \cup B(O_{i,k})$ with $O_{j,k} \in O(V_j)$ and $O_{i,k} \in O(V_i)$, $k = 1, 2, ..., |O(V_j)|$.

R' is the same as R except that each $<O_{i,k}, OP> \in R(S)$ with $O_{i,k} \in O(V_i)$ is replaced by $<O_{j,k}, OP> \in R(S)$ with $O_{j,k} \in O(V_j)$.

A' is the same as A except that each $<O_{i,k}, I>$ with $O_{i,k} \in O(V_i)$ is replaced by $<O_{j,k}, I>$ with $O_{j,k} \in O(V_j)$, and each $<O, I_{i,l}>$ with $I_{i,l} \in I(V_i)$ replaced by $<O, I_{j,l}>$ with $I_{j,l} \in I(V_j)$.

G' is the same as G except that each $T \in G(O_{i,k})$ with $O_{i,k} \in O(V_i)$ is replaced by $T \in G(O_{j,k})$ with $O_{j,k} \in O(V_j)$.

The vertex merger is a transformation used to share hardware resources by operations so as to improve the design in terms of cost. For example, as illustrated in Figure 6.3, the two comparison operations can be implemented by the same comparator by merging the two vertices together. A vertex merger should only be performed when the two vertices' associated control places will never

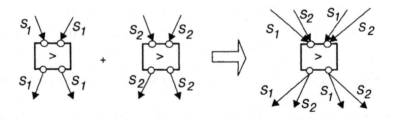

Figure 6.3: An example of vertex merger.

possess tokens simultaneously. In this way, the two sets of operations will not be performed at the same time and thus can share the same operators safely. In this example, the two vertices perform the same operation, which will become the operation performed by the new vertex. In the cases when the merged vertices perform different operations, the generated vertex will perform the corresponding operations according to the new selection mapping **R'**. Since a merger will change neither the order of operations nor the operations themselves, it will not change the semantics of a design [Pen88].

6.1.4   Timing

The data/control flow systems are defined based on the notation of Petri nets. Since normal Petri net models do not contain an explicit notion of execution time, it is impossible to determine, for example, any performance measure of the modelled system. To solve this problem, we have introduced timing into the data/control flow system by requesting a token to reside in a place for some period of time before it can be used to fire a transition. This time interval represents the time required to finish the associated operations in the data path. To reduce the analysis complexity, we introduce, at the same time, a restriction that requires each transition to be fired *immediately* when it is enabled and the guarding condition is true instead of allowing arbitrary firing time of the transitions.

Definition 6.17   A timed data/control flow system (the ETPN model) is a two-tuple, $<\Gamma, L>$, where $\Gamma = (D, S, T, F, C, G, R, M_o)$, with $S = \{S_1, S_2, ..., S_n\}$, is a data/control flow system, and $L = \{l_1, l_2, ..., l_n\}$ is a finite set of time intervals each of which is assigned to a control place. The behavior of a timed data/control flow system is the same as its corresponding data/flow system except that:

1) A token must stay in place S_i for a time period of length l_i before it can be used to enable its output transitions, $i = 1, 2, ..., n$; and

2) A transition will be fired *immediately* when it is enabled and its guarding condition is true.

When the data path vertices are bound to the basic RTL components stored in the module library and their placement is fixed, the time needed for the data path operations associated with each place can be accurately estimated (see Section 6.3 for more details). Usually, the time intervals associated with different places are different, and the ETPN model is inherently *asynchronous*. A particular implementation, such as in the case of CAMAD, however, can assume that a synchronous clocking scheme is used and the time intervals of Definition 6.17 equal the time of one or multiple clock cycles.

6.2   Mapping VHDL Specifications to ETPN

In this section we briefly outline how VHDL specifications are translated into the ETPN design representation. We focus here mainly on representation of the basic control structures. Representation of higher level structures, like processes and subprograms, as well as issues concerning process interaction, will be discussed in Chapter 7.

6.2.1   Compiling VHDL to ETPN

Compilation of VHDL to ETPN is performed by passing through two interme-diate forms: the *program graph* and the *data/control flow description* [EKPM92]. Figure 6.4 shows how a VHDL specification is successively trans-lated to ETPN. The translation process is performed in four steps:

1. The first step of the compiler does a complete syntactic and semantic analysis of the VHDL specification and transforms it into an internal form called program graph [ASU88]. A program graph consists of trees corresponding to the syntactic structure of the statement parts in the program (statement parts of architecture bodies, blocks, subprograms and processes) and additional data nodes pointed from these trees. These data nodes represent declared objects on which the state-ments operate.

2. The second step performs structural transformations on the program graph and, at the same time, some code optimization. The transformations performed during this step are aimed at reducing some high-level constructs, still present in the program graph, to a level which is suitable for the final generation of the ETPN representation. Examples of such transformations are: concurrent signal assignment statements and concurrent procedure calls are transformed into equiv-alent process statements; sensitivity lists are eliminated from processes and the equivalent wait statements are introduced; processes are expanded to the equiv-alent loops; and a loop with a *for* iteration scheme is transformed into an equiv-alent sequence based on a *while* scheme. Several transformations are also performed during this step. They are constant folding, elimination of common subexpressions, code motion, and height-reduction for expression syntax trees. Some of these transformations are discussed in Section 6.4.

3. The data/control flow description generated in the second step reflects already the flow of data and control through the program at the level corresponding to constructs supported by ETPN. The main task of the third step is to perform parallelization of the internal representation. The initial VHDL specification typically consists of a set of concurrent, interacting processes. This coarse grain, process level parallelism is specified by the designer. Computations inside each process are specified as a set of statements to be executed sequentially, one after the other. These sequences of statements inside each process have to be paral-lelized automatically by taking advantage of the potential parallelism intrinsic to

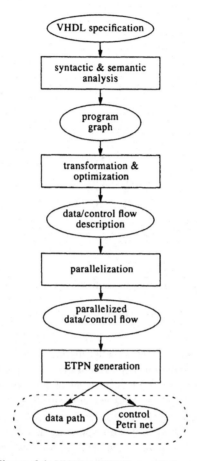

Figure 6.4: VHDL to ETPN compilation.

the input specification [GEKP96]. The functional effect of executing statements according to the generated representation has to be the same as for the initial specification. In Figure 6.5 we illustrate this step by presenting the ETPN representation corresponding to the following VHDL sequence, with and without parallelization:

```
          . . .
S1:       A:=1;
S2:       X:=1;
S3:       while 100>=X loop
S4:                  V:=V*2;
S5:                  X:=X+1;
          end loop;
S6:       R:=V+A;
          . . .
```

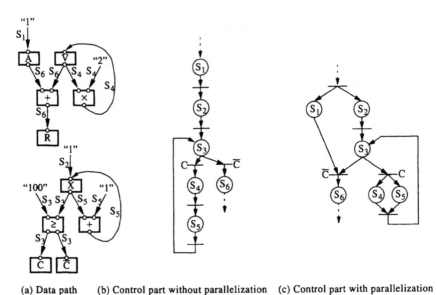

(a) Data path   (b) Control part without parallelization   (c) Control part with parallelization

Figure 6.5: ETPN representations with and without parallelization.

The ETPN representation generated after the following step will thus correspond to a hardware implementation with a high degree of parallelism. The synthesis algorithm will decide on an adequate level of implemented parallelism which represents the optimal trade-off between hardware cost and required performance.

4. The last step transforms the data/control flow description generated in the previous step into ETPN with its corresponding data path and control Petri net. Basically, each statement node in the data/control flow description has its corresponding control place in the control Petri net. An arc connecting two nodes is modeled as a transition. The construction of the ETPN data path is carried out according to the convention that each variable or signal will be mapped into a vertex (signals can require a more complex representation, as will be discussed in Chapter 7). All references to the variable or signal will be to this vertex. Each operation will also be represented by a vertex, dedicated exclusively to that operation. Thus, if there are, for example, ten instances of a "+" operation in the data/control flow description, they will be represented as ten different data manipulation vertices in ETPN. According to the flow of data between the data path nodes, arcs are generated to connect the output port of one node to an input port of another. These arcs are guarded by the corresponding control places in the control Petri net. In the next subsection and Chapter 7, we will give more details concerning the ETPN representation of some VHDL constructs.

In theory, the ETPN representation generated by the compiler can be viewed as a structural specification which could be implemented directly. It would correspond to a very expensive and fast solution, as no sharing of resources is introduced. Practically, however, the generated ETPN representation does not indicate how the design is to be implemented in hardware. In the following stages of the synthesis process, transformations will be performed on the ETPN representation to produce a structure with a reasonable sharing of resources, an optimized scheduling, and an efficient allocation and binding of storage elements.

6.2.2   ETPN Representation of Behavioral VHDL

A behavioral VHDL specification to be translated into ETPN consists of package declarations, package bodies, entity declaration and architecture body. The architecture body contains processes, concurrent signal assignments and concurrent procedure calls. The latter two are treated at compilation as their equivalent process statements (see Section 2.6).

The ETPN representation generated by compilation has an overall structure like that presented in Figure 6.6. The initial marking of the control Petri net consists of a single token in a place (S_0) called the starting place. The output transition of this place (T_0) is an input transition for the starting places of all processes. The control part corresponding to these processes consists of places which control operations described as sequential statements inside each process. As will be discussed in Chapter 7, a single control Petri net can be generated for a group of processes, or the control Petri nets corresponding to each process can be completely disjoint.

According to the VHDL language specification constants and variables are

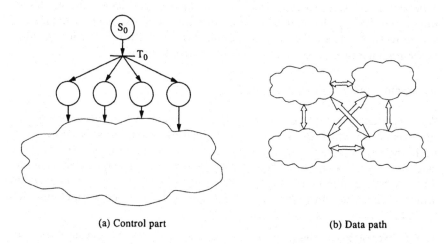

(a) Control part   (b) Data path

Figure 6.6: ETPN representation resulting from compilation of behavioral VHDL.

declared locally to processes. Thus, data path nodes corresponding to these items are grouped in chunks, locally accessed by each process. The generated data path also contains communication elements which support process inter-action and correspond to the VHDL signals. Representation of signals will be discussed in Chapter 7.

Variables of elementary type are mapped onto data nodes which have one single input and one single output port. Array variables are mapped to nodes which have an additional input port for the index value. Bit-string variables which are accessed also for slices are mapped to nodes which have two additional input ports for the slice limits. For variables of record type separate data nodes are generated for each field.

All activities inside the data path, related to evaluation of an expression, are controlled by one place in the control part[1]. Storage of the resulting value in a data node is controlled by the same place. Thus, for example, in Figure 6.5 place S_4 controls all activities connected to the variable assignment statement $V:=V*2$.

6.2.3   Representation of VHDL Control Structures

In this section we discuss the ETPN representations corresponding to the VHDL *if*, *case*, and *loop* statements.

Representation of the *if* statement

The mapping of an *if* statement (see Subsection 2.3.3) to ETPN is straight-forward. The data path nodes and arcs corresponding to the conditions are constructed. As illustrated in Figure 6.7, for each condition one place is generated to control its evaluation. Each of these places has two output transi-tions, guarded by the corresponding condition and its complementary one respectively. These are the input transitions for the control parts corresponding to the statement sequences on the two alternative branches.

Representation of the *case* statement

A *case* statement (see Subsection 2.3.4) can be mapped to ETPN as a sequence of decisions in which the value of the expression is successively compared to each choice. This implies practically the implementation of the equivalent if statement. Such a representation leads often to inefficient implementations. Thus, a second alternative for mapping a case statement has been provided, which is illustrated in Figure 6.8. The mapping is based on a functional data

---

1. If the expression contains a function call, several control places can contribute to its evaluation. Subprograms are discussed in Chapter 7.

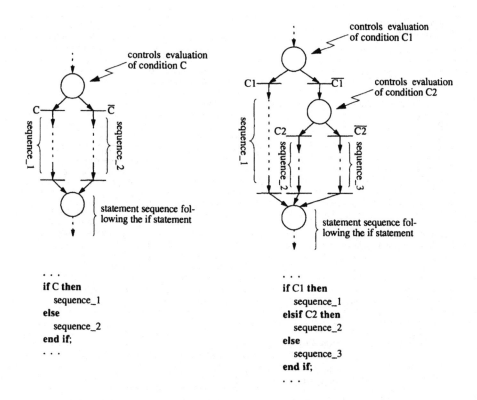

: Control part for the ETPN representation of the *if* statement.

path node which implements a decoder. This node gets as input the value of the
expression and has one output port associated to each possible choice. An
output, interpreted as a condition, will have the value *true* if the corresponding
choice has to be selected and *false* otherwise.

The Petri net place that loads the value of the expression into the decoder (S_E
in Figure 6.8) has output transitions corresponding to each possible choice,
which are guarded by the respective conditions. These are the input transitions
to the control parts corresponding to the statement sequences for the respective
alternative.

In order to reduce the number of conditions to be produced by the decoder a
modified implementation, like the one presented in Figure 6.9, can be generated
for the same case statement. This representation will be generated if the
expression is of the predefined type INTEGER, or if the range of the expression
type exceeds the actual range of the choices. This means that there are much
more possible values for the expression than explicit choices specified in the

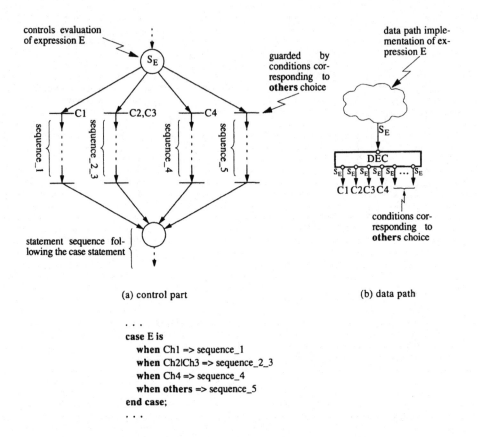

(a) control part   (b) data path

```
. . .
case E is
    when Ch1 => sequence_1
    when Ch2|Ch3 => sequence_2_3
    when Ch4 => sequence_4
    when others => sequence_5
end case;
. . .
```

Figure 6.8: ETPN representation of the *case* statement.

case statement. In such a case it is possible to filter out the situations which correspond to expression values that are smaller than the smallest choice, Ch_min, or greater than the greatest choice, Ch_max. For the case statement which we use as an example in Figure 6.8 and Figure 6.9 these limits are:

$$Ch\_min = \min(Ch1, Ch2, Ch3, Ch4)$$
$$Ch\_max = \max(Ch1, Ch2, Ch3, Ch4)$$

If the value of the expression is outside these limits, execution has to be directed to the sequence corresponding to the **others** choice. If the expression value is inside the limits, execution proceeds like in the previous implementation. The decision on which way to proceed is controlled, in Figure 6.9, by place S_1.

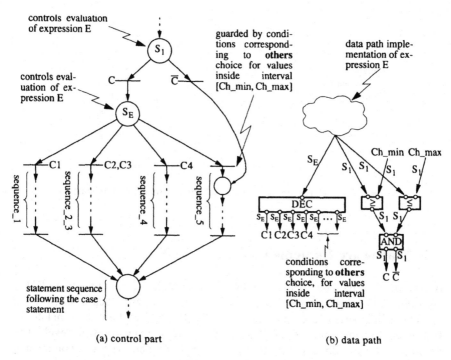

(a) control part   (b) data path

Figure 6.9: Modified representation of the *case* statement.
(for the same example like in Figure 6.8).

Representation of loops

Repetitive structures are built in VHDL using the loop statement (see Subsection 2.3.5). They can be created using one of the two iteration schemes: *while* or *for*. If no iteration scheme is mentioned the loop is executed indefinitely. The ETPN mapping of such an indefinite loop is straightforward: the statement sequence inside the loop body is generated and the output transition of the last place is connected to the first place of the sequence (Figure 6.10).

The ETPN mapping of a loop with a *while* scheme is illustrated in Figure 6.11. The place S_1 controls evaluation of the condition and has two output transitions guarded by the condition and the complementary one respectively. The first transition is connected to the control part of the loop body while the other one to the next statement after the loop. The output transition of the last statement in the loop body is connected to place S_1.

A loop with a *for* scheme is mapped to ETPN by transforming it into the equivalent *while* loop preceded by the initialization of the loop parameter, as depicted in Figure 6.12.

```
. . .
loop
    sequence
end loop;
```

Figure 6.10: ETPN representation of a loop with no iteration scheme.

6.3 Hardware Implementation of ETPN

Before we describe how to select and apply a set of transformations to an ETPN representation so as to reach a satisfactory design, we will discuss the problem of how to implement the final design in hardware, i.e., how to translate an ETPN description into digital hardware at the register-transfer level. The implementation is divided into two parts. First, the data path is implemented using RTL components, such as registers, functional units and multiplexers and then the control Petri net is converted into a finite state machine.

6.3.1 Data Path Implementation

The transformation of an ETPN data path to a RTL design is carried out by a net-list generation procedure. It is assumed that each ETPN data path vertex has at least one module in a module library which directly implements its function.

```
. . .
while C loop
    sequence
end loop;
. . .
```

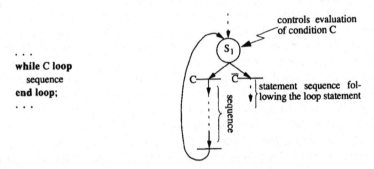

Figure 6.11: Control part for ETPN representation of a *while* loop.

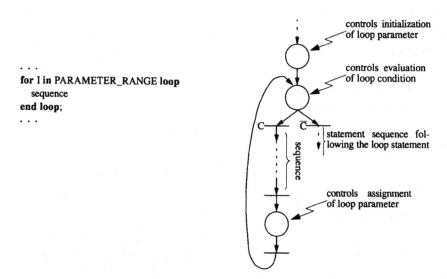

for I **in** PARAMETER_RANGE **loop**
   sequence
end loop;

controls initialization of loop parameter

controls evaluation of loop condition

statement sequence following the loop statement

controls assignment of loop parameter

sequence

Figure 6.12: Control part for ETPN representation of a *for* loop.

The module library consists of a fixed number of module types which can be instantiated with various parameters. With each type of modules, a set of functions which it can implement is specified together with some attributes concerning the physical dimensions and propagation delay of its implementation. With a given data path function, there could exist several module types which can implement it and several different sets of parameter values to be selected. The net-list generation procedure makes both the final selection of the module types and the decision on the appropriate parameters.

6.3.2 Petri net Implementation

The timed Petri net description of control logic cannot be directly mapped into hardware. It is first transformed into a finite state machine (FSM) description and then implemented as microprograms, PLAs or dedicated circuits [KuPe87].

The method used for FSM generation is based on the reachability-marking generation algorithm [Pet81]. However, to follow the restricted execution rules as specified by Definition 6.5, firing multiple transitions simultaneously is enforced and an as-soon-as-possible (ASAP) firing rule is assumed. This means that the ASAP scheduling is used as the default scheduling. Note, however, many different types of rescheduling transformations are applied to the ETPN design during the synthesis process. Therefore, the general scheduling strategy is not inherently of ASAP nature.

The FSM description is captured as a directed graph with nodes representing

FSM states and arcs representing state transitions. It is formally defined by the following definition.

Definition 6.18   The *FSM graph*, **FG**, is a four-tuple **FG=(FS, FA, FC, CC)** where

> **FS** $\subseteq 2^{\mathbf{P}}$ is a finite set of states; each state consists of a collection of places of the timed Petri net that can possess tokens simultaneously;

> **FA** $\subseteq$ **FS** $\times$ **FS** is a finite set of arcs representing possible control flow between two states;

> **FC** = $\{FC_1, FC_2, ..., FC_3\}$ is a finite set of conditions that guard the arcs;

> **CC** : **FC** $\rightarrow 2^{\mathbf{FA}}$ is mapping from conditions to sets of arcs; an arc FA_i is guarded by a condition FC_j if $FA_i \in$ **CC**(FC_j).

A state corresponds to a control step and denotes a really reachable marking of the control Petri net, i.e., a set of Petri net places which hold tokens at the same time and consequently their associated operations will be performed simultaneously. When these operations have been completed, a next state will be entered depending on the conditions guarding the arcs coming from the current state.

The FSM graph generation algorithm is based on the reachability-marking algorithm. Every state generated by this algorithm can be either frontier, interior or terminal state. A frontier state is the state which is used to find next states. An interior state denotes a state which has already been used for generation of next states and a terminal state is a state which has no next states. Only the frontier states are used for generation of next states. After generation they become terminal, if they have not any next states, or interior if they have next states. In some cases, the frontier state is removed if it is the same as an already existing interior or terminal state. The algorithm starts generation of states from an initial frontier state which represents and initial marking of the Petri net. It then uses frontier states to generate new frontier, and terminal states. The basic FSM generation algorithm is illustrated by the pseudo-code depicted in Figure 6.13.

> **Example 6.8:** Consider the Petri net depicted in Figure 6.14a. The FSM generation algorithm generates the FSM graph given in Figure 6.14b, which is equivalent to the current default scheduling. The FSM code is represented in Figure 6.14c. The generated states contain usually several places. For example, state M_1 contains three places S_1, S_2 and S_3 which control data path activities simultaneously. State transition can be done by firing several transitions simultaneously. For instance, transition from state M_1 to state M_2 requires firing of three transitions. Note, that there are states such as M_3, M_4 and M_5, that have places which were already used in the previous state to control operations in the data path and

procedure FSMgeneration;
  begin
    M_0 := the set of initial places defined by the initial marking;
    Node_type(M_0) := frontier;
    while $\exists M_i$, Node_type(M_i) = frontier **do**
      if $\exists M_j$, (Node_type(M_j) ≠ frontier) ∧ (M_i and M_j contain exactly the same places) **then**
        Redirect the arc from predecessor(M_i) to M_i so that it connects predecessor(M_i) to M_j
        instead, and then remove M_i
      else
        if Enabled_transitions(M_i) = ∅   (* M_i does not enable *) **then**
          Node_type(M_i) := terminal
        else (* M_i is not duplicate and there exist enabled transitions *)
        begin
          Node_type(M_i) := interior;
          for each maximally firable set of transitions, T_m, enabled by M_i **do**
          begin
            Create a new node M_k as a set of places that will hold
              tokens when all transitions in T_m are fired;
            Create a new arc A_l from M_i to M_k (M_i = predecessor(M_k));
            Guarding_condition(A_l) := intersection of conditions
              guarding the transitions in T_m;
            Node_type(M_k) := frontier;
          end;
        end;
  end.

Figure 6.13: The FSM generation algorithm.

should not be present in these states. They are present in the figure only
to explain the basic algorithm, and are encircled to indicate that they are
not present in the final FSM states.

                                                                     □

The main difference between the above algorithm and a reachability graph
generation procedure [Pet81] is that when generating the reachable states from
an interior state, a set of transitions is fired in our case instead of one single
transition. Further, only states which are reached by firing maximally firable
sets of transitions are included in the FSM graph. A *firable* transition set is a set
of transitions which are enabled by the current state and the intersection of
their guarding conditions is not always false. A *maximally* firable transition set
is defined as a firable transition set that is not a proper subset of any of the
firable transition sets of the current state. The maximally firable transition set
concept is introduced to get rid of redundant states in the generated FSM
graphs so as to reduce their complexity and implementation cost.

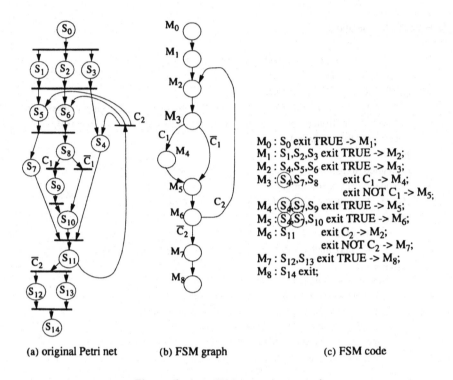

Mo : So exit TRUE -> M1;
M1 : S1,S2,S3 exit TRUE -> M2;
M2 : S4,S5,S6 exit TRUE -> M3;
M3 : S4,S7,S8   exit C1 -> M4;
                  exit NOT C1 -> M5;
M4 : S4,S7,S9 exit TRUE -> M5;
M5 : S4,S7,S10 exit TRUE -> M6;
M6 : S11   exit C2 -> M2;
                  exit NOT C2 -> M7;
M7 : S12,S13 exit TRUE -> M8;
M8 : S14 exit;

(a) original Petri net   (b) FSM graph   (c) FSM code

Figure 6.14: A FSM generation example.

The FSM graph is generated for a timed Petri net assuming synchronous implementation and a clock cycle time execution for every state. Every place in this net has assigned a time interval which represents the time needed to carry on related operations in the data path. Thus the complete algorithm takes the place time interval into account when generating states. A state contains only those places which have not consumed their time interval. A place will consume one clock cycle time when staying in one state. The clock cycle time will then be reduced from the place time interval, and if there is still time left the place will stay in the next state. This process will continue until no time left for the place. This strategy for FSM generation makes it possible to use the algorithm for multi-cycle operations (chaining).

6.4   Basic Transformations

In this section, we will discuss design transformations which can be applied during the high-level synthesis process. This kind of transformations is applied early in the design process to improve certain design characteristics, such as resource utilization, degree of parallelism or power consumption. The high-level synthesis transformations differ from the low level transformation in that they are usually independent of the selected target technology and can be applied regardless of the later design steps.

For the purpose of this book we will classify design transformations, based on their primary function, into the following four groups: compiler oriented, scheduling oriented, data path oriented and control oriented transformations. This classification is based on the primary function of every transformation. Some transformations change also other aspects of the design in addition to the primary function. For example, a scheduling oriented transformation can change the data path allocation.

The transformations carry out local changes in the design. Based on the preconditions for a transformation a subpart of the design is identified and the transformation is carried out. The selected subpart is transformed to a new semantically equivalent structure.

Design transformations can be applied to different representations. Typically compiler oriented transformations are defined for parse trees or directed acyclic graphs (DAGs) which are used by compilers [ASU88]. Other design transformations are defined for some kind of internal design representations which are used by synthesis tools. In our case, we will use the ETPN design representation [PeKu94].

The selection of appropriate transformations can be done in different ways. Obviously the transformation can be selected and applied manually to the part of the design which needs improvements. The designer's knowledge can also be systematically represented in a rule-based system which can later be used for selection of transformations. Finally, transformation can be selected by an optimization algorithm which implements a search of the design space toward an optimal solution. The sequence of selected transformations in this case, follows a certain design space search strategy which is typical for a given optimization algorithm (see Chapter 5).

6.4.1   Compiler Oriented Transformations

This category includes code-improving transformations which are used in compilers for programming languages [ASU88]. These transformations are used to make the generated code run faster and/or occupy less memory. They can be applied in system synthesis to optimize hardware resources as well as make a design faster [BhLe90, PoRa94]. Of course, these transformations should preserve the original semantics of the design and be easy to apply.

Another group of compiler oriented transformations has originally been proposed for compilation of programs to be executed on parallel machines. These transformations use special techniques to parallelize and schedule parts of programs. They are usually applied to the parts of programs which are the most promising candidates for parallelization, such as loops.

In this section, we will consider machine-independent transformations that can improve the target code without taking into account the special properties of the target machine. Such transformations are of great importance for system synthesis since they can easily be implemented and incorporated into an automatic synthesis algorithm.

Compiler oriented transformations can be used in different stages of the system synthesis process. The natural way is to apply them directly to the behavioral VHDL specification or the program graph generated during the compilation step in the same way as they are used during compilation of programming languages. For example, common subexpression elimination, dead code elimination, code motion and constant and variable propagation are such transformations. Their application will always improve a design. Such transformations can be performed on the program graph during compilation of VHDL into ETPN (see Subsection 6.2.1). Another group of transformations, such as in-line procedure expansion and loop unrolling, can also be performed during the HLS process. By applying them, different designs in a design space can be explored and the optimal one, in respect to the given design criteria, can be selected.

Algebraic transformations

Algebraic transformations are applied to transform expressions. They make use of basic algebraic laws, such as *associativity, commutativity* and *distribution*, to simplify expressions. The simplification of an expression means either the reduction of operations or the replacement of expensive operations by cheaper ones. In some cases, algebraic transformations change the structure of an expression in the way that further synthesis steps find an implementation which can be considered as a better one. For example, tree-height reduction intro-duces more parallelism and gives rise to more design alternatives.

Associativity of operations gives possibility to group operands in different ways. For example, a+b+c can be grouped either as (a+b)+c or a+(b+c) assuming the availability of two input adders. Depending on the surrounding structure, selection of one of them can be beneficial for the design. The most common use of associativity is the so called tree-height reduction transformation.

Example 6.9: Consider the expression a+b+c+d which usually is compiled to the structure depicted on Figure 6.15a. However, after applying associativity transformations we get the structure represented on Figure 6.15b. It is obvious that the first structure needs 3 steps for evaluation while the second one can be executed in two steps because of the parallelism between the two first additions: a+b and c+d.

□

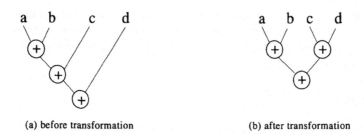

(a) before transformation   (b) after transformation

Figure 6.15: Tree-height reduction transformation example.

The algorithms for constructing a minimum-height syntax tree are presented, for example, in [BaBo69], [Kuck76], [Mura71].

Commutativity makes it possible to change the order of operands. For example, a*b=b*a. It can be used together with other transformations to create more efficient designs.

Distributivity can be used in many cases to perform factorization of multiplication in respect to addition and reduce the number of relatively expensive multipliers in the design. For example, since $x^2+a*x=x*(x+a)$ the first expression uses two multipliers and one adder while the second one only one multiplier and one adder.

Common subexpression elimination

The common subexpression elimination transformation has been introduced to speed-up the execution by storing commonly used expressions and reusing their values in later computations. This technique can be directly used in synthesis since we can save computation time while introducing, in some cases, an additional register for storing intermediate results.

Example 6.10: Consider the following program part:

    e := b-d;
    a := b+c-d;
    f := c+e;

It can be transformed into the equivalent one:

    e := b-d;
    a := e+c;
    f := a;

Note that in this case no additional variables are needed.

□

Dead code elimination

Dead code elimination copes with the problem of removing statements which produce results never used subsequently in the program. For example, if a program contains a statement which assigns a value to a variable which is never used in the subsequent execution of this program then this statement can be safely removed.

Code motion

The code motion transformation tries to rearrange program statements to improve the execution time of the entire program. Since moving statements out of the most computation intensive parts gives execution time improvements, this transformation tries to move loop invariants out of a loop. This is possible since loop invariants do not change their value throughout subsequent iterations of the loop. This transformation improves the execution time since the loop invariants do not need to be unnecessarily computed several times inside a loop. Instead they are computed once before the loop execution starts.

Example 6.11: Consider the following fragment of a VHDL program
```
for i in 63 downto 0 loop
    k := a * b;
    x(i) := y(i) + k;
end loop;
```
This loop contains a statement which assigns always the same value to k since neither a nor b changes their value during the execution of the loop. This statement can be moved out of the loop and the transformed program will be the following one:
```
k := a * b;
for i in 63 downto 0 loop
    x(i) := y(i) + k;
end loop;
```
□

Constant and variable propagation

Defined constants can often be used to compute the value of expressions or parts of expressions before run-time execution of the program. By this, we achieve a better performance of the design since some expressions are computed only once during compilation.

Example 6.12: Consider the following part of a program:
```
i := 0;
j := i + 2;
k := 2 * j;
```
It can be transformed into an equivalent sequence of instructions:
```
i := 0;
```

```
j := 2;
k := 4;
```
                                                                    □

Variable propagation, on the other hand tries to reduce the number of variables by eliminating copies of variables. The assignment of a variable, like v := z will permit, as a result of further analysis, to replace v in some expressions by z.

Example 6.13: Consider the following part of a program:
```
a := b;
c := a + d;
e := f * a;
```
It can be transformed into an equivalent program part:
```
a := b;
c := b + d;
e := f * b;
```
The statement a := b can be removed by a subsequent dead-code elimination transformation.

                                                                    □

In-line procedure expansion

A procedure can be treated as a macro and expended in a calling program. In this case, the body of the procedure substitutes a procedure call statement and actual parameters substitute formal ones. In a hardware implementation this transformation can give opportunity for further transformations, such as resource sharing.

Example 6.14: Consider the following part of a program:
```
procedure Max (X, Y : in integer; Z : out integer) is
    begin
        if X > Y then Z := X else Z := Y end if;
    end Max;

    :

a := b + c;
Max(a, f, i);
```
It can be transformed into an equivalent program part:
```
a := b + c;
if a > f then i := a else i := f end if;
```

                                                                    □

It can be noted that in some cases functions can also be extended in line in a calling program.

Loop unrolling

This transformation applies to iterative statements (loops) which are data independent. It instantiates iterations and replaces them by equivalent statements. In case of a full unrolling every statement of the loop body is instantiated and replaced by an equivalent statement.

Example 6.15: Consider the following for loop:

```
for i in 0 to 7 loop
    ack := ack + (x(i) + x(15-i)) * h(i);
end loop;
```

The possible ETPN structure for this loop is depicted in Figure 6.16. A number of assumptions is made. The controller is sequential, even if some operations, such as S_1 and S_2, can be executed in parallel. To select a realistic clock cycle additional registers, such as t1 and t2, were introduced to cut long combinational paths. All operations, including memory accesses, need one clock cycle execution. The memory access operations fetches data from a location indicated by an address register (e.g., i for memory h) and deposit them into a register (e.g., R3 for memory h).

This loop can be fully unrolled:

```
ack := ack + (x(0) + x(15)) * h(0);          -- instr. 0
ack := ack + (x(1) + x(14)) * h(1);          -- instr. 1
ack := ack + (x(2) + x(13)) * h(2);          -- instr. 2
ack := ack + (x(3) + x(12)) * h(3);          -- instr. 3
ack := ack + (x(4) + x(11)) * h(4);          -- instr. 4
ack := ack + (x(5) + x(10)) * h(5);          -- instr. 5
ack := ack + (x(6) + x(9)) * h(6);           -- instr. 6
ack := ack + (x(7) + x(8)) * h(7);           -- instr. 7
```

The fully unrolled loop has 8 statements, however, the loop construct and the index calculation are not needed any longer. In case of hardware implementation, the full loop unrolling makes the controller more complex. The data path has a similar structure as depicted in Figure 6.17. All functional units are shared by the instructions in this example. It should be mentioned that this is not the only possible implementation of the fully unrolled loop. The inherent parallelism of the computations can be exploited. However, in this case, the memories represent a potential bottleneck which can be solved by a careful scheduling of memory accesses.

□

Full loop unrolling with a large number of iterations leads to a big amount of generated code. In the case of hardware synthesis, this usually leads to larger controllers while a data path units can be shared between computations.

In practice, partial unrolling of loops is used very often. This can reduce the

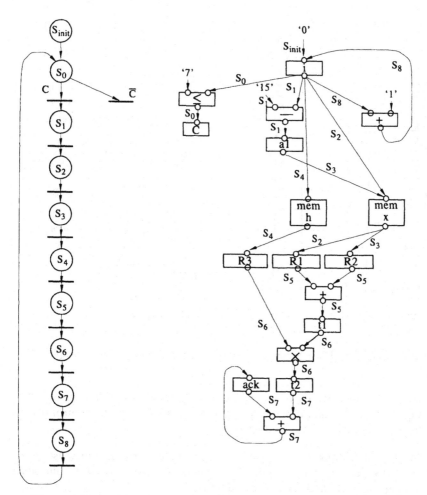

Figure 6.16: The ETPN representation for the original loop for the Example 6.15.

number of iterations and increase possible parallelism of loop execution.

Example 6.16: The loop unrolling with the factor 2 for the loop of the Example 6.15 results in the following program fragment:

> **for** i **in** 0 **to** 3 **loop**
>   ack := ack + (x(2*i) + x(15-2*i)) * h(2*i);
>   ack := ack + (x(2*i+1) + x(15-(2*i+1))) * h(2*i+1);
>   **end loop**;

Note that after the loop unrolling transformation, parts of the computation inside the loop can be executed in parallel. Computation of (x(i0) + x(15-i0)) * h(i0) where i0:=2*i, and (x(i1) + x(15-i1)) * h(i1) where i1:=2*i+1

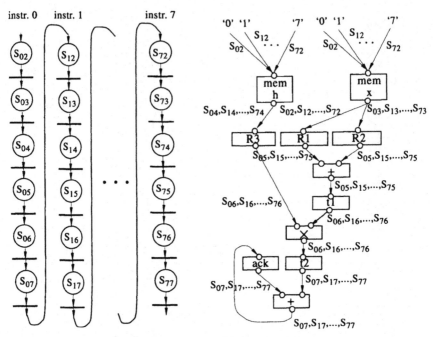

Figure 6.17: The ETPN representation for the fully unrolled loop for the Example 6.15.

are data independent and potentially can be parallelized as depicted in Figure 6.18. The additional empty places are added into the control part to avoid memory access conflicts.

□

The loop unrolling transformation, which potentially can increase parallelism between statements within the loop, is mainly used to introduce pipelining as it will be discussed in Section 6.5.

6.4.2   Operation Scheduling Oriented Transformations

While the previously described transformations were originally proposed for code optimization, the operation scheduling oriented transformations have been specially developed to improve hardware synthesis results. They change in first place scheduling oriented properties of a design.

Scheduling oriented transformations deal with three main aspects of operation scheduling:
1. determination of the serial/parallel nature of the design,
2. division or grouping of operations into time steps, and
3. change of the order of operations.

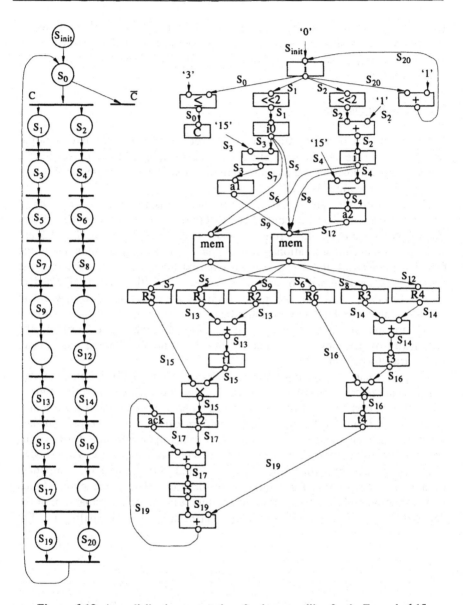

Figure 6.18: A parallelized representation after loop unrolling for the Example 6.15.

The first group of operation scheduling oriented transformations changes the degree of parallelism in the design. For example, *place-stretch* transformations make two parallel places sequential ones while *parallelization* performs a reverse transformation. The second group of these transformations deal with the division or grouping of operations into time steps. *Place splitting* and *place folding* transformations are used for this purpose. The last set of operation scheduling oriented transformations changes the order of operations. This is achieved by *rescheduling* transformations.

Place-stretch transformation

The place-stretch transformations can be used to sequentialize two parallel, thus decreasing the degree of parallelism. It can be applied if two places have a common input transition and a common output transition. An example of place-stretch transformations is given in Figure 6.19.

The place-stretch transformation is semantics-preserving, if at least one of the stretched places does not control any external arc. The reason for this is that since S_i and S_j are in parallel order in the original control Petri net, S_i and S_j are not in a data-dependent relation unless the data-dependent relation is due to the fact that both $C(S_i)$ and $C(S_j)$ contain external arcs. Therefore, whether the operations associated with S_i and those associated with S_j are performed in parallel or in sequence does not matter from the semantics point of view. It can be shown that the ETPN representations before and after the place-stretch transformation are data-invariantly equivalent to each other. Further, the complexity of the control Petri net will not be radically affected by a place-stretch transformation.

The performance of the design can, however, be affected if the place-stretch transformation stretches the critical path, i.e., the path whose operation time is critical to the performance. In other cases, the place-stretch transformation may affect the performance by stretching a non-critical path into the critical one. Whether a stretch should be done or not depends upon whether it allows the improvement of certain design parameters. For example, if the two sets of operations associated with S_i and S_j can be bound to the same RTL component by a vertex merger transformation (to be described later), the place-stretch transformation makes it possible to do so, thus resulting in a compaction of the data path. This is an example of trading performance for data path cost. The

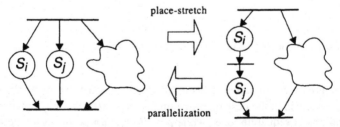

Figure 6.19: An example of place-stretch and parallelization transformation.

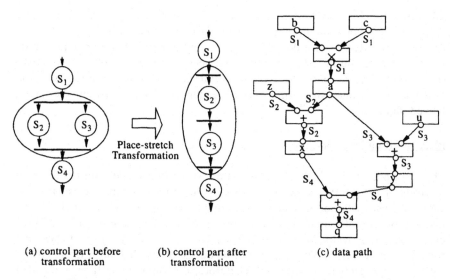

(a) control part before   (b) control part after   (c) data path
     transformation   transformation

Figure 6.20: Place-stretch transformation applied to Example 6.17.

place-stretch transformation may also have no effect at all on performance if the stretched path does not constitute the critical path. For example, there may exist a complicated set of operations associated with another path between the input and output transitions of S_i and S_j, as depicted in Figure 6.19 as a cloud.

Example 6.17: Consider the following fragment of a VHDL program:

```
a := b * c;        -- S1
x := z + a;        -- S2
y := a + u;        -- S3
q := x + y;        -- S4
```

This sequence of statements can be translated into the ETPN representation depicted in Figure 6.20. Before place-stretch transformation, the design needs two adders and three computational steps, while after applying the transformation it could be synthesized using one adder and four computation steps, assuming a subsequent vertex merger transformations (see Example 6.20). It is obvious that after transformation we got a cheaper but slower design.

$\square$

The place-stretch transformation can also be applied to two places that have a common input transition but do not have a common output transition. They can also be extended to stretch subparts of the Petri net graph instead of places.

Parallelization transformation

The transformation set includes also the inverse transformations of the different stretch transformations discussed above (see Figure 6.19). The purpose of the corresponding *parallelization* transformation is to increase the degree of parallelism in a design by, for example, parallelizing a sequence of Petri net places and introducing additional operation vertices into the data path (if needed). The decision to apply a parallelization transformation must be based upon the analysis of the data dependence relation between the parallelized places.

Place-splitting transformation

The other operation scheduling oriented transformations deal with the division or grouping of operations into time steps. As mentioned before, we assume that each Petri net transition is synchronized by a clock signal and each place holds a token for normally one clock cycle. It can also hold a token for several clock cycles only if it waits for synchronization or for some conditions to become true. This, however, is not the result of an operation's delay but it is the consequence of the global control flow. Therefore, the clock cycle time is decided taking into consideration the place which controls the most time-consuming set of operations.

In many cases, it can happen that one place, which controls a long computation path, determines the clock cycle. This long clock cycle has to be used by all computations controlled by the other places. These computations very often use only a small fraction of the selected clock cycle. On the other hand, a design's critical path is determined by the selected clock cycle. One way to make such a design faster is to divide the operation sets that require too much time into sequences of more primitive operations, each of which consumes approximately the same unit of time. This time unit can be given by the designers as a design constraint. A *place-splitting* transformation transforms a place into a sequence of places and introduces additional registers in the data path, as illustrated in Figure 6.21. Additional registers have to be introduced since primitive operations, obtained after splitting an operation set, are controlled by separate places. Thus data which are produced in the data path by one control place and used by another one have to be stored in registers.

Example 6.18: Consider the VHDL assignment statement
ack := ack + (x(0) + x(15)) * h(0);
By default, it is compiled into an ETPN representation where one place controls all three arithmetic operation: two additions and one multiplication (Figure 6.22a). Such a structure requires a very long clock cycle which is equivalent to the sum of the delays of two adders' and a multiplier. As a result of applying a place-splitting transformation two times the assignment statement will be controlled by three consecutive places. One place will control the addition of x(0) and x(15), the next place the multiplication of the partial result and h(0) and the final one the addition

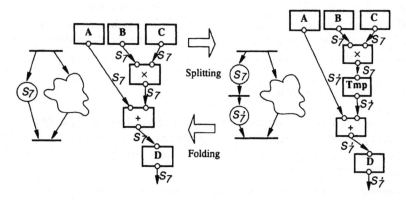

Figure 6.21: An example of place-splitting and place-folding transformations.

of the created partial result and ack. The design after two place-split transformations will need one multiplier and one adder (assuming subsequent vertex-merger transformation) but also two temporal registers. It will be computed during three clock cycles.

□

Place-folding transformation

The inverse transformation of place-splitting is *place-folding* which folds a sequence of places into a single one, as illustrated in Figure 6.21. Note that the set of operations associated with the folded places will still be executed in the

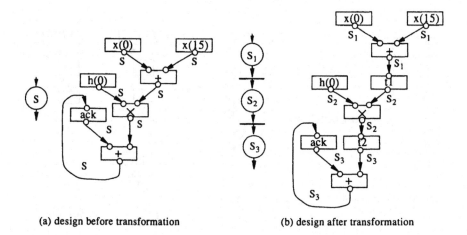

(a) design before transformation   (b) design after transformation

Figure 6.22: Place-splitting transformations for the Example 6.18.

same order if these places are data dependent upon each other.

This transformation is usually used to group together several operations which are very fast. For example, comparison and logical operations can be folded together to speed-up a design.

Rescheduling transformation

The *rescheduling* transformations introduce dummy places into the Petri net so as to change the default scheduling of operations. The dummy places are introduced in parallel paths to influence the way places will be grouped together for controller generation.

Example 6.19: Consider the following fragment of a VHDL program:

                a := b + c;   -- S1
                d := s - a;   -- S2
                e := d * f;   -- S3
                x := y + z;   -- S4

Figure 6.23a depicts the initial control part representation for this design while Figure 6.23c shows its related data path. In Figure 6.23b the result of applying a rescheduling transformation is shown. Two dummy places are introduced to delay place S_4 so that its associated operation, namely addition, can eventually share the same RTL component as the addition associated with S_1 and the subtraction associated with S_2.

□

Rescheduling transformations are mainly used to balance the load of each control step, as illustrated with this example. It creates the opportunity for vertex merge transformation which can be applied later.

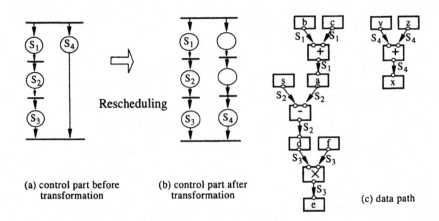

(a) control part before   (b) control part after   (c) data path
    transformation   transformation

Figure 6.23: An example of rescheduling transformation.

6.4.3   Data Path Oriented Transformations

In general, data path allocation determines the number and type of hardware modules to be included in the physical implementation and the mapping of data path operations into available hardware modules. It decides also on the communication channels between the allocated hardware modules. In a transformational synthesis approach allocation is basically done through *merger* transformations. These transformations merge together data path units until a one-to-one mapping to available physical hardware is feasible.

<u>Vertex merger transformation</u>

The *vertex merger* transformation introduced by Definition 6.16 folds two combinatorial vertices into one, as illustrated in Figure 6.3, and represents the decision to share a hardware module by two operations.Two vertices will be considered as candidates for merger, if their functions can be implemented by the same RTL component. When two vertices with different operations, for example, a subtraction and an addition, are merged, the generated vertex will perform a set of operations. A module-binding procedure will be carried out to map it to an available hardware module, such as an ALU, in the RTL module library.

According to Definition 6.16, their associated control places should not hold tokens simultaneously in any of the reachable markings (**RM**), before two vertices can be merged. The restricted transition firing rule enforced by the ETPN timing model, however, reduces the set of *really reachable* markings to a much smaller one than **RM**. Therefore, CAMAD assumes that two vertices can be merged if their associated places are scheduled into different control steps according to the current default scheduling (see Subsection 6.3.2). When stretch or rescheduling transformations are performed, however, the default scheduling will be changed and the associated places of the merged vertices could end up in the same control steps, resulting in a resource conflict. Checking for these conflicts must consequently be carried out when stretch or rescheduling transformations are to be performed.

> **Example 6.20:** Consider the data path of the Example 6.17 which is depicted in Figure 6.20c. After the place-stretch transformation of places S_2 and S_3, the related addition operation can share the same adder as depicted in Figure 6.24. Further sharing of the adder is also possible between operations S_2, S_3 and S_4. This is, however, not depicted in the figure.
>
> □

We should also note that a vertex merger may cause the introduction of multiplexors (as in the example given in Figure 6.3) or the increase of multiplexor inputs, the

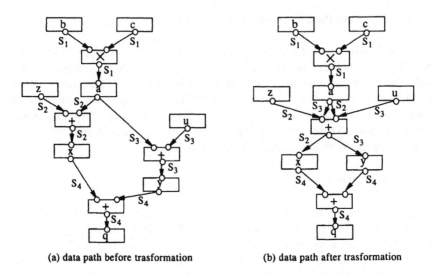

(a) data path before trasformation   (b) data path after trasformation

Figure 6.24: An example of vertex merger transformation.

impact of which, in terms of timing delay and implementation difficulty, must be taken into account before making a decision to carry out such a transformation.

Vertex splitter transformation

The *vertex splitter* transformation is an inverse transformation to the vertex merger. Vertices shared by more than one operation are split into two nodes. The arcs together with the controlling place(s) follow their related vertices.

The transformation increases the area of the design while it has no direct influence on the performance. It can, however, enable further transformations, such as parallelization.

6.4.4   Control Oriented Transformations

Control synthesis deals with the design of the controller and the selection of a clocking scheme. The implementation of the Petri net control description has been discussed in Subsection 6.3. We present in the following only transformations which will contribute to the simplification of the control logic.

Control merger

Control merger folds a set of Petri net places, that are strictly parallel (they all have the same input transition), into one place. This transformation should be done only if all the output transitions of the places are guarded by always true

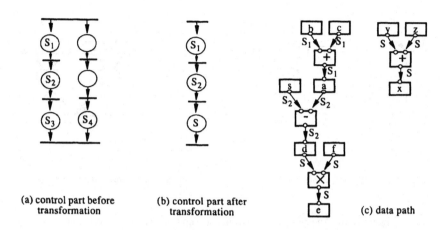

(a) control part before   (b) control part after
    transformation   transformation   (c) data path

Figure 6.25: An example of control merger transformation for the Example 6.19.

conditions. After a control merger, all of the arcs in the data path that are controlled by the original places will be changed to be controlled by the new one.

Example 6.21: Consider the ETPN structure discussed in the Example 6.19. After the rescheduling transformation, the new ETPN contains two parallel paths which can be easily merged together by three consecutive control merger transformations as depicted in Figure 6.25.

□

Control/data path exchange

Conditional statements execute different statement sequences depending on a computed condition. A typical if statement executes statements in one of two conditional branches based on a condition. This statement is usually compiled into a representation that after computing the condition selects one of two possible execution paths (see Subsection 6.2.3). This is a typical sequential software view on the execution model. The condition is computed and after that a decision is made on which statement sequence is to be selected for execution. Hardware, however, is inherently parallel and it is possible to transform this structure into a different one which executes in parallel both statement sequences and later, based on the condition, selects the results from one of them.

Example 6.22: Consider the following VHDL conditional statement:
    if cond **then** a := b+c; **else** a := d + e; **end if**;
This statement can be implemented in two ways: controller based or data-path based as depicted in Figure 6.26. When it is implemented

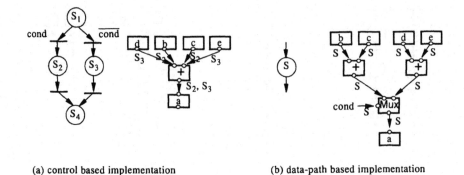

(a) control based implementation   (b) data-path based implementation

Figure 6.26: An example of control/data-path exchange transformation.

using control based style the control part has two paths depending on the condition cond. It executes either operations controlled by S_2 or S_3. The data-path based implementation has only one control place but, in addition, it has to use a multiplexer which selects the result from either the left or the right path to written into the register a.

□

Transformations, like the one presented above, can also be applied to other conditional structures, such those described by a case statement.

Depending on the selected method for conditional statement implementation different design features are influenced. The controller based implementation creates a more complex controller but data path is simpler and consumes less power. The data-path oriented implementation, on the other hand, will require a more complex data path and more power since more parallelism is implemented while the controller becomes simpler.

6.5   Pipeline Transformations

Pipelining is a special processing method which makes it possible to execute simultaneously several computations on the same hardware by using its different parts. In this case, the computation is divided into execution steps and different steps are executed by different pipeline stages. A new computation is started before the old one finished its execution. Several computations co-exist in the same pipeline, each of them being in a different execution phase.

Figure 6.27 compares sequential, parallel and pipelined execution of N computations. Each computation contains three execution steps which are performed on different data and produce different results. It is assumed that the execution time for a single computation step is the same for all three execution methods. Thus to produce results for a single computation all methods need the

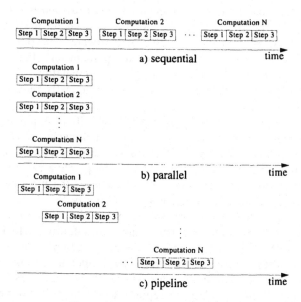

Figure 6.27: Different execution methods.

same time. However, the total execution time for N computations varies very much between methods. The sequential execution is the slowest while the parallel is the fastest one. Pipelining is the intermediate solution. It offers a good trade-off between performance and cost. The total execution time is very much reduced, comparing to sequential execution time, but is higher than for parallel execution while the cost of the implementation is similar to the sequential implementation (with some overhead for pipeline implementation). The parallel implementation is N times more expensive because it requires N copies of the same hardware. All these features make pipelining an efficient architecture for implementation of many computational processes.

To make an evaluation of different sequential, parallel and pipeline implementations we need to have an area and a performance measures for them. An area measure is defined for the ETPN representation (see Section 6.6) and can be directly applied here. The performance measures have to be re-defined for pipelines. For the purpose of this section, we will use execution time defined in clock cycles. We define sequential (T_{seq}), parallel (T_{par}) and pipeline (T_{pipe}) total execution times. We assume that we have N computations and every computation is computed during k steps. Additionally, we assume that a new computation is initiated in a pipeline every clock cycle. Then the total execution time, measured in execution steps, is represented by the following three formulas:

$$T_{seq} = N \cdot k, \qquad T_{par} = k, \qquad T_{pipe} = k + (N-1)$$

The execution time is the greatest for sequential execution and the lowest for the parallel one. The parallel execution does not depend on the number of computations as long as we have N copies of hardware. The pipeline total execution time is between the sequential and the parallel one. For very high values of N and small k, it is close to the number of computations. This means that the execution time of a single computation is practically reduced to the execution time of a stage.

In this book, we assume the following pipeline definition. A pipeline has n stages and every stage has execution time of one clock cycle. Since in some situations, due to conflicts, the pipeline can not be initiated every clock cycle we define a so called initiation rate. For example, the pipeline can be initiated every second clock cycle. In this case, we still consider the pipeline to be n stages long but the execution time is calculated differently.

To calculate the pipeline execution time we define the initiation rate for a pipeline as $IR = \frac{I}{C}$, where I is the number of initiations during C clock cycles. For the situation when every clock cycle a new computation is initiated $IR=1$ and represents the best possible initiation rate. If the pipeline, because of collisions, can accept only one initiation every two clock cycles the $IR = \frac{1}{2}$. The total execution time for such pipelines is represented by the following formula:

$$T_{pipe} = k + \left\lceil \frac{1}{IR}(N-1) \right\rceil$$

Other models have been also used. For example, a pipeline stage can be defined as several clock cycles long [JGB 96]. In this case, an n stage pipeline with two clock cycles initiation rate defined in accordance to our definitions will be defined as an $n/2$ stage pipeline with a two clock cycle stages as depicted in Figure 6.28.

The definition adopted in this book is, in our opinion, more flexible and makes it possible to consider more complex initiation schemes. The pipeline can be initiated in this case using an initiation schedule which defines time slots for initiation. For example, we can start a pipeline computation at time slots 0, 2 and 5, and repeated initiations afterwards in time slots 2 and 5 until an exit condition will become true. This definition offers also more flexibility in analyzing resource utilization and possible resource sharing. In our definition, the resources will be shared if the analysis of the reservation table will allow this. In other approaches, the resources can only be shared within a single multi-cycle pipeline stage.

6.5.1   Analysis and synthesis methods

In this section, we discuss pipeline analysis and synthesis methods, which can directly be applied to build pipelines, using the transformational design method. The main idea is to build a simple controller, using the ETPN

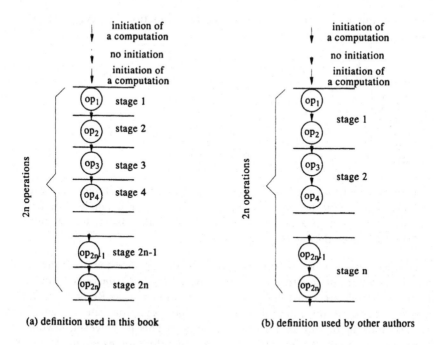

(a) definition used in this book   (b) definition used by other authors

Figure 6.28: Different methods to define pipeline execution schemes.

formalism and later synthesize it, using available controller synthesis tools (see Subsection 6.3.2). The general goal is to find a controller which can provide a *maximum throughput* of the pipeline while guaranteeing absence of *collisions* between different operations. The maximal throughput can be obtained when a new computation is started in the pipeline every clock cycle. This ideal situation is very often disturbed by collisions. The collision occurs when two or more operations would like to use the same pipeline stage simultaneously. In this case the start of a new computation has to be delayed and the throughput decreases. To analyze these situations and build efficient controllers methods, described below, are used.

The *reservation table* is the basic method used to analyze dynamic properties of pipelines. The reservation table is, in principle, a timing diagram which shows the flow of data through data path units. It is formulated as a two dimensional table. Each row of this table represents a data path unit and each column represents a time step. When the mark is set in a cell of the table it means that the pipeline stage represented by this row is used during the execution step indicated by the column. A set of marks filling the reservation table represents the execution pattern for a given computation.

Example 6.23: Consider the hardware structure depicted in Figure 6.16 representing the loop computation of the Example 6.15. The following

reservation table is created to depict usage of functional units (we use Xs
to represent markings of the table):

	1	2	3	4	5	6	7	8
SUB	X							
Mem_x		X	X					
Mem_h				X				
ADD_1					X			
MUL						X		
ADD_2							X	
ADD_3								X

It is assumed that we have 8 execution steps and 7 different hardware
units which will be used as pipeline stages. For example, the unit
ADD_1 is used in execution step 5. On the other hand, the memory
Mem_x is used during execution steps 2 and 3.

□

The reservation table is primarily used to design pipeline controllers. We
will discuss here the controller synthesis based on *collision vectors* and *state
diagrams* which are constructed using reservation tables. Reservation tables
can also be used for improving the pipeline execution rate by analyzing
possible introduction of additional delay stages. This is, however, beyond the
scope of this book. An interested reader can refer, for example, to [Sto87,
Hwa93].

To implement a pipeline execution model, we would like to start a next
computation on the same data path hardware in the way that both computations
will co-exist in the same pipeline. All computations co-exist in the same
pipeline if they do not use the same hardware at any possible execution time. It
means that, in principle, a next computation can be started in a pipeline if it
will not collide in any future time with the previously initiated computations.
The collision can be easily checked using reservation tables. We use the reservation table filled with the current execution pattern represented by X marks,
for example. To check if the next computation will collide with the current one
we simply overlay another copy of the execution pattern, marked for example
with Y, shifted to the right by the certain number of clock cycles. The execution
pattern Y represents the next computation initiated in the pipeline. As a result,
we have two identical execution patterns in our reservation table; one, marked
with X, representing the current computation and one, marked by Y, representing the next computation. If at least one cell of the reservation table
contains both X and Y, a collision exist between the current and the next
computation. Start of the next computation at this execution step is thus
forbidden.

The information on possible collisions is represented in a compact way by a
collision vector. The collision vector is a binary vector of length k, where k
represents the number of execution steps of the considered computation. Every

position *i* in this vector contains a bit indicating whether the new computation can be started *i* clock cycles after the current computation. A 1 represents collision and 0 indicates that no collision occurs.

Example 6.24: Consider the reservation table from the Example 6.23. We represent a current computation by X and the new one by Y. It is also assumed that the new computation starts at the next clock cycle. The reservation table below illustrates this situation.

	1	2	3	4	5	6	7	8	9
SUB	X	Y							
Mem_x		X	XY	Y					
Mem_h				X	Y				
ADD_1					X	Y			
MUL						X	Y		
ADD_2							X	Y	
ADD_3								X	Y

The collision vector for this reservation table is 1000000. The first 1 represents the collision depicted in the reservation table as a shaded cell. The other cases of initiation of the pipeline are not shown in the table but they do not produce any collisions.

□

Finally, to build a controller for our pipeline we need to make decisions when to start the next computation in the pipeline. We will call it an *initiation schedule* since it assigns starting points for the next computations. To determine the initiation schedule, the analysis of all possible states of the pipeline for all possible initiations is needed. This analysis leads finally to selection of the optimal initiation schedule. The selection of the optimal initiation schedule is usually done based on the analysis of a so called, *state diagram*. The state diagram is built based on the collision vector and it shows possible states of the pipeline as well as all possible initiations.

The state diagram is a directed graph. The nodes in this graph represent pipeline states derived from the collision vector by initiating next computations. The arcs represent transitions between two consecutive initiations. Every node is labeled by a binary vector of length *k*, where *k* represents the number of execution steps of the considered computation. An initial node is labeled by the collision vector and the other nodes are labeled by a binary vector of the same length representing possible collisions existing in the current pipeline state. The label of the arc is an integer number representing a number of clock cycles between two consecutive initiations of the pipeline represented by the connected nodes.

As mentioned before, the initial node of the state diagram is created for an original collision vector and consequently labeled by the collision vector. A new node of the state diagram is created for every *i'th* position of this vector which is 0. The arcs are labeled by the value *i* indicating the distance between two initiations when moving from one pipeline state into another as indicated by state diagram nodes. The new label of the node is obtained by shifting the current label *i* positions to the left, appending with *i* zeros on the right and performing OR operation on this vector and the collision vector. The last *k* positions of the resulting vector are the new node label. This method is applied iteratively to state diagram nodes, however, if the newly computed node already exists in the state diagram the node is not created but instead an arc to the existing node is generated. The detail algorithm can be found, for example, in [Sto87].

> **Example 6.25:** Consider the collision vector 01101101. The initial node labeled by this collision vector is created. Since the collision vector has three 0's on positions 1, 4 and 7, the three arcs are created labeled by 1, 4 and 7. The new state label is computed based on the method presented above. This computation is depicted in the Figure 6.29 for the initial node and the arc labeled by 1. The new node labeled by the vector 11111111 is created. The same procedure is repeated for the other two alternatives and later applied to newly created nodes. The corresponding state diagram is depicted in Figure 6.29. On the state diagram, there is always a possible connection from every state to the initial state after 9 execution steps. This represents a simple non-pipeline execution and these arcs are not shown in the figure.
>
> □

In the next section, the introduced formalism will be used to build an initiation schedule and a related controller.

6.5.2   Pipeline transformations in ETPN

In high-level synthesis, two types of pipelining are usually identified: functional and algorithmic. *Functional pipelining* means that pipelined functional units are used. *Algorithmic pipelining* assumes pipelining of the

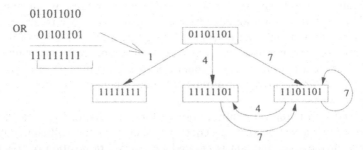

Figure 6.29: The state diagram for the collision vector 01101101.

computations of a whole algorithm or a part of the algorithm. Algorithmic pipelining is usually applied to loops, since there we can identify a number of similar computations which have to be performed successively on different loop iterations.

Functional pipelining transformation

Functional pipelining requires availability of pipelined functional units in the module library. Pipelined functional units are capable of initiating a new computation before the previously initiated computation is finished. Multipliers are typical examples of such units. They have usually a higher latency and, if not pipelined, they can become a performance bottleneck.

The ETPN transformation for pipelined functional units simply performs place-stretch transformation on the place controlling the considered functional unit.

> **Example 6.26:** Consider that a pipelined multiplier is available in the module library and it replaces the ordinary multiplier. Figure 6.30 represents this situation. We assume that the multiplier needs two clock cycles to produce results. The pipelined multiplier has the additional registers/logic to pipeline computations.
>
> □

The use of functional pipelining gives the same advantages as place-stretch transformation. Basically, it makes it possible to use a lower clock cycle for a design and consequently improves its performance.

Controller transformation for algorithmic pipelining

The main idea of algorithmic pipelining is to design a pipeline which will be used to execute computations of a whole algorithm or of some of its parts. In this case, the analysis of the algorithm is based on the previously described reservation tables, collision vectors and state diagrams (see, for example [Sto87]). The result of this analysis provides information which is later used to

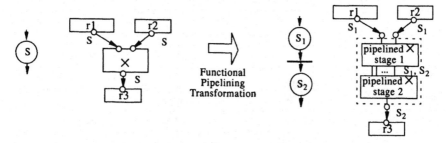

Figure 6.30: Functional pipelining with two stage pipelined multiplier.

synthesize a controller for a given data path structure. In our approach, the
controller will still be represented as a Petri net.

Basically the controller has to implement the pipeline initiation schedule. It
starts a new computation at the scheduled time step when the collision can not
occur. Later it needs to control data path activities for all active and co-existing
computations. The controller is designed based on the analysis of the state
diagram. To achieve the best performance, the highest initiation rate is selected
from the diagram and a related controller which implements this initiation
schedule is then designed.

Example 6.27: Consider the state diagram from Figure 6.29. The highest
initiation rate is represented by the left hand branch where after initiating an initial computation we initiate the next computation after 1
clock cycle and the following one after 9 clock cycles. Then we have a
loop initiating computations after 1 clock cycle and 9 clock cycle delays.
The related Petri net controller is depicted in Figure 6.31. The first
computation in the pipeline is started after firing the transition T_3. Then
the place S_1 implements one clock cycle delay and a next computation is
started by firing the transition T_2. After that the controller has to delay
an initiation of a new computation by 9 clock cycles which is implemented by places S_2, ..., S_{10}. When this computation is started, the
consecutive computations will be started after 1 and 9 clock cycles in a
loop implemented by places S_1, S_2, ... and S_{10}. Every place which
initiates a new computation, i.e., S_0, S_1 and S_{10}, selects, based on the

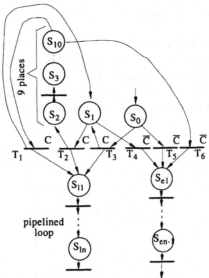

Figure 6.31: The controller for the pipeline represented by the state diagram in Figure 6.29.

value of the condition C, a new computation or finishes pipeline execution by selecting one of the transitions guarded by $\overline{C}$. Note, that the loop contains the same places (S_{11}, ..., S_{1n}), however, a feedback arc is cut since the loop executions are right now initiated by the specially created control structure (places S_0, ..., S_{10}).

□

The controller transformation, as illustrated in the previous example, consist of the following steps: (1) generate a reservation table for a given computation, (2) create a reduced state diagram and find a maximum rate initiation cycle, and (3) restructure the related Petri net. The restructuring of the Petri net can be divided into two main modifications. First, a loop closing arc is removed and *n-1* dummy places are added in the exit branch, where *n* is the pipeline length. Adding *n-1* dummy places is required to synchronize the loop exit with the computation termination which in a pipeline case requires additional clock cycles to empty the pipeline. Second, the initiation pattern which provides a proper initiation scheme obtained from the reduced state diagram analysis is added. The Petri net skeleton of the initiation pattern depends on the application but, in principle, it is a loop structure with several pipeline initiations and exits. The transformation is schematically depicted in Figure 6.32.

Example 6.28: Consider the original loop of the Example 6.15. The reservation table for the pipeline is presented in the Example 6.23 and the collision vector is 1000000. The state diagram is depicted in Figure 6.33. Assuming that all functional units require 1 clock cycle execution

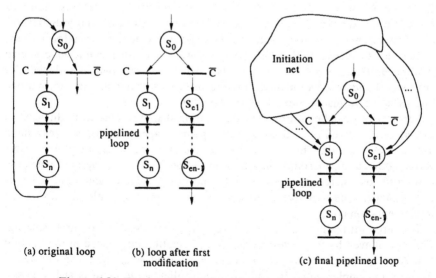

(a) original loop   (b) loop after first modification   (c) final pipelined loop

Figure 6.32: The cotroller transformation for algorithmic pipelining.

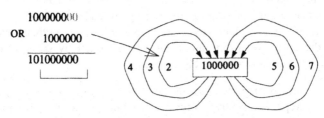

Figure 6.33: The state diagram for the pipeline implementation of the Example 6.15.

the actual representation of the pipeline is depicted in Figure 6.34. Since the initiations of the pipeline are possible in execution steps greater than 2, we select the lowest possible value 2. The control Petri net implements this pipeline controlling strategy. When the computations of the loop are finished (condition C becomes false) we need to wait 7 clock cycles for the pipeline to be emptied. This is implemented as 7 places on the false branch of the controlling Petri net.

The number of clock cycles to execute this loop sequentially is 7*8=56 while the parallel execution requires 8 clock cycles. This pipeline will need 8+⌈2*(7-1)⌉=20 clock cycles to execute the same computation.

□

Vertex merger/splitter for algorithmic pipelining transformation

The direct implementation of a pipeline, as described in the previous examples is, in many cases, inefficient. It can be further improved by applying several data-path oriented pipeline transformations. By applying these transformations, we can try to rearrange the data path in a way that either main performance limiting bottlenecks are removed or the cost of the data path is reduced. An analysis of the reservation table can be used to select both transformations improving performance and resource utilization.

Performance can be improved by duplicating shared resources. In this case, a shared vertex which is used by several pipeline stages is split and every stage uses a separate unit. This transformation changes the structure of the reservation table and furthermore the collision vector. It should be applied when the new collision vector provides a better initiation scheme and thus a better performance. This transformation improves performance while increasing the implementation cost.

Resource utilization can be improved in some cases by resource sharing. This is achieved by the vertex merger transformation introduced in the Section 6.4. A vertex merger transformation can be applied when the initiation rate is decided and implemented in a pipeline controller while there is still a possibility to share some resources without changing the collision vector. The vertex

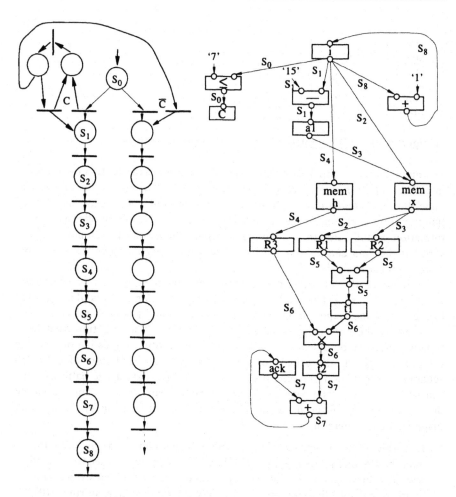

Figure 6.34: A pipelined representation for the Example 6.15.

merger transformation will introduce resource sharing but it will not affect the pipeline performance since the collision vector will still be the same. This transformation will reduce the implementation cost.

Example 6.29: Consider the pipeline from Example 6.28 (Figure 6.34). The original reservation table for this pipeline is presented in the Example 6.23. It can be further optimized by merging adders 2 and 3. The resulting reservation table showed below implies the same collision vector as before and the same controller structure. In this case, the pipeline performance is unchanged but two adders are used instead of three.

□

	1	2	3	4	5	6	7	8
SUB	X							
Mem_x		X	X					
Mem_h				X				
ADD_1					X			
MUL						X		
ADD_2							X	X

Memory mapping for algorithmic pipelining transformation

The main source of pipeline performance degradation is reduction of initiation rate. The initiation rate is reduced because of collisions caused by sharing the same data path resources, such as functional units and memories. Sharing of functional units can easily be removed by applying a vertex duplication transformation, as indicated in the previous section. The memories can not easily be split but the data can be mapped into several memory modules. This requires more complex analysis.

To improve the pipeline performance, the memory, which is shared by several pipeline stages has to be replaced by several memory modules assigned to these pipeline stages. Every memory module should only keep data which are used by the current pipeline stage, if possible. If distribution of data to separate memory modules assigned to different pipeline stages is not possible the pipeline performance can not be improved. The distribution of data into separate memory modules requires an analysis of the data usage patterns. This can not always be determined statically, which means that in some cases the data mapping into different memory modules can not be done by the memory mapping transformation.

> **Example 6.30:** In the case of the pipeline discussed in Example 6.28, the main bottleneck is the memory x. It is accessed in two consecutive steps which limits pipeline initiation rate to every second clock cycle. This memory can be divided into two memory modules: one storing $x(0)$, $x(1)$, ..., $x(7)$ and the other one storing $x(8)$, $x(9)$, ..., $x(15)$. In this case, the reservation table will be different and the collision vector will become 00000000 which allows initiation every clock cycle. The pipeline will need $8+(7-1)=14$ clock cycles to execute the computation.
>
> □

6.6   Selection of Transformations

This section will discuss another important issue of a transformational approach, namely how to select a sequence of transformations to be applied to the initial solution in such a way that the final implementation will be a near-optimal one. This selection can be carried out either manually by the designer or automatically by an algorithm.

In the case of automatic selection of transformation, we have an optimization problem, and by the nature of the transformational approach, algorithms of neighborhood search can be used. A neighborhood can be defined, in this case, to comprise all the designs which can be generated by applying one of the pre-defined transformations. Selection of a transformation in each search step can be guided by problem-specific schemes or general heuristics, such as simulated annealing and tabu search techniques.

6.6.1   Problem-Specific Selection Scheme

A problem-specific selection scheme will make use of the high-level synthesis specific information to guide the selection of design transformations. One of such schemes has been implemented in CAMAD and described in [PeKu93].

<u>Optimization Criteria</u>

The main optimization criteria considered by CAMAD are estimated performance and implementation cost. Other design factors such as pin-out count, power-consumption, and testability can be formulated as design constraints. Given an ETPN design representation, its implementation cost is estimated as the area taken up by the hardware implementation generated according to the ETPN data path implementation method and the Petri net implementation algorithm described in the previous sections. The cost measure is based upon calculation of the cost of the data path and the cost of the control structure. The cost of the data path is in turn calculated according to the cost of each data path unit. The cost of data path units which perform logic, arithmetic, or storage operations is given by the corresponding module parameters stored in the module library. The cost of data path units for communication (i.e., arcs and bus vertices) depends, however, upon the placement of the components as well. To make a more accurate estimation, a rough floorplan is made each time the cost is to be calculated (as well as when accurate timing information is needed). In this way, CAMAD takes into account the geometrical information of a design.

The floorplanning algorithm makes use of a simple heuristic based on the connectivity between the data path vertices. All vertices are first sorted by their connectivity with the other nodes. The node with maximal connectivity is placed in the middle of the silicon plane. Subsequent nodes are then placed around the already placed nodes in the sorted order. This heuristic will not give an optimized floorplan. It provides only a rough indication of how much it will cost to route the wires in the final implementation and indicates an upper bound of the wiring delay.

Given a floorplan, the *area cost* for an ETPN data path $D = (V, I, O, A, B)$, is estimated by the following formula.

$$\text{Area}(\mathbf{D}) = \sum_i Area(V_i) + \sum_j [Len(A_j) \times Wid(A_j)]$$

where

$Area(V_i)$ = area cost of the module corresponding to V_i;
$Len(A_j)$ = length of the connections represented as A_j;
$Wid(A_j)$ = width of the connections represented as A_j, which is the bit width of the connection multiplied by a given implementation factor.

The cost of the control structure will be estimated by an abstract measure which indicates the complexity of the Petri net $\mathbf{PN} = (\mathbf{S}, \mathbf{T}, \mathbf{F})$ in terms of its static structure defined as follows.

$$\text{Cost}(\mathbf{PN}) = \sum_i |{}^\circ S_i| + \sum_j |{}^\circ T_j|$$

where

${}^\circ S_i = \{T \mid (T, S_i) \in F\}$ the pre-set of $S_i \in \mathbf{S}$;
${}^\circ T_j = \{S \mid (S, T_j) \in \mathbf{F}\}$, the pre-set of $T_j \in \mathbf{T}$.

This complexity measure is to be multiplied with an implementation factor to get an estimation of the layout area taken by the controller. Once the implementation style decision for the controller is made, for example, to generate the control signals by a microprogram, more accurate estimation can be made by first running the FSM generation algorithm and then measuring the size of the generated FSM.

The performance of a design can be measured in different ways. If the design is basically a large loop representing, for example, the fetch and execution cycle of a microprocessor, we can break this loop and base the performance measure on the average time needed to complete such a cycle. Sometimes, however, the speed of a particular sequence of operations of the design is more important, therefore, the performance measure must be based upon how fast the specified operation sequence can be executed. Both of these two cases can be captured by the following definition.

The *performance* of an ETPN design is defined as the *average* time needed for a token to travel from an initial place to a final place specified by the designer. The average is taken over the alternative paths between the initial place and the final place, taking into account the estimated execution frequencies of the different paths as well as estimated loop repetition. That is, the time for a token to travel from the initial place to the final place equals the sum of times of each path multiplied by the probabilities of the path being executed. The dynamic execution information such as the percentage of times when a condition is true and the average loop repetition count can be given by designers or extracted from simulation. The simulator *executes* the ETPN design representation with typical input data and collects statistics about operation utilization and control flow choices in the given design.

<u>General Optimization Strategy</u>

With the definition of the implementation cost and performance of an ETPN design, the optimization algorithm can be used to guide the transformation process towards: (1) a design with minimal implementation cost whose performance is within a given constraint; or (2) a design with minimal execution time between two given control places whose implementation cost is less than a given upper-bound. We will concentrate on the cost-minimization strategy (case 1) in the following discussion.

The cost-minimization strategy is based upon a concept of critical path. A *critical path* consists of a sequence of control places which dominates the time needed for a token to flow from the initial place to the final place. For example, in Figure 6.14a, if the initial place is selected to be S_0 and the final place S_{14}, the critical path consists of S_0, S_1, S_5, S_8, S_9, S_{10}, S_{11}, S_{12}, and S_{14}[1] (we assume here that each place will hold a token for normally one clock cycle time). The critical path of a design is used as the first criterion to select a part of the design to apply transformations. This is similar to the approach taken by the MAHA system [PPM86]. However, CAMAD can handle Petri net graphs with nested loops and make use of dynamic execution frequency when analyzing the critical path.

The second selection criterion used by CAMAD is the *repetition count* of the control places which belong to the critical path. The repetition count of a place indicates how many times *on average* the set of operations associated with the place will be executed when the token travels from the initial place to the final place. As in the example of Figure 6.14a, the repetition count for place S_0, S_1, S_{12}, and S_{14} is 1; for S_5, S_8, S_{10}, and S_{11}, it equals the loop repetition number which is determined by the probability when condition C_2 is true; and for S_9, it equals the loop repetition number multiplied by the probability when condition C_1 is true. Those places with the highest repetition count will be selected for consideration first.

The above two criteria are used to select a small set of control places to apply transformations. Since the primary goal of the cost minimization strategy is to minimize the cost of the final design, the main transformations are used to share hardware resource as much as possible. In order that operations having as

---

1. Note, in this example, we can identify another path, consisting of the same places except that place S_5 is substituted by place S_6, which is as critical, because the operation associated with S_5 and that associated with S_6 take the same time to execute. In such situations, one of the paths is arbitrarily selected. Since these paths usually overlap with a very high degree, it makes little difference which one is selected.

much similarity as possible are first grouped together to shares the same hardware resource, a concept of "temperature" similar to that in annealing is introduced. At the beginning, when the temperature is high, the operation vertices have a high movability, and therefore, will be merged to each other only if they perform very similar operations. When the temperature decreases, the vertices' movability is reduced, and they will be merged to each other even if they perform quite different operations but can be implemented by the same RTL component stored in the module library. For example, an addition vertex will only be merged into other addition vertices with the same word-length (100 percent similarity) when the temperature is the highest (99 degrees). With the decreasing of the temperature, it can be merged into addition vertices with different word-lengths, then into subtraction vertices, and eventually into vertices whose operations can be implemented by the same ALU. The similarity between two vertices which cannot be implemented by the same RTL component is defined to be zero. This strategy avoids the generation of ALUs that are capable of performing many different operations unless absolutely needed.

The Algorithm

A sketch of the cost-minimization algorithm is illustrated by the pseudo-code given in Figure 6.35.

The first major step of each optimization iteration is to calculate the time required for the set of operations associated with each Petri net place. This information is then used to determine the minimal clock cycle time, which equals the time required for the most time-consuming operation set. If the minimal clock cycle time cannot meet the clock constraint (if given) or the times needed for the places are very diverse, place-splitting and/or place-folding transformations are applied so as to balance the required times of the different places and to meet the constraint. These transformations divide or group *operation sets* into control steps and invalidate the default scheduling. A new default scheduling is, therefore, generated using the FSM generation algorithm described in Section 6.3. The algorithm then finds the current critical path.

If the given performance constraint is met, the current default scheduling is acceptable and the algorithm proceeds to the next major step. It finds the set of control steps which the places with the maximal repetition count in the critical path are assigned to and examines them one by one in the order of decreasing implementation cost. The implementation cost of a control step is defined as the hardware resources needed to execute the operations associated with the places assigned to it. For each control step, the vertices which are associated with any place assigned to it and are not marked as allocated are considered for merger. If vertex V_i is similar enough, under the current temperature, to that of

begin
    Temperature := 99;
    while not Satisfied **do**
    begin
        Compute the required time for each place;
        Balance the place times and determine the clock cycle time;
        Generate the current default scheduling using the FSM generation algorithm;
        Find the current critical path, let it be C_p;
        Mark all data path vertices as un-allocated;
        if performance within design constraint **then**
        begin
            while $C_p \neq \varnothing$ **do**
            begin
                MP := the set of maximal repetition count places in C_p;
                MS := the set of control steps (FSM states) which the places in MP are
                    assigned to;
                while $MS \neq \varnothing$ **do**
                begin
                    M_i := the control step in MS which has the maximal implementation cost;
                    for each un-allocated $V_i \in \mathrm{ASS}(S_i)$ and S_i is assigned to M_i **do**
                    begin
                        for each allocated vertex V_j such that Similarity(V_i, V_j) > Temperature
                            and V_i is not yet merged **do**
                        begin
                            if all associated places of V_i and V_j are not in the same control step
                          then Apply vertex merger transformation to verge V_i into V_j
                          else if the associated places of V_i which are in the same control step as
                              those of V_j can be rescheduled or stretched **then**
                          begin
                            Apply rescheduling or place-stretch transformation;
                            Apply vertex merger transformation to merge V_i into V_j;
                        end;
                      end;
                    if V_i is not merged **then** Mark V_i as allocated;
                  end;
                Remove M_i from MS;
                end;
                Remove MP from C_p;
            end;
            Compact data path vertices not associated with the critical path;
           Satisfied := (other constraints met) $\wedge$ (Temperature = 0) $\wedge$
                (the current C_p = the previous C_p);
           Temperature := Temperature/Reduction_factor;
           if Temperature < 10 **then** Temperature := 0;
        end else
        begin (* Performance has to be improved *)
           Find a set of parallelizable places in C_p using repetition counts as search priority,
               and apply the place-parallelization transformations;
           otherwise
           begin
               Decide on a new clock cycle time to meet the performance constraint;
               Apply place-splitting transformations to places whose required time is larger
                  than the new clock cycle time;
           end;
        end;
    end;
end.

Figure 6.35: The problem-specific optimization heuristic of CAMAD.

a data path vertex already allocated, say V_j, they will be merged provided that the associated places of V_i and V_j are not in the same control step.

If some of the associated places are in the same control steps, they must be rescheduled or stretched if possible. That is the outcome of the transformation must not violate the performance constraint. If neither rescheduling nor stretch is possible and there exists no allocated vertex which V_i can be merged to, V_i is marked as allocated resource. The vertices marked as allocated by the *last* optimization iteration will constitute the final data path design, while in other iterations the allocated vertices will be marked as un-allocated at the beginning of the next iteration. This heuristic for merging data path units is used to map off-critical vertices[1] on top of the critical ones. (In the case when pre-specified partial structures are given in the input specification, the relevant ETPN parts are marked as permanently allocated. In this way, CAMAD can also handle synthesis with pre-specified partial structure.)

The above synthesis step will be repeated for all the control places in the critical path and then transformations will be applied to the off-critical path control places. The off-critical path synthesis uses a similar transformation selection scheme. But it will try to reduce cost as much as possible, when trade-off between cost and performance must be made. That is, place stretch and rescheduling transformations will be performed as much as possible to enable vertex mergers. In this way, an off-critical path may become the critical path.

If the performance constraint cannot be satisfied, the algorithm tries to shorten the critical path by, for example, parallelizing some control places. The set of the places in the critical path will be checked to see whether two of them can be parallelized, i.e., they are not in a data dependent relation. A set of parallelization transformations will be applied until the constraint is met or no more parallelization transformations can be carried out. In the latter case, the only way to satisfy the performance requirement is to increase the clock frequency. A new clock cycle time will be calculated and place-splitting transformations are applied to the places whose required times are longer than the clock cycle time.

The optimization cycle will be repeated for each decreasing temperature step until the temperature reaches zero degree, other design constraints have been met, and the new critical path is the same as the previous one, which means that no more optimization can be done.

6.6.2   General Heuristics

The selection of a sequence of transformations to be applied to an ETPN design can also be carried out using general heuristic algorithms described in Chapter

---

1. Off-critical vertices are data path vertices that are associated with places not in the critical path.

5. For example, a tabu search technique is proposed in [HaPe95] to solve the problem of multicycle scheduling with a multiple-component library (i.e., a library that contains several alternative implementations of the same operation and modules that can implement several different operation types). It is assumed that the delay of operations can be an arbitrary number of clock cycles. Since many designs contain operations with a wide spectrum of delays this approach allows designs with a much better performance. A neighborhood can be defined by the set of ETPNs generated by changing the implementation of one of the data path nodes. During each iteration all neighbors are evaluated based on a cost function. The goal of the tabu search algorithm is to find the best, according to the cost function, feasible solution. Thus, when one or several of the neighbors are feasible (and not tabu classified) the best one of these neighbors is chosen to be the next solution. The search will stop after a predetermined number of iterations. The tabu search algorithm can also be used to deal with synthesis under timing constraints, which will be discussed in the next chapter.

7

Synthesis of Advanced Features

In this chapter several issues concerning system-level synthesis starting from VHDL specifications are discussed. We concentrate here on aspects which are characteristic to the synthesis of hardware components. Specifications containing subprograms and interacting processes are of main interest in this context. The last section is devoted to the problem of specifying timing constraints and of hardware synthesis under restrictions imposed by such constraints.

Further issues concerning system-level synthesis will be elaborated in the following chapter. There we will discuss how to decide on the specific part of the system specification which has to be synthesized into hardware. For the discussion in this chapter we consider that such a decision has already been made.

7.1 Synthesis of Subprograms

Procedures and functions are powerful constructs which have been used for a long time by programmers and are now available in VHDL (see Section 2.5) as well as other modern hardware description languages. The advantages of using subprograms in system specifications are the same as those already known in software development; namely structuring of the specification, reduction of the specification size, improvement of readability, flexibility, maintainability and reusability.

However, from the synthesis point of view a subprogram can be also interpreted as a structural entity [CST91]. This means that a subprogram is viewed not only as the specification of a certain behavior but also as a partition which can be implemented as a separate entity. Assuming such an interpretation, the designer has the opportunity to influence, by using subprograms, the final structure of the implementation.

The structural meaning of a subprogram can also be ignored during synthesis. In this case, the final implementation is not influenced at all by the fact that some activities have been specified by the designer as subprograms. The synthesized structure is identical with the one resulting from a specification which does not contain any subprogram, but all the activities are explicitly described at the point where they have to be executed.

There are strong arguments for considering subprograms as structural entities [RNVG93]. A flattened specification with none of its parts encapsulated as separate modules (except maybe some processes) can be too large to be accepted for synthesis by a design automation tool. And, even if the tool is able to perform the synthesis, this can take a very long time and the quality of the resulting implementation can be lower compared to a synthesis performed for certain modules separately. Sometimes there are technological limitations which require that a certain design should be synthesized to a set of modules and not to a single piece of hardware. In order to facilitate partitioning of the design, for reasons like those presented above, new subprograms can automatically be created in addition to those specified by the designer. In this case, sequences of statements inside a process are automatically extracted and replaced by a call to a newly created subprogram [Vah95].

If the design is not too large there can also be arguments for flattening the structure by inline expansion of subprograms. As will be discussed in the next section, one main gain resulting from such a synthesis approach can be an improvement of execution speed.

In the following sections we often will use procedures in order to illustrate the discussion. This is because procedures are more general than functions and thus, our conclusions are valid for both kinds of subprograms. A specific aspect concerning functions is that they very often are simple and imply only evaluation of expressions. They also are usually side effect free (pure functions, see Subsection 2.5.1). In such situations the function can be synthesized to a pure combinational structure which is used as part of the data path of the calling process.

Several alternatives for hardware synthesis of subprograms are discussed in the following section. Then, ETPN design representations corresponding to different synthesis alternatives are presented. In Subsection 7.1.3 we show how the synthesis of subprograms is influenced by specific features of VHDL concerning concurrent processes and signals.

7.1.1 Implementation Alternatives

Three implementation alternatives for subprograms are presented in this section [EKPM92]. One of them is based on a flattened representation while the other two are preserving, to a certain degree, the structural identity of the subprogram.

<u>Inline Expansion</u>

For inline expansion, the subprogram hierarchy is flattened by the expansion of each subprogram call into the calling process. Thus, the subprograms, as individual entities, disappear from the resulting structure and are dissolved in the calling processes, by each call to a subprogram being replaced with the body of the subprogram. In Figure 7.1 we show the architecture resulting after synthesis of a specification consisting of a process which contains two calls to a given subprogram. After inline expansion the synthesized architecture consists of a single control unit and a data path.

By performing scheduling on a unique, flattened representation, the synthesis tool can take advantage of all the potential parallelism intrinsic to the specification. Thus, it can produce a high-performance implementation by overlapping in an optimal way operations from the subprogram and the calling process. From this point of view, this approach is superior to both implementation alternatives which will be further discussed.

Another advantage of this approach is that sharing of data path resources between subprogram and process as well as between different subprograms is possible, which can reduce the cost of the final implementation.

The main disadvantage of inline expansion is the high complexity of the resulting control logic. By expanding the subprograms, the number of control steps in the synthesized control unit can be very large especially when there is a high number of subprogram calls in the synthesized process.

Another aspect is the complexity of the synthesis process itself. The design

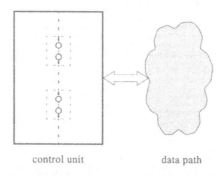

control unit   data path

Figure 7.1: Subprogram synthesis with inline expansion.

representation resulting after inline expansion can be so complex that it can not be accepted for synthesis assuming reasonable computation resources, or, if synthesis can be performed, it takes a prohibitively long time. Both complexity of the resulting control logic and complexity of the synthesis process are reduced in the following two approaches by synthesizing a single copy of the subprogram into a separate control unit (independent control unit/data path) or into a single control unit together with the calling process (subroutine implementation).

Independent Control Unit/Data Path

As opposed to inline expansion, this approach results in the synthesis of a separate control unit and data path for the subprogram. As shown in Figure 7.2, the calling process and the subprogram communicate by a handshaking protocol for activation of the subprogram and transmission of the input parameters as well as for reception of the output parameters after completion of the subprogram. For this approach the subprogram is practically transformed into another process which communicates with the calling process by passing messages in both directions. Implementation strategies for VHDL processes communicating through messages will be further discussed in Section 7.4.

The main advantage of this approach is the reduced complexity of the control logic. However, no sharing of data path resources between subprogram and calling process as well as between subprograms can be provided in this case.

Another advantage of generating an independent control unit and data path is the reusability of the resulting module. It can be instantiated several times, in the same or in other designs, whenever needed. This is not the case for the other two approaches discussed here.

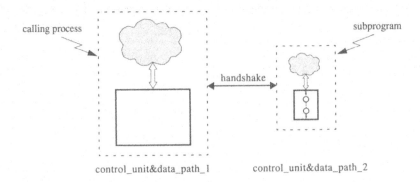

Figure 7.2: Synthesis of an independent control unit/data path.

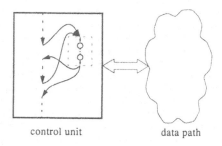

Figure 7.3: Subroutine implementation.

Subroutine Implementation

Subroutine implementation is situated somewhere between the two previous approaches. On the one side, we have a single control unit and data path resulting from synthesis, like for inline expansion. But, on the other side, for each subprogram the resulting control unit has a certain sequence of states, which is entered whenever the respective subprogram is activated. After leaving the last state corresponding to the subprogram, control is transferred to the state following the subprogram call. This synthesis approach is illustrated in Figure 7.3.

The resulting control unit is simpler than that produced with inline expansion (this difference growth with the number of calls to the same subprogram). However, as will be discussed in the next section, additional control logic and registers have to be provided for parameter passing and for the return mechanism.

Like inline expansion, this approach also allows a good sharing of data path resources between subprogram and calling process.

The complexity of the synthesis process is reduced, compared to inline expansion, as the design representation contains only one copy of the subprogram. However, process and subprogram are still synthesized together, which entails a higher complexity than separate synthesis, as is the case at generation of an independent control unit/data path.

7.1.2   Representation and Transformation

The ETPN design representation corresponding to a VHDL specification which contains subprograms, will differ in certain aspects depending on which implementation alternative has been selected [EKPM92].

For implementation as independent control unit/data path there is a completely separate ETPN representation of the subprogram, which is straightforward. A message passing mechanism has to be established, based on a handshaking protocol, in order to implement parameter passing. In this way, as already mentioned, the subprogram is transformed into a process which is

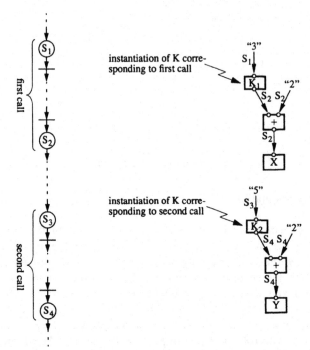

Figure 7.4: ETPN representation with inline expansion.

activated after receipt of a message, and which finally answers by sending back values for the output parameters. ETPN representation for VHDL processes interacting through messages, will be discussed in Section 7.4.

We will use the following VHDL example to illustrate ETPN representation for inline expansion and subroutine implementation:

```
. . .
procedure P(A: in INTEGER; B: out INTEGER) is
   variable K: INTEGER;
begin
   K:=A;
   . . .
   B:=K+2;
end P;
. . .
process
   variable X,Y: INTEGER;
begin
   . . .
   P(3,X);
   . . .
   P(5,Y);
   . . .
end process;
```

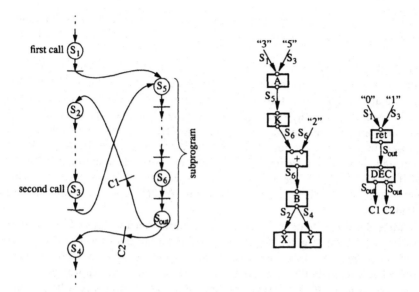

Figure 7.5: ETPN representation with subroutine implementation.

If an implementation with inline expansion is preferred, synthesis has to proceed from an ETPN representation like that presented in Figure 7.4. Control places corresponding to operations inside the subprogram are moved into the main control part of the calling process and are copied as many times as there are calls to the respective subprogram. Local variables of the subprogram (like K in our example) are instantiated one time for each call. This is consistent with the handling of variables at ETPN generation and with the semantics of local variables in VHDL. Further transformations will optimize, during synthesis, allocation of registers and functional units by sharing. As the subprogram appears like an ordinary sequence inside the process body, no parameter passing or return mechanism has to be provided. This also means that no data nodes have to be allocated for the formal parameters.

For subroutine implementation, as shown in Figure 7.5, we have a single, separate control sequence corresponding to the subprogram (places S_5 to S_{out}). This sequence is entered each time there is a call to the respective subprogram. After the subprogram has been left, control returns to the place after the subprogram call. Transfer of the argument value to the input parameter A is controlled by the places S_1 and S_3 respectively. After return, the value of the output parameter B is copied into the variables X and Y respectively, under the control of places S_2 and S_4.

The return mechanism is modeled by the conditional transitions at the output of place S_{out}, which controls return from the subprogram. Before entering the subprogram, the identity of the call is stored into the local node *ret*. Based on this value, before leaving the subprogram, a decoder node generates the conditions which direct the control to the right place (conditions *C1* and *C2* in Figure 7.5)[1].

At VHDL compilation, the ETPN representation for subprograms is generated according to user options and can be any of the three alternatives discussed above. The default option is subroutine implementation, except when there is a single call to the subprogram (the subprogram is expanded inline) or when there are certain limitations concerning VHDL semantics which will be discussed in the next section. During the synthesis process transformations concerning subprogram representation can be performed automatically, or guided by the designer. Depending on the optimization goals, the following transformations, which are defined for the ETPN representation, can be applied:

1. *inline expansion*: one or several calls of a subprogram are replaced by its body;
2. *exlining* [Vah95]: a given sequence (which can but does not need to be the result of a previous inline expansion) is extracted as a procedure and represented according to the independent control unit/data path approach;
3. *process generation*: the representation of a subprogram according to the subroutine implementation approach is transformed into the representation corresponding to the independent control unit/data path alternative;
4. *subroutine generation*: representation of a subprogram according to the independent control unit/data path approach is transformed into the representation corresponding to the subroutine implementation alternative.

7.1.3   Processes, Signals and Subprogram Synthesis

In the previous discussion, we assumed that a subprogram is called from a single process only. However, in VHDL it is possible that subprograms are shared by two or more processes. If a subprogram is called by more than one process and it is expanded inline, each process will execute its own copy of the subprogram and there are no problems related to sharing. It is also possible to implement shared subprograms with the independent control unit/data path approach, using an adequate communication interface. As mentioned, the subprogram behaves in this case like a process which becomes active after executing a *receive* operation for the message containing the input parameters, and returns a message containing the values of the output parameters. Thus, a subprogram implemented in this way implicitly accepts requests sequentially, one at a time.

If a subprogram which is shared by several processes is synthesized using subroutine implementation, it is necessary to protect it against being entered by

---

1. In our particular example, when we have only two calls to the subprogram, the conditions can be directly generated without the need of a decoder.

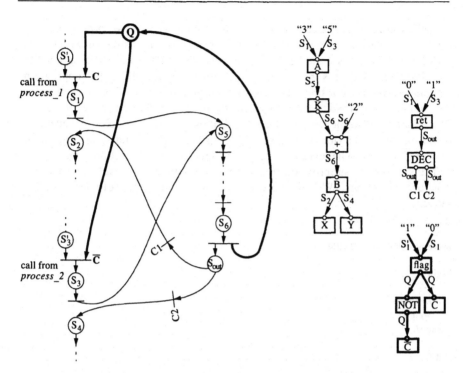

Figure 7.6: ETPN representation with subroutine implementation for shared subprograms.

more than one process at a time [EKPM92, EKPM93a]. This is a basic techno-logical requirement and, at the same time, it is a formal requirement of our internal design representation. If a subprogram is called from two or more processes, the ETPN representation corresponding to the subroutine implemen-tation approach will be unsafe (see Subsection 6.1.2) as long as no protection mechanism is provided. This is illustrated in Figure 7.6, which shows the ETPN representation for a similar example like in Figure 7.5, but considering that the two procedure calls are coming from parallel processes. The representation is unsafe, if we do not consider the protection mechanism represented with thick lines, because the places between S_5 and S_{out} (which correspond to the subprogram) potentially could hold two tokens.

Protection of a subprogram, if synthesized according to the subroutine implementation approach, can be solved by using mutual exclusion mecha-nisms and transforming the respective subprogram into a critical resource. Such a solution is illustrated in Figure 7.6, where the part responsible for the protection mechanism is emphasized by thick lines. If place Q holds a token[1]

---

1. Place Q gets a token with the initial marking of the system; similarly data node *flag* is initialized with 0.

the subprogram is free and can be activated by one of the two processes. It is easy to observe that no token will be present in Q as long as the subprogram is active and thus no second activation is accepted before the subprogram has been left. What happens, however, if both processes are calling the subprogram simultaneously? According to the solution shown in Figure 7.6, protection for this situation is provided by giving priority to the call issued from *process_1*. This strategy is implemented using conditions C and $\overline{C}$ which are generated based on the value stored in node *flag*. This node gets (under the control of place *S1'*) value 1 whenever *process_1* attempts to enter the subprogram.

Transforming a subprogram into a critical resource can result, however, in deadlock situations. We illustrate this problem with the following VHDL sequence:

```
signal S: INTEGER:=0;
 . . .
procedure P(X: in INTEGER) is
 . . .
begin
 . . .
  if X=1 then
    wait on S;
  end if;
 . . .
end P;
 . . .
process
begin
 . . .
  P(1);
 . . .
end process;

process
begin
 . . .
  P(2);
  S <= . . .
 . . .
end process;
```

During simulation there is no problem with the above example. According to the language definition, a VHDL simulator accepts multiple activations of a subprogram. However, after hardware synthesis, if the procedure is not expanded inline but is implemented as a hardware structure protected from multiple activations, the potential of deadlock exists when it executes the wait on signal S. If the call P(1) is executed first, the two processes will deadlock: the first process will wait, inside procedure P, for an event on signal S which has to be produced as result of the signal assignment in the second process, while the second process will wait for procedure P to become free. Such a behavior, which does not correspond to the semantics of the specification language, is not acceptable. Hence, *a subprogram called by more processes will be expanded inline if it contains a wait on a signal or it calls another subprogram that contains such a wait.*

Additional care must be devoted to subprograms which have parameters of class **signal**. According to the definition of VHDL, the actual signal argument is associated to the formal parameter at the beginning of each call; thereafter, during the execution of the subprogram body, a reference or an assignment to the formal parameter is equivalent to a reference or an assignment to the actual signal. However, both for the independent control unit/data path and for the subroutine implementation (Figure 7.5) approach, parameter passing is based on the initial transfer to the subprogram of the values for parameters of mode **in** and **inout** and the final transfer back of the values for the parameters of mode **out** and **inout**. There is no direct access from the subprogram to the actual argument, as would be needed for signal parameters. Even if such a direct access can be practically solved, it implies an important additional complexity of the resulting hardware.

However, the problems connected to parameters of class **signal**, as presented above, are of practical importance only when the subprogram contains at least a wait statement or it calls another subprogram that contains one. As discussed in Section 2.4, according to the VHDL simulation cycle, the value of a signal is updated only after *all* processes are suspended as result of executing a wait statement. Thus, if the subprogram does not execute any wait statement, no signal argument will have to change its value during the subprogram execution. The following conclusion has to be drawn for the synthesis of subprograms: *a subprogram with parameters of class signal will be expanded inline if it contains a wait statement or it calls another subprogram that contains a wait statement.*

7.2 Synthesis Strategies for Interacting VHDL Processes

In this and the following two sections we will focus on the synthesis of interacting processes which are specified in VHDL. More precisely, we consider that there is a set of processes specified in VHDL which has to be implemented in hardware. This set of processes represents a complete specification of a system or a part of a system specification which has been allocated to hardware as result of previous partitioning steps.

In Section 2.7 we had a discussion on VHDL as a synthesis language. We mentioned there that in the context of logic and high-level synthesis certain restrictions to the VHDL language subset and some restrictive modeling rules have been accepted, in order to facilitate synthesis. At the same time we emphasized that some of those restrictions are not acceptable in the context of system level synthesis. Thus, at this level it must be possible to specify systems consisting of several interacting processes. The explicit use of a clock signal and prescheduling of operations in terms of clock cycles can also not be imposed at the system level.

However, accepting interacting VHDL processes for synthesis introduces several complex problems. Subection 7.2.1 is devoted to the discussion of these problems, while Subsection 7.2.2 outlines our approaches to the synthesis of interacting processes.

7.2.1   Simulation/Synthesis Correspondence

As our presentation in Chapter 2 revealed, VHDL has been defined as a simulation language. This creates several difficulties when the language is used for synthesis purposes, since semantics of some central features of VHDL is explicitly defined in terms of simulation. The most difficult issues in this context originate from the way signal assignments and wait statements are defined. Unlike variables, which are updated as soon as they are assigned a value, signals are only updated at the start of a new simulation cycle (see Section 2.4). This means that the update of signal values must be synchronized with the execution of a wait statement by *every* process in the system and has to be performed simultaneously for *all* signals that are active in that simulation cycle.

To illustrate the main difficulties with VHDL semantics for signal assignment and wait statement, and to draw some conclusions concerning synthesis, we refer to the example in Figure 7.7. We assume that there is no wait statement in processes P1 and P2, except those given explicitly. Let us consider first only process P1. The value assigned to variable Z is the value that has been given to A in the *previous* execution iteration of the process, because the value of signal A will be updated only when executing the wait statement. At synthesis, this behavior has to be captured, regardless if signal assignment to A and reference to the signal value for assignment to Z will be scheduled into the same clock cycle or not. In the general case this has to be solved by latching the value assigned to the signal and by updating the signal value only when a wait statement is executed.

Latching of the signal value is enough to preserve at synthesis the computational effects of the simulation cycle, only if we consider an isolated process, without any interaction to other processes. To illustrate this we look now at both processes, P1 and P2, in the example above. At simulation, the two values for signal A referred in process P2 will be the same for one execution iteration of the process. This is because the value of the signal cannot change unless both processes, P1 and P2, execute a wait statement. But, if we consider the two processes separately and isolate them for synthesis, and then let the resulting

```
signal A, D: INTEGER:=0;
. . .
P1: process
   variable Z: INTEGER;
   . . .
begin
   wait on D;
   . . .
   A<=D+1;
   . . .
   Z:=A;
end process P1;

P2: process
   variable Y1, Y2: INTEGER;
   . . .
begin
   wait on A;
   . . .
   Y1:=A*3;
   . . .
   Y2:=A;
end process P2;
```

Figure 7.7: VHDL example with interacting processes.

structures work together, the value of A could be changed between the two assignments to variables Y1 and Y2. Thus, it is not sufficient at synthesis to latch values assigned to signals and to update them when wait statements are executed by the given process. This update has to be carried out only when *all* processes are executing a wait statement. This behavior has to be implemented in the synthesized hardware, as long as we are not restricting the use of signals and wait statement in the synthesizable VHDL subset.

Another important issue concerning VHDL semantics is the modeling of time. According to the language definition, strict timing is specified by *after* clauses in signal assignment statements and wait statements with time clauses. From the point of view of high-level and system-level synthesis, the decision on when a certain operation will be executed is left to the synthesis tool and is not strictly specified by the designer. That is the reason why synthesis tools working at these levels do not consider strict timing specifications[1] [CST91, RKDV92, BeKu93].

At the system level we consider time as a notion of causality and hence we are interested in the partial ordering relation of operations on signals and ports.

---

1. In Section 7.5 we will discuss how VHDL timing facilities are used for specification of timing constraints, and how these constraints can be handled during synthesis.

Providing a certain partial order of operations on signals and ports in a VHDL specification entails introducing some synchronization between processes. According to VHDL semantics, the designer can enforce this synchronization by using signal assignments and wait statements.

The synthesis strategies presented in the next sections preserve this temporal relationship between simulation model and the synthesized hardware structure. Thus we achieve *simulation/synthesis correspondence*, which means that both the simulation model and the synthesized hardware react with the same values (sequences of values) of the signals and ports to identical sequences of stimuli applied to the inputs. Considering correspondence at the level of signals and ports is sufficient, since we are only interested in the external behavior of the resulting hardware.

7.2.2   Implementation Alternatives

Our approach for implementation of interacting VHDL processes supports two main specification and synthesis styles [EKPM94, EKP96]: one using Signal

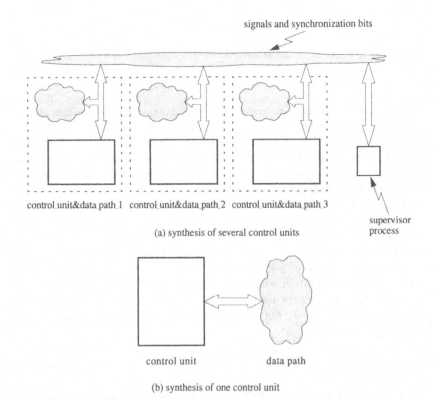

Figure 7.8: Implementation of processes interacting at the signal level.

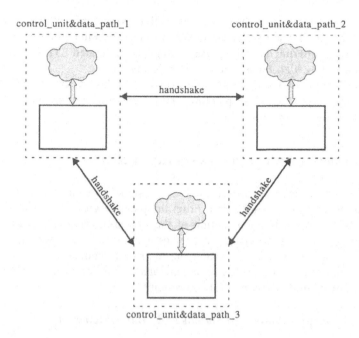

Figure 7.9: Implementation of processes interacting through message passing.

Assignment and Wait statements for process interaction (SAW for short), and the other one based on synchronous Send/Receive message passing primitives (SR for short).

The SAW style makes it possible for the designer to express process interaction using signals and wait statements, as described in the VHDL language definition. The hardware that will be synthesized according to this style is controlled either by a single control unit (Figure 7.8b) or by several control units working together (Figure 7.8a). For the implementation alternative with several control units, overall synchronization, as required for a correct synthesis according to VHDL semantics, is solved by an additional *supervisor process* generated automatically during synthesis (Figure 7.8a).

Specification of process interaction using signals and wait statements, assumed by the SAW style, can be regarded as too low level for system specifications. Hence, the SR specification and synthesis style is based on higher level communication primitives for synchronous message passing. Hardware implementation according to this style is illustrated in Figure 7.9. It results in several independent controllers, one corresponding to each process, which are communicating by means of a handshaking protocol.

Hardware synthesis according to the alternatives presented above is based on translation of the VHDL system specification into an ETPN structure which depends on the selected synthesis style. To achieve simulation/synthesis correspondence, as discussed in the precedent section, we compile VHDL in such a way that its essential semantics with respect to process interaction is explicitly captured in the internal design representation. The generated ETPN structure is then optimized by high-level synthesis transformations.

The SAW and SR specification and synthesis styles as well as the ETPN representation for the corresponding implementation alternatives will be discussed in the following sections.

7.3   Synthesis with Signal Level Interaction

The SAW style accepts an unrestricted use of signals and wait statements for specification of process interaction. From the point of view of synthesis SAW implies the hardware implementation of the synchronization imposed by the simulation cycle. This means that processes have to wait for each other, until all of them are executing a wait statement, in order to update the signal values. As shown in Figure 7.8, the hardware generated according to this style can be controlled by a single or by several control units.

7.3.1   Representation of Signals and Wait Statements

A wait on a signal is represented in ETPN by a transition with an associated condition. This transition is located in the control part corresponding to the process executing the wait statement [EKPM92, EKP96]. The condition will be produced as the result of an assignment to the respective signal. In Figure 7.10 we show the ETPN representation of a signal and the control part corresponding to a wait statement. According to ETPN semantics, the process will wait on transition T, until condition C_S generated in the data path becomes *true*.

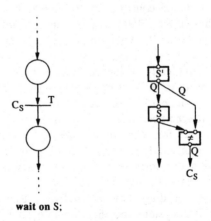

wait on S;

Figure 7.10: Representation of signals and wait statements.

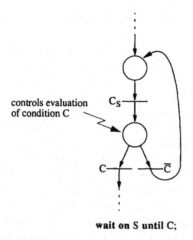

wait on S until C;

Figure 7.11: Representation of wait statement with condition clause.

Figure 7.11 shows the representation corresponding to a more general wait statement which consists of both a sensitivity and a condition clause. The process first waits for an event on signal S and then evaluates condition C. If the condition is false the process resuspends on the signal.

Signals are modeled by two data path nodes like S and S' in Figure 7.10. The value referred by the processes accessing the signal is that in node S while the node S' stores the last assigned value. Referring to the discussion in Section 2.4 and to Figure 2.7, we can say that node S holds the current signal value while node S' represents the driver corresponding to the signal. A driver consisting of a single data node is sufficient under the circumstances that, according to Subsection 7.2.1, no *after* clauses in signal assignments are considered for synthesis.

Condition C_S, as depicted in Figure 7.10, indicates an *event* on signal S, and hence will be true only when the new value of the signal differs from the old one. The proposed structure for signal representation can also be extended to produce a condition corresponding to a *transaction* on the signal (see Subsection 2.4.5 and [EKPM92]).

Updating the signal, by passing the value from node S' to node S, is controlled by a place, called Q in Figure 7.10, *that will hold a token only when all processes are executing a wait statement.* The control structure containing this place consists of a single control unit or a collection of control units, as will be discussed in the following two sections. For reasons of simplicity, in our figures, we will use a compressed data path representation for signals depicting only the two register nodes (as, for example, nodes A, A', and D, D' in Figure 7.12).

7.3.2   Synthesis of Several Control Units

Synthesis of several control units, one for each process, is performed on a
design representation containing several independent control Petri nets
synchronized through shared data path nodes. In Figure 7.12 we show the
ETPN structure corresponding to the VHDL example introduced in Figure 7.7
(portions related to process interaction are emphasized with thick lines). This
representation is generated if the synthesis of several control units is required.
The *supervisor process* is automatically generated during compilation of the
VHDL description, and is responsible for the synchronized updating of signals,
under the control of place Q. This place will hold a token only when *all*
processes in the system are executing a wait statement.

The required synchronization is achieved by using the one-bit data nodes x_1,
x_2, ..., x_k, one for each process $P1$, $P2$, ..., Pk in the design. Each node x_i is
initially reset by the starting place S_0. Setting of x_i is controlled by any of the
places in the respective set of places Ω_i, where Ω_i consists of all control places
corresponding to the wait statements that belong to a process Pi. Node x_i will
be reset under the control of any place in the set Ω_i', where Ω_i' consists of the
places that are direct successors of those in Ω_i. In our particular example x_1 is
set under control of place W_{P1}, and is reset under control of place W_{P1}' and S_0;
W_{P2} controls setting of node x_2, and W_{P2}' with S_0 reset the same node.

Place α in the supervisor process does not control any arc. This dummy
place indicates that a certain delay has to be introduced after signal update,
which is performed under control of place Q, and before a new evaluation of
condition C, controlled by place Q' (both in the supervisor process). This delay
is necessary to complete the resetting (controlled by places in the sets Ω_i') of
the one-bit nodes x_i, corresponding to those processes that are leaving the wait
state after signal update. In this way the supervisor process is forced to stay in
state α until all processes leaving a wait state have executed actions controlled
by places in sets Ω_i'. Thus, all nodes x_i, which had to be reset, got their new
value before condition C is reevaluated. This means that all processes which
were waiting for an event that has been produced, had the necessary time to
leave the wait state before the supervisor process reaches place Q' in which it
will stay until all processes are executing again a wait statement. The delay
assigned to place α will be equal to the clock cycle time if we use a single
global clock; if the individual control units use their own clock the delay will
be equal to the maximum clock cycle time.

Since the control Petri nets corresponding to the processes are disjoint,
synthesis of a separate control unit for each process is possible. The set of
control units is synchronized by the control unit corresponding to the supervisor process (Figure 7.8a). With the help of the nodes x_1, x_2, ..., x_k in the data
path, the supervisor process coordinates the global synchronization so that
signals are updated only when all processes are in a wait state.

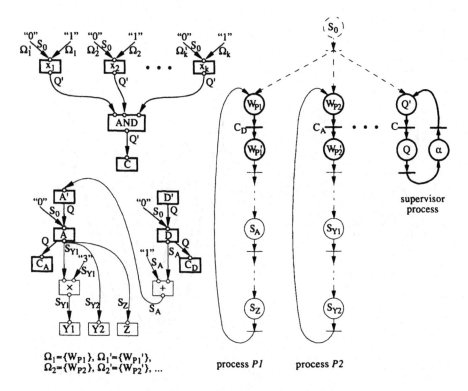

$\Omega_1 = \{W_{P1}\}, \ \Omega_1' = \{W_{P1}'\},$
$\Omega_2 = \{W_{P2}\}, \ \Omega_2' = \{W_{P2}'\}, \ ...$

process *P1*   process *P2*

Figure 7.12: ETPN representation for implementation with several control units.

The main advantage of this solution is the reduced complexity of the control logic. However, in this case, no data path resources are shared between processes, which may result in a certain hardware overhead.

Another advantage is the reduced complexity of the synthesis process, compared to generation of a single control unit. For this reason, generation of several control units can be the only practical alternative for more complex designs.

7.3.3   Synthesis of a Single Control Unit

If the complexity and/or the number of processes are not very large the system specification can be synthesized to a single control unit and a data path (Figure 7.8b). This structure is generated using a design representation that differs from that in Figure 7.12. If the designer requests the synthesis of the example presented in Figure 7.7 using a single control unit, the VHDL compiler generates the ETPN structure represented in Figure 7.13 (portions related to process interaction are emphasized with thick lines). Synchronization between

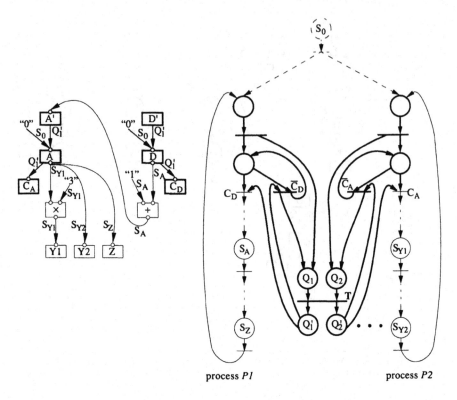

Figure 7.13: ETPN representation for implementation with a single control unit.

processes in order to update signals is moved entirely into the control part where it becomes explicit. The control places Q_1, Q_2, ..., Q_k, one for each process, hold a token only when the corresponding process executes a wait statement. When *all* processes are waiting, the transition T can be fired and the places Q_1', Q_2', ..., Q_k' will get (all at the same time) tokens. Any of these places can be used to control the update of signals (in Figure 7.13 we used place Q_1' to control the update of the two signals A and D). If the condition (C_A or C_D in our particular example) associated to a signal on which a process Pi is waiting becomes true, process Pi will continue (the token is passed from Q_i' to the output place of the transition on which the process was waiting). If the condition associated to the signal is false (the expected event did not happen) Pi enters again its waiting state (the token is passed back from Q_i' to Q_i).

For the example illustrated in Figure 7.13 we considered only one wait statement in a process. The solution for a more general situation is depicted in Figure 7.14. Here we show a part of the control Petri net corresponding to a process which contains two wait statements (on signals called *S1* and *S2* respectively). Generalization for several wait statements is straightforward.

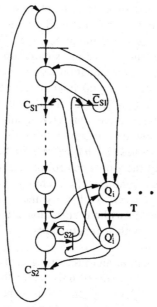

Figure 7.14: Control Petri net for a process containing two wait statements.

Moving synchronization entirely into the control part, and considering all processes together as a single structure, increases the complexity of the control Petri net and that of the resulting control logic. But this does not necessarily mean an explosion of control states. The additional constraints, introduced into the control part, are used at controller generation to eliminate unreachable states and thus to avoid state explosion (see Subsection 6.3.2).

For this solution no supervisor process has to be generated and there is no need for the data nodes $x_1, x_2, ..., x_k$ in the data path, as is the case at generation of several control units. Handling the whole representation globally at synthesis, for the design of a single control unit, offers at the same time the possibility of sharing data path resources between different processes. This can result in a reduction of implementation cost.

One serious disadvantage of this approach is the high complexity of the synthesis process. Synthesis has to be performed for a representation consisting of all processes together and, thus, the time needed for generation of the hardware structure can be prohibitively large. This is why larger designs have to be synthesized using several control units.

7.4   Synthesis with System Level Interaction

The SAW synthesis style, presented in the previous section, entails the implementation in hardware of the global control imposed by the VHDL simulation cycle. Thus, it results in a strong synchronization of the processes, that very often exceeds the level needed for the intended functionality of the synthesized system. This *over-synchronization* is the price payed for an unrestricted use of signals while preserving at the same time simulation semantics at synthesis. The question is if this over-synchronization can be relaxed, allowing a higher degree of parallelism, without giving up simulation/synthesis correspondence.

When describing a system in VHDL for simulation, the designer can rely on the definition of the simulation cycle. Our SAW synthesis style guarantees, by implementing the synchronization imposed through the simulation cycle, a similar behavior of the synthesized hardware and the simulation model. If hardware implementation of the global control imposed by the simulation cycle is not supported, a correct synthesis (as defined by simulation semantics) cannot be guaranteed *in general*. Such a synthesis approach would update signals and schedule processes without enforcing global synchronization, and thus can produce hardware with a functionality that differs from simulation behavior.

Relaxing over-synchronization and preserving at the same time simulation/ synthesis correspondence can be solved only under the following assumption: the correct behavior of the VHDL specification to be synthesized should not rely on the *implicit* synchronization enforced by the simulation cycle. We call a description that conforms to this requirement *well synchronized*. In a well synchronized VHDL description all the assumptions that provide the proper synchronization and communication between processes are *explicitly* stated by operations on signals.

We will now present a synthesis strategy that does not reproduce the simulation cycle in hardware while maintaining simulation/synthesis correspondence [EKPM94, EKP96]. It accepts designs specified according to a certain description style and produces independent control units which work in parallel. The descriptions conforming to this style are implicitly well synchronized. When defining this specification style we started from the following main considerations:

- It is possible to produce an implementation with a high degree of parallelism and asynchrony when the circuit can be described as a set of loosely connected processes. This means that the amount of communication between these processes is relatively low and that a process communicates usually with a relatively small number of other processes. The enforcement of the simulation cycle for this class of hardware can result in a considerable reduction of the potential parallelism (and consequently of the performances) and produces at the same time an increase in implementation complexity.
- For the designer, the specification style must be defined through a small number of *simple* rules that have to be respected for circuit description in VHDL.

7.4.1   The Designer's View

With the message passing based SR specification style the designer describes the system as a set of VHDL processes communicating through signals. Any number of processes can communicate through a given signal (we say that these processes are connected to the signal) but only one of these processes is allowed to assign values to it. Assignment of a value to a signal is done by a *send* command. Processes that refer to the signal will wait until a value is assigned to it, by calling a *receive* command. Both *send* an *receive* have the syntax of ordinary procedure calls.

A *send* command, denoted as *send(X, e)*, where X is a signal, e is an expression, and e and X are type compatible, is executed by a process P in two steps:

1. process P waits until all other processes connected to signal X are executing a *receive* on this signal (if all these processes are already waiting on a *receive* for X, then process P enters directly step 2);
2. expression e is evaluated and its value is assigned to signal X. This value becomes the new value of X. After that, process P continues its execution.

A *receive* command, denoted as *receive(X)*, where X is a signal, causes the executing process to wait until a send on signal X is executed.

Communication with send and receive can be achieved also through several signals:

send(X,e1,Y,e2, ...), where X, Y, ... are signals, and $e1$, $e2$, ... are expressions type compatible to the respective signals, solves communication between the executing process and the other processes connected to signal X (that will get the value resulting from evaluation of expression $e1$), to signal Y (that will get the value resulting from evaluation of expression $e2$), etc. This communication implies successive synchronization with the respective groups of processes, before assigning the new value to the corresponding signal (according to the rules for simple *send* and *receive*); the order of successive synchronizations is not predefined. Execution of the command terminates after communication on all signal arguments succeeded.

receive(X,Y, ...), where all the arguments are signals, causes the executing process to wait until a *send* on all arguments is executed; synchronization with the sending processes can be realized in any order.

The definition of the *send/receive* commands ensures that between the execution by a process of two consecutive receives on a given signal X, the value of this signal remains unchanged. This is due to the fact that in this interval no *send* on that signal can become active. This property is very important from the point of view of synthesis (see Subsection 7.4.3).

To avoid undesired blocking of a process on a *receive* command (and

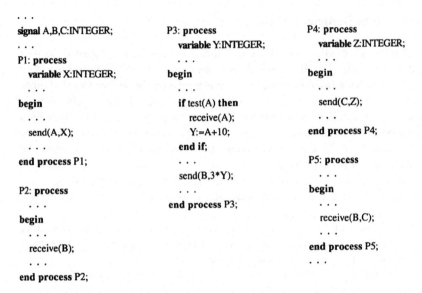

Figure 7.15: Example of VHDL processes interacting through message passing.

possible deadlock situations), the boolean function *test* is provided; *test(X)*, where *X* is a signal, returns *true* if there is a process waiting to execute *send* on X; otherwise the function returns *false*.

Communication with *send* and *receive* requires synchronization between processes. However, it is important to note that this synchronization does not necessarily affect *all* processes but only those involved in a specific communication (the processes connected to a given signal). We illustrate this with the example in Figure 7.15. Processes P1 and P3 are connected to signal A; processes P2, P3, and P5 are connected to signal B; Processes P4 and P5 are connected to signal C. When P1 executes the *send* on A it will synchronize with process P3 that executes a *receive* on the same signal. P3, for executing *send* on B, has to synchronize with P2 and P5 that have to execute *receive* on B. P4 and P5 are synchronized for *send* respectively *receive* on signal C. Except these restrictions, no other synchronization is required for the correct behavior of the system.

A VHDL description corresponding to this style is transformed by a preprocessor into an equivalent standard VHDL model for simulation, by expanding *send*, *receive,* and *test* commands into equivalent sequences containing signal assignments and wait statements. Starting from the same initial description, the VHDL synthesis-compiler generates the ETPN internal representation that is synthesized using high-level synthesis transformations. The resulting architecture consists of several interacting control units (Figure 7.9). Simulation/synthesis correspondence will be preserved during the synthesis process, without providing any synchronization of processes additional to the explicit synchronization required by the *send/receive* commands.

7.4.2   The Simulation Model

A VHDL description based on the SR style is translated by a preprocessor to a standard VHDL program for simulation. The generation of the simulation model is solved in two main steps:

1. A package that exports (resolved) bit signals is generated. For each signal *X* declared in the VHDL description, the package will export the bit signals *P1_X*, *P2_X*, ..., *Pn_X*. Each of these signals corresponds to one of the processes labeled *P1*, *P2*, ..., *Pn*, that execute *receive* on signal *X*. The generated bit signals will be used for implementation of the handshaking protocol between processes connected to a given signal. In order to implement the *test* function, for each signal *X* that is used as argument of a *test*, a bit signal *T_X* will be exported by the generated package. Considering, for instance, the example in Figure 7.15, the following package declaration is generated:

```
package P_GEN is
    function RES(S: BIT_VECTOR) return BIT;
    -- function RES implements wired or
    signal P2_B, P3_A, P5_B,P5_C: RES BIT:='0';
    signal T_A: BIT:='0';
end P_GEN;
```

2. *send* and *receive* commands are expanded, based on predefined templates, to VHDL sequences that implement the handshaking protocol for synchronous message passing between processes connected to the same signals. A reference to function *test(X)* will be expanded to the boolean expression *(T_X = '1')*, and to a *wait for 0* statement preceding the statement that contains the reference.

The generation of standard VHDL sequences corresponding to operations *send* and *receive*, both on one signal and on several signals, is presented in [EKPM93b]. We will illustrate here three such sequences in the context of the example in Figure 7.15:

a) send(B, 3*Y) in process P3 will be expanded to:

```
if P2_B/='1' or P5_B/='1' then          -- P2 and P5 receiving?
    wait until P2_B='1' and P5_B='1';    -- wait for receive on B;
end if;
P2_B<='0'; P5_B<='0';
B<=3*Y;                                  -- update signal B;
wait for 0;
```

b) send(A, X) in process P1 will be expanded to:

```
if P3_A/='1' then                    -- P3 receiving?
    T_A<='1';                        -- for the test(A);
    wait until P3_A='1';             -- wait for receive on A;
end if;
P3_A<='0';
T_A<='0';                            -- for the test(A);
A<=X;                                -- update signal A;
wait for 0;
```

c) receive(B, C) in process P5 will be expanded to:

```
P5_B<='1'; P5_C<='1';
wait until P5_B='0' and P5_C='0';
```

7.4.3   A Synthesis Strategy

For the synthesis of interacting processes according to the SR style, a signal can be represented as a simple data path node (nodes A, B, C, for instance, in Figure 7.16). After an assignment (as result of a *send* executed on the respective signal) the value of the node is updated directly. Synchronization between the process assigning to a signal and all those accessing it, imposed by the *send/receive* mechanism, makes an assignment in two steps (like the latch/ update mechanism in the SAW synthesis style) unnecessary.

The handshaking protocol between the process executing a *send* and those executing *receive* on a certain signal can be implemented using one-bit data path nodes, similar to the bit signals generated for the simulation model discussed in the previous section. For illustration, in Figure 7.16 we show the ETPN representation generated by the synthesis compiler for the example introduced in Subsection 7.4.1 (Figure 7.15).

For synchronization between process P3, executing *send* on B, and the other two processes, P2 and P5, connected to the same signal B, condition C_B is used. After both P2 and P5 are waiting on *receive* for B (and consequently both data path nodes $P2\_B$ and $P5\_B$ are set), C_B will be *true* and process P3 can go on to update the value of signal B, under control of place $S3_7$. Place $S3_8$ controls the resetting of nodes $P2\_B$ and $P5\_B$, and thus the condition for process P2 to continue is produced ($\overline{C}_{P2\_B}$ is *true*). Process P5, that executes *receive* on B and C, is allowed to continue only after process P4 has also executed the *send* on C (and thus $\overline{C}_{P5\_BC}$ becomes *true*). Synchronization between P1 and P3 for *send* and *receive* on signal A, and between P4 and P5 for *send* and *receive* on signal C has been implemented in a similar way.

The decision based on *test* with signal A, in the control part corresponding to process P3, has been implemented using conditions C_{T_A} and $\overline{C}_{T\_A}$ produced in the data path with the one-bit data node T_A.

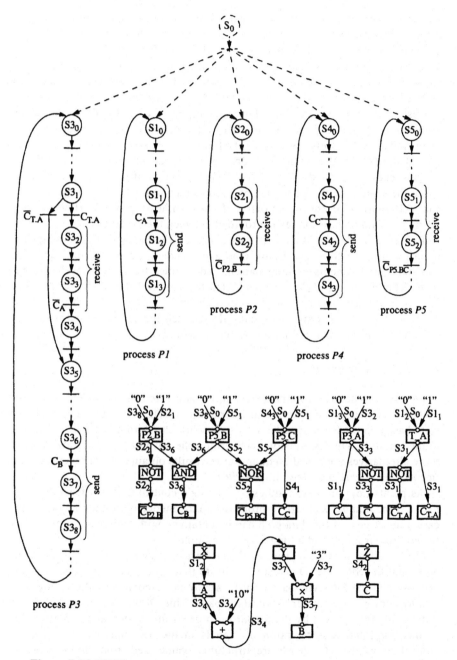

Figure 7.16: ETPN representation for implementation of interacting processes according to the SR style.

The SR synthesis style leads to hardware structures that work at a higher degree of parallelism than those synthesized according to the SAW style presented in Section 7.3. It does not require a global synchronization of *all* processes. Signals need not be implemented by special data nodes with additional functional elements in the data path. They are represented and updated exactly like ordinary variables. As a consequence of the fact that process interaction is based on a handshaking protocol and does not need any global control, the Petri net controllers of the processes can be implemented as independent and loosely coupled control units working in parallel. This is the typical strategy for the synthesis of designs described according to this style.

As already mentioned for the SAW style, synthesis of separate control units has at the same time the advantage of a reduced complexity of the control logic and a limitation of the complexity of the synthesis process. However, no sharing of data path resources is provided between the processes.

To take advantage of this synthesis approach the designer has to adapt his description to the SR style. Using *send* and *receive* instead of wait statements and signal assignments is more natural and simpler for system level descriptions. This modeling style facilitates also the organization of the design into loosely coupled processes with a well structured interface. If such an organization is possible, this style is a natural approach for synthesis and results in efficient and highly parallel hardware implementations.

7.4.4   Discussion

In the previous sections, we presented three hardware synthesis alternatives for system specifications consisting of interacting VHDL processes. The first two were based on the SAW specification style while the third one works according to the SR style. Our conclusions concerning the advantages and drawbacks of these alternatives are based on quantitative results obtained from several experiments which were presented in [EKP96].

As pointed out already, synthesis of a single control unit is a feasible alternative if the number and size of processes in the model are not too large. In this case a lower cost of the data path can be obtained, with a higher complexity of the control structure and a longer synthesis time.

Synthesis of several control units can be performed according to both the SAW and the SR style. Our results indicate that synthesis of models specified according to the SR style, based on the send/receive communication paradigm, requires lower cost in terms of data path. This difference grows with the number of signals used by the model for process interaction and is mainly due to the simplified representation of signals in the SR approach. In terms of control complexity, the hardware structures synthesized from the two specification styles are very similar. This leads to the conclusion that the SR synthesis style produces a hardware of lower cost which, as discussed in the previous sections, works at a higher degree of parallelism.

At the same time, system specification using the SR style is adequate for the design of medium size and large systems. For this kind of specifications process interaction at the level of signal assignment and wait statement, as defined in the VHDL standard and available with the SAW style, is at an unacceptably low level. With the SR style a higher level process interaction mechanism has been adapted for VHDL. It is very important that specifications using this mechanism are implicitly well synchronized and, thus, they can be synthesized without enforcing the strong synchronization required by the VHDL simulation cycle.

A process interaction mechanism based on simple synchronous message passing primitives has the advantage of a well defined semantics which can be used to reason about the communication behavior of the system with the possible support of verification tools. Other communication models, such as buffered communication, can be easily modeled by describing intermediate components as additional processes.

In Subsection 7.4.3 we presented a concrete solution for hardware implementation of processes with the SR style. It is only one of the possible alternatives and is based on an implementation of the handshaking protocol using shared registers. Several other implementations can be produced, depending on the particularities of the communication interface. Automatic generation of interfaces which implement communication channels for synchronous message passing is described in [LVM96, VLM96a, VLM96b]. The components communicating through these interfaces can operate at different clock frequencies derived from the same system clock, or the clocks used by the different components can be unrelated. Integration of asynchronous components is also supported. Such communication interfaces can be implemented both as hardware and in software.

Systems specified in VHDL according to the SR style can be implemented on architectures using different hardware and programmable components. In Section 7.2, 7.3, and 7.4 we focused on the synthesis of the hardware components of the system. In Chapter 8 we discuss how the system specification can be efficiently analyzed and partitioned for implementation as a mixed hardware/software system.

7.5 Specification and Synthesis with Timing Requirements

At system-level and high-level synthesis the time schedule of operations to be executed by the implemented system is determined as result of optimizing decisions made during the synthesis process. Therefore, a system specification accepted for synthesis at these levels, usually does not contain timing restrictions which strictly determine the schedule of implemented operations.

However, the prescribed behavior of the implemented system often imposes certain constraints on timing aspects which have to be considered as additional requirements by the synthesis tool.

Requirements on the timing aspects of a system can be incorporated as *timing constraints* (TC) in the specification submitted to the synthesis system. They are considered during all phases of the synthesis process, from allocation of architectural components and system partitioning until the final generation of the implemented hardware and software.

In this section we first present a notation for specification of timing constraints in VHDL descriptions. Then we discuss how the high-level synthesis of the hardware component of a system can be performed when TCs have to be considered.

The specification of timing constraints and the synthesis algorithms that will be presented support the generation of a hardware structure that not only fulfills the strict functionality requirements in terms of input/output values but also corresponds to the imposed timing restrictions. They allow a top-down design process with iterative improvements, as shown in Figure 7.17. One essential feature, which supports this methodology, is that the notation for specification of timing constraints is recognized and properly interpreted not only by the synthesis tool, but also by the VHDL simulator.

The initial specification defining functionality and timing requirements is first simulated using a standard VHDL simulation environment. At this stage estimated values are used for simulation of timing (usually, timing constraints are given as an interval between an upper and lower limit and, at this point, the designer does not know yet what concrete delays on the constrained sequences will result after synthesis). During this step mainly functional errors and some basic errors concerning timing are detected and the initial specification is updated until the simulation results are acceptable.

The specification is then synthesized to an RT-level structure. Taking into consideration design goals and time constraints, the synthesis algorithm produces a time schedule for the operations and thus, the actual values of the synthesized times can be back-annotated into the initial VHDL specification [EKPD97]. This means that values for time intervals, which have been initially estimated by the designer, are now automatically replaced with the concrete delays that have been produced as result of high-level synthesis. The back-annotated specification is used for behavioral simulation to check design correctness, which allows the detection of not only functional errors, but also errors concerning the timing requirements of the designed hardware. The simulation of the back-annotated behavioral VHDL specification has several advantages comparing to simulation of the generated RT-level model. First, it corresponds directly to the original specification and, thus, is readable for the designer which facilitates rectification of the model before starting another iteration of the design process. Second, behavioral simulation is much faster than RT-level simulation. Thus, by performing iterations at the behavioral level

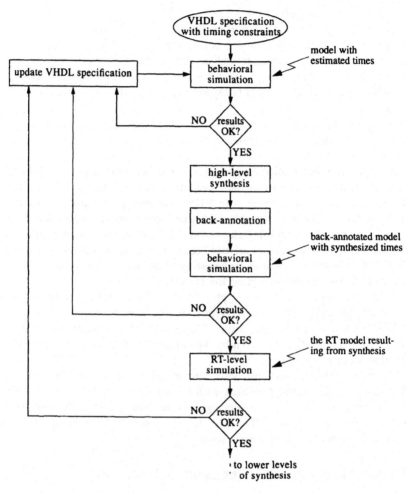

Figure 7.17: Design flow for hardware synthesis with timing constraints.

in order to fix functionality and timing of the design, the number of RT-level simulations, which are more accurate but very slow, can be reduced.

7.5.1   Specification of Timing Constraints in VHDL

The notation for specification of TCs in VHDL, presented here, has been defined considering the following main requirements:
- specification of minimal, maximal, exact, and range constraints between arbitrary statements should be possible (typically constraints are specified between the first and last statement of a process or between two arbitrary

operations on signals or ports);
- it should allow specification of nested TCs;
- simulation of the VHDL design before synthesis, with some estimated values for the delays on the constrained sequences, has to be performed;
- high-level synthesis with TCs, preserving correspondence between the behavior of the synthesized hardware and that of the simulation model has to be performed;
- back-annotation of the synthesized times for post-synthesis behavioral simulation should be possible;

Some hardware description languages, such as HardwareC [KuDM92] and DSL [CaKu86], include as part of their definitions a notation for TC specification. There is, however, no provision in VHDL for the specification of a minimum, maximum or range of acceptable TC values for synthesis. According to VHDL standard semantics both *after* clauses in signal assignment statements and *wait* statements with time clauses are used to express strict simulation timing. Thus, if VHDL is used as an input language for high-level synthesis, some conventional notation has to be adopted for TC specification.

Some synthesis systems accepting VHDL as an input language, such as CALLAS [Bie93], rely on *exact* synthesis of a user specified number of clock cycles between certain operations. The CALLAS approach in fact requires scheduling decisions to be taken by the designer and thus limits the freedom for synthesis optimization.

Synopsis VHDL tools use both commented lines and attributes to specify TCs [Syn92]. The main drawback with these two methods is that no simulation of the behavioral model, considering also the timing aspect, is possible, either before nor after synthesis. The TCs are ignored by VHDL simulators and are recognized only by the synthesis tools.

Predefined Procedures for Timing Constraint Specification

The conventional notation we have introduced for specification of TCs in VHDL is based on a set of predefined subprograms [EKPD95]. According to the first requirement formulated before, the following restrictions need to be expressed on the execution time of a sequence of operations:
- *minimal delay*: the execution of the sequence has to take at least a certain time;
- *maximal delay*: the execution of the sequence has to take at most a certain time;
- *range delay*: the execution of the sequence has to take a time between two given limits;
- *exact delay*: the execution of the sequence has to take exactly a given time.

```
package CONSTRAINTS is
    subtype TIME_1 is TIME range 150 ns to 150 ns;
    subtype TIME_2 is TIME range 0 ns to 200 ns;
    subtype TIME_3 is TIME range 100 ns to TIME'HIGH;
    subtype TIME_4 is TIME range 100 ns to 1200 ns;
    constant CONSTR_1:TIME_1;
    constant CONSTR_2:TIME_2;
    constant CONSTR_3:TIME_3;
    constant CONSTR_4,CONSTR_5:TIME_4;
end CONSTRAINTS;

package body CONSTRAINTS is
    -- the values are estimated for presynthesis simulation;
    -- they are back-annotated after synthesis.
    constant CONSTR_1:TIME_1:=150 ns;
    constant CONSTR_2:TIME_2:=100 ns;
    constant CONSTR_3:TIME_3:=130 ns;
    constant CONSTR_4:TIME_4:=500 ns;
    constant CONSTR_5:TIME_4:=650 ns;
end CONSTRAINTS;
. . .
process
    variable ANCHOR1, ANCHOR2: TIME;
. . .
begin
. . .
    anchor(ANCHOR1);
        . . .
    exact_time(CONSTR_1, ANCHOR1);
        . . .
    if COND then
        anchor(ANCHOR2);
            . . .
        max_time(CONSTR_2, ANCHOR2);
    else
        . . .
        anchor(ANCHOR2);
            . . .
        min_time(CONSTR_3, ANCHOR2);
        . . .
    end if;
        . . .
    range_time(CONSTR_4, ANCHOR1);
    . . .
end process;
. . .
```

Figure 7.18: VHDL specification with timing constraints.

For each of the above restrictions a predefined procedure has been provided in our design environment. To specify a certain restriction the user calls the corresponding predefined timing procedure passing as an argument the value of the time interval associated with the constraint. The location of the respective call determines the end point (the *sink*) of the constrained sequence. The starting point of the statement sequence is defined by the so called *anchor*. To specify the anchor, each call to a predefined timing procedure contains an argument representing a variable of type TIME. The anchor point, representing the start of the constrained sequence, is defined by the precedent call to a timing procedure that contains as an argument the same time variable. Figure 7.18 gives an example of VHDL specification with timing constraints, where the constrained sequences are emphasized by arrows.

The predefined package TIME_RESTRICT, implemented as part of the design environment, exports the timing procedures *anchor*, *range_time*, *min_time*, *max_time*, and *exact_time*. The procedure *anchor* is used to set a starting point for a subsequent timing constraint. The other procedures specify a time restriction and set an anchor for a possible subsequent constraint referring to the same time variable. The anchor and the sink of a time constraint have to be located in the same branch of an if statement, variant of a case statement, body of a loop, procedure, or process.

The time values associated with the constraints are specified as constants of a subtype of type TIME. At synthesis the ranges of these subtypes are identified as the constraint limits. The values of the time constants are ignored by the synthesis tool. Thus, for the example, in Figure 7.18 the synthesis tool will consider the following constraint limits as being imposed by the designer:
- *exact* constraint: 150 ns (type TIME_1 of constant CONSTR_1);
- *max* constraint: 200 ns (type TIME_2 of constant CONSTR_2);
- *min* constraint: 100 ns (type TIME_3 of constant CONSTR_3);
- *range* constraint: 100 ns .. 1200 ns (type TIME_4 of constant CONSTR_4).

For simulation, the constraint associated with a time constant (the range corresponding to its type) is not relevant, but its value is considered. This value is passed as a parameter to the procedure exported by the package TIME_RESTRICT and is considered by the VHDL simulator. In our example these values are specified in the body of package CONSTRAINTS. For pre-synthesis simulation the values are estimated by the designer; after synthesis they are automatically replaced at back-annotation with the synthesized times which are then considered for post-synthesis simulation [EKPD97].

At simulation the four procedures *range_time*, *min_time*, *max_time*, and *exact_time* act in an identical way. In Figure 7.19 this is illustrated with a sequence from the package body TIME_RESTRICT. Simulation of the delay on the constrained sequence is solved by the **wait for** statement in the procedure *wait_delay*. The actual wait is for the amount of time (TO_WAIT) left after the

```
package body TIME_RESTRICT is
  procedure wait_delay(DELAY: TIME; T_ANCHOR: inout TIME) is
    variable TO_WAIT: TIME;
  begin
    TO_WAIT:=DELAY-(NOW-T_ANCHOR);           -- time left to wait
    if TO_WAIT>=0 then
      wait for TO_WAIT;
    else
      assert FALSE                 --already more time spent than expected
        report "timing restriction error"
          severity WARNING;
    end if;
    T_ANCHOR:=NOW;               -- set time for anchor
  end wait_delay;

  procedure range_time(DELAY: TIME; T_ANCHOR: inout TIME) is
  begin
    wait_delay(DELAY, T_ANCHOR);
  end range_time;
  . . .
  procedure anchor(T_ANCHOR: out TIME) is
  begin
    T_ANCHOR:=NOW;               -- set time for anchor
  end anchor;
  . . .
end TIME_RESTRICT;
```

Figure 7.19: Part of the package body TIME_RESTRICT.

previous waits executed inside the constrained sequence corresponding to the specified anchor[1]. This solution supports, according to our requirements, nested timing constraints.

The synthesis tool ignores the body of the predefined timing procedures. The procedures are recognized by their names and the respective constraints are translated into the internal design representation, with time limits corresponding to the ranges of the subtype associated to the time constants.

By specifying the constraining values as constants of a subtype of type TIME, and allowing the subtype ranges to express the constraint limits, we get a VHDL model that implies for synthesis a set of possible hardware implementations. The estimated values for pre-synthesis simulation define one of these allowed implementations. After synthesis the actual synthesized delays are back-annotated and the resulting simulation model thus corresponds to the behavior of that particular hardware structure that has been generated.

---

1. In Figure 7.19 the function NOW is used which is predefined according to the standard VHDL language definition; it returns the current simulation time.

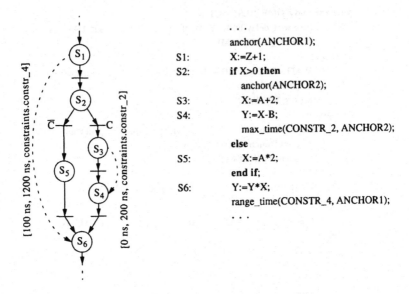

```
                              . . .
                              anchor(ANCHOR1);
                     S1:      X:=Z+1;
                     S2:      if X>0 then
                                  anchor(ANCHOR2);
                     S3:      X:=A+2;
                     S4:      Y:=X-B;
                              max_time(CONSTR_2, ANCHOR2);
                              else
                     S5:          X:=A*2;
                              end if;
                     S6:      Y:=Y*X;
                              range_time(CONSTR_4, ANCHOR1);
                              . . .
```

Figure 7.20: Control Petri net with timing constraints.

ETPN representation with timing constraints

TCs are captured by the ETPN representation as additional arcs in the control
part. These arcs are attributed with the time limits associated to the constraint
and with an identifier corresponding to the respective time constant (the
identifier is needed for back-annotation of the synthesized times in the VHDL
specification). Such an arc is placed between the control places corresponding
to the start and to the end of the constrained sequence. The VHDL compiler for
synthesis recognizes the predefined timing procedures by their name and
generates an ETPN representation with the additional attributed arcs for TCs.
In Figure 7.20 we show the control Petri net of the ETPN representation corresponding to a VHDL sequence that contains TCs. The time constants
CONSTR_2 and CONSTR_4 are imported from the same package
CONSTRAINTS defined in Figure 7.18. The additional arcs expressing TCs are
depicted as dotted lines.

Specification of timing constraints across process borders

Until now we discussed only specifications of TCs on sequences of statements
inside a process. However, design restrictions can also be expressed as
constraints on the time interval between certain operations on signals and it is
possible that these operations are performed by different processes.

Constraints on the time interval between two signal operations[1], which can be also executed by different processes, are specified using the concurrent assertion statement (see Subsection 2.6.3), like follows:

assert condition
   report "timing constraint violation"
     severity WARNING;

The *condition* in the statement above consists of a single function call, to one of the four predefined boolean functions exported by package TIME_RESTRICT: *min_assert, max_assert, range_assert*, or *exact_assert*. The function will be selected according to the kind of the constraint that has to be specified (a minimum, maximum, range or respectively exact constraint). The two signals affected by the constraint and the constraint limits are specified by the function arguments.

In the example shown in Figure 7.21, the concurrent assertion statement specifies that a transaction on signal B has to come not earlier than 100 ns and not later than 800 ns after a transaction on signal A. We remember that in the SR specification style (see Section 7.4) a signal event or transaction can be produced only as result of a *send* operation.

During simulation the concurrent assertion statement will be triggered at each *send* executed on signal A or B, and an assert violation occurs if the constraint is not satisfied. We present, for illustration, the predefined function *range_assert*, as defined in the body of package TIME_RESTRICT:

function range_assert (**signal** A, B: **in** BIT; T1, T2: **in** TIME) **return** BOOLEAN **is**
begin
  return not B'ACTIVE **or**
    (NOW-A'LAST_ACTIVE>=T1 **and** NOW-A'LAST_ACTIVE<=T2);
end range_assert;

The function above uses the signal attributes ACTIVE and LAST_ACTIVE which are predefined by the standard VHDL language definition. B'ACTIVE returns *true* if signal B is active during the current simulation cycle, and *false* otherwise; A'LAST_ACTIVE is the amount of simulation time that has elapsed since the last time at which signal A was active. Thus, the function *range_assert* returns *false*, which produces an assertion violation, when B is active and the time interval from the previous event on A is outside the requested limits.

The VHDL synthesis compiler translates the assertion statement into a constraint represented as one or several edges in the internal design representation. If a constraint has been expressed between signals *S1* and *S2* then an arc has to be generated between each control place corresponding to a *send* on *S1* and each place corresponding to a *send* on *S2*. In Figure 7.22 we show the

---

1. Such constraints are often called *global*, in opposition to those on sequences of statements, which are called *local*.

```
architecture SYNTH of SYNTHESIS is
    use CONSTRAINTS.all;                    -- the package is defined in figure 7.18
    use TIME_RESTRICT.all;
    signal A, B: INTEGER;
begin
    assert range_assert(A'TRANSACTION, B'TRANSACTION, 100 ns, 800 ns)
        report "timing constraint violation"
            severity WARNING;

P1:process
    variable T: TIME; variable X: INTEGER;
begin
    anchor(T);
    . . .
    range_time(CONSTR_4, T);
    . . .
    send(A, X);
    . . .
    anchor(T);
    . . .
    max_time(CONSTR_2, T);
end process P1;

P2:process
    variable T: TIME; variable Y: INTEGER;
begin
    receive(A);
    . . .
    anchor(T);
    . . .
    range_time(CONSTR_5, T);
    . . .
    send(B, Y);
    . . .
    anchor(T);
    . . .
    exact_time(CONSTR_1, T);
    end process P2;
end SYNTH;
```

Figure 7.21: VHDL specification with timing constraints across process borders.

control Petri net corresponding to the example in Figure 7.21.

The analysis needed to determine if a given constraint can be satisfied by a feasible schedule is more complex if interacting processes are considered and constraints across process borders are specified [KuDM92]. Scheduling of the representation depicted in Figure 7.22, for example, will be based on the fact that there is a synchronization point between the two processes which have to handshake for performing *send/receive* on signal A. Thus, the requirement specified on the time interval between the operations on signals A and B is

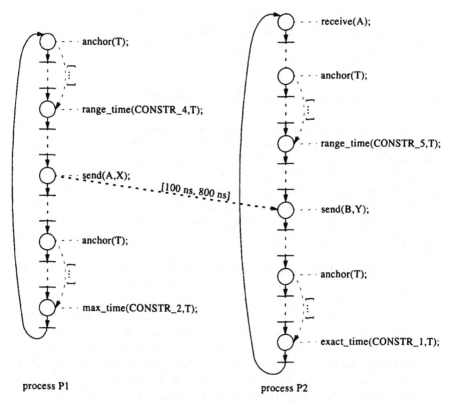

anchor(T);

range_time(CONSTR_4,T);

send(A,X);

anchor(T);

max_time(CONSTR_2,T);

receive(A);

anchor(T);

range_time(CONSTR_5,T);

send(B,Y);

anchor(T);

exact_time(CONSTR_1,T);

[100 ns, 800 ns]

process P1   process P2

Figure 7.22: Control Petri net with timing constraint across process borders.

solved at synthesis as a constraint local to process P2, concerning the delay on
the operations between the place corresponding to *receive* on A and that corresponding to *send* on B.

A VHDL design that contains no other TCs than global ones, specified using
concurrent assertion statements, results in a correct synthesis, according to the
specified timing requirements (if timing analysis and operation scheduling are
feasible), but no simulation which is relevant from the point of view of TCs can
be performed. During such a simulation, the assertion statement would report a
violation after each *send* on the constrained signals, because there is no
operation in the VHDL model that produces any progress of the simulation
time.

Such a situation is eliminated with a complete design, like that in Figure
7.21. Both TCs on sequences of statements inside the processes and constraints
on signal operations have been provided. The global constraints, expressing
timing requirements on signal operations, are significant in this context both at
simulation and synthesis. Simulation verifies if local constraints satisfy global

requirements on signals, expressed by assertion statements. At synthesis the global constraints are interpreted as additional restrictions that have to be considered when exploring the design space spanned by the constraints on sequences of statements.

7.5.2   Synthesis Considering Timing Constraints

Different approaches have been proposed to take into account timing constraints in the high-level synthesis process. Most of these approaches have some limitations. For example, an approach given in [StDu92] is used to solve the interface timing problem in high-level synthesis and requires that the algorithmic description of the design determines the interface timing exactly on the basis of clock cycles. In [KuDe92], methods for scheduling of the operations in a design relative to anchors (i.e., operations with unbounded delays) are described. The problem of resource binding is assumed to have been solved before the scheduling starts.

To avoid some of the limitations, Nestor et al. propose an algorithm that tries to minimize the resource cost during the scheduling process [Ne93]. First, an initial scheduling of the operations is performed by a simple algorithm. Second, the initial schedule is iteratively improved under the control of an optimization algorithm. It is shown that every legal schedule can be reached using this method. One advantage of using an iterative improvement technique to high-level synthesis under timing constraints is that since a completed design is available all the time, the accurate estimation of its timing characteristics is possible. Another advantage is that powerful optimization heuristics such as those described in Chapter 5 can be used to guide the selection of improvement transformations.

An iterative improvement technique based on timing constraints captured in ETPN to drive the high-level synthesis process has been proposed in [HaPe95]. It consists of two basic steps. In the first step, a multicycle scheduling technique is used to find a clock period and an initial scheduling for the design that fulfills the specified timing constraints. Trade-offs are made in this step between speed and cost when deciding on which physical module should be used to implement a particular operation in the data path (we assume here that there are several modules with different delays and areas which can be used to implement an operation). The multicycle scheduling problem is formulated as a combinatorial optimization problem, and a tabu search heuristic is used to solve it. In the second step, a resource allocation and binding technique is used to allocate physical modules and perform rescheduling whenever needed. The technique only makes transformations on the design that do not cause any timing constraint violation.

Multicycle scheduling assumes that the delay of operations (i.e., ETPN data path nodes) can be an arbitrary number of clock cycles (see Section 3.2.4). Each timing constraint covers one or several paths of places. The delays of each path of places, called a constrained path, has to fulfill the corresponding timing constraint. We assume that the designs are synchronous. Therefore, the delay, D_x, of a place P_x is a multiple of the clock cycle,

$$D_x = T \cdot \left\lceil \frac{d_x}{T} \right\rceil$$

where T is the clock period of the design and d_x the time needed to execute the data path operations controlled by P_x. The clock period can be calculated as the maximal value for which the timing constraints will be fulfilled. It has to be recalculated whenever the execution time of a constrained path is changed (i.e., the implementation of a data path node is changed in such a way that the corresponding place requires more or fewer clock periods to execute).

The tabu search technique is then used to solve the combinatorial optimization problem of finding the best design which satisfies all timing constraints. The neighborhood is defined by the set of ETPN representations generated by changing the implementation of one of the data path nodes. During each iteration all neighbors are evaluated based on a cost function. The goal of the tabu search algorithm (TSA) is to find the best, according to the cost function, feasible solution. Thus, when one or several of the neighbors are feasible (and not tabu classified) the best one of these neighbors is chosen to be the next solution. The search will stop after a predetermined number of iterations.

The cost function considers the estimated area and the clock period of the design. These are calculated from data given by the ETPN representation of the design. The cost function, f_c, is given by the following equation

$$f_c = A - C_{infl} \cdot T$$

where A is the area and T is the clock cycle time of the current solution, and C_{infl} is the clock influence factor which is given to the TSA as a parameter. The goal is to minimize the cost, f_c, which promotes longer clock periods, if possible. A long clock period gives several advantages, e.g., lower power consumption (see Chapter 10).

The tabu classification is based on the changes of the implementation of the data path nodes. When the implementation of a data path node is changed, the implementation used before the change is classified as tabu, and will not to be used, for a number of iterations proportional to the number of neighbors. However, when the aspiration criterion is satisfied the implementation of the data path node can be changed to a tabu classified data path node implementation. The aspiration criterion is given as that the solution is feasible and the best so far, i.e., the cost function returns a better value than for any other feasible solution found.

Diversification will be carried out if no new best solution has been found in the last N_{div} iterations. The implementation of each data path node is then changed, with the probability p_{div}, to the physical module that has been used the least number of iterations to implement that node (N_{div} and p_{div} are given as parameters to the TSA). In this way randomness is introduced in the TSA and the probability of finding the optimal solution is increased (see the discussion in Section 5.4).

When a scheduling of the ETPN places, which satisfies the timing constraints, has been found the resource allocation and binding task has to be performed to complete the high-level synthesis process. Before the resource allocation and binding process there is no sharing of physical modules in the ETPN as discussed in the previous chapter. The allocation and binding algorithm (ABA) merges data path nodes to decrease the total number of physical modules in the design. For each physical module type (e.g., 16-bit adder or 32-bit multiplier) the maximal number of that type used in any control step determines how many instances that has to be allocated. Hence, to achieve the goal of minimizing the number of physical modules it might be necessary to re-schedule some of the places in the ETPN. For example, a design that contains two additions can be implemented with one single adder. The scheduling algorithm assumes that the additions are implemented with dedicated adders, but the ABA will merge these two adders. However, if the additions are performed in the same control step one of the places controlling an addition has to be delayed by rescheduling to enable the merge of the data path nodes.

Both place rescheduling and data path node merging may change the timing of the design; rescheduling by introducing dummy places (see Section 6.4.3) and thus increasing the length of the execution paths inside the timing constraints, and node merging by introducing multiplexors. Introduction of multiplexors can lead to the increase of delay of the associated places. The change of the delay of a place can cause a timing constraint violation if it requires the place to use more clock cycles to complete. When this happens the clock cycle time is adjusted so that the number of clock cycles spent on the place does not increase. If this leads to a timing constraint violation then the merge of these data path nodes is prohibited.

During each iteration of the ABA the place rescheduling transformation is repeatedly applied to the most expensive control step (the cost of a control step equals the hardware resources needed to execute the operations controlled by the places assigned to it, as defined in Section 6.6.1) as long as any places can be re-assigned to other control steps and no timing constraints are violated. When the cost can not be reduced further the physical modules required by the most expensive control step are allocated. The rest of the control steps are examined and the data path nodes are merged into other nodes for which physical modules already have been allocated. If this cannot be done, additional physical modules are allocated for the data path nodes.

8

Hardware/Software Partitioning

One of the main tasks to be performed during system synthesis, as we discussed in Chapter 4, is partitioning. At the system level this implies that the behaviors captured by the processes in the system specification are partitioned among system components. One typical problem at this level is to decide which parts of the specification are to be implemented as hardware and software respectively. This is a typical partitioning problem at the system level, known as the "hardware/software partitioning problem" (see also Section 4.6).

Some approaches to partitioning in general and to hardware/software partitioning in particular were discussed in Chapter 4. In the approach presented in this chapter [EPKD96a, EPKD96b, EPKD97] we accept as input a system specification formulated in VHDL using the SR specification style. The synthesized system has to produce maximal performance (in terms of execution speed) using a given amount of hardware and software resources. In order to achieve such an implementation, the partitioning strategy which will be presented entails four successive steps. The first two are operating on the VHDL source specification and finally produce an internal system level representation called the process graph. The effective partitioning is then performed on this graph representation and produces two sets of processes:

1. the set of processes assigned for hardware, which will be further synthesized according to the strategies discussed in the previous chapters;
2. the set of processes assigned to the software domain, which will be further compiled to C in order to be implemented on the target processor which is

part of the system architecture.

Two algorithms will be presented for partitioning of the process graph structure. Both are based on iterative improvement heuristics and thus operate with successive transformations starting from an initial partitioning. One algorithm is based on simulated annealing and the other one on tabu search. In its attempt to find the implementation producing the highest performance at the imposed cost limit, the partitioning strategy is based on metric values derived from profiling (simulation), static analysis of the specification, and cost estimations.

We will present this hardware/software partitioning strategy considering a simple implementation architecture which is typical for embedded systems with hardware and software components. The following assumptions have been made:
1. There is a single programmable component (microprocessor) executing the software processes;
2. The microprocessor and the hardware coprocessor are working in parallel (the architecture does not enforce a mutual exclusion between the software and hardware);
3. Reducing the amount of communication between the microprocessor (software partition) and the hardware coprocessor (hardware partition) improves the overall performance of the application.

8.1   The Partitioning Strategy

The main goal pursued during partitioning is to assign processes to hardware and software so that the performance of the implemented system, in terms of execution speed, is maximized and the implementation cost is within a user imposed limit. In order to achieve this we distribute functionality between the software and the hardware partitions taking into account communication cost and overall parallelism of the synthesized system. The following three objectives are considered during partitioning as vehicles for performance improvement:
1. To identify *basic regions* (processes, subprograms, loops, and blocks of statements) which are responsible for most of the execution time in order to be assigned to the hardware partition;
2. To minimize communication between the hardware and software domains;
3. To increase parallelism within the resulting system at the following three levels:
   • internal parallelism of each hardware process (during high-level synthesis, operations are scheduled to be executed in parallel by the available functional units);
   • parallelism between processes assigned to the hardware partition;
   • parallelism between the hardware coprocessor and the microprocessor executing the software processes.

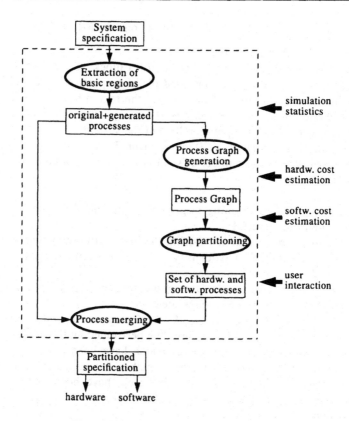

Figure 8.1: The partitioning steps.

The partitioning goals are achieved taking into account simulation statistics, information from static analysis of the source specification, and cost estimations. At the same time, all major decisions taken by the design tool can be influenced through user interaction.

Statistics data are collected from simulation of an internal representation generated by a front-end compiler, running the system with sets of typical input stimuli. Two types of simulation statistics are used by the partitioning algorithm:

1. *Computation load* (CL) of a basic region is a quantitative measure of the total computation executed by that region, considering all its activations during the simulation process. It is expressed as the total number of operations (at the level of internal representation) executed inside that region, where each operation is weighted with a coefficient depending on its relative complexity:

$$CL_i = \sum_{op_j \in BR_i} N\_act_j \cdot \varphi_{op_j}; \quad N\_act_j \text{ is the number of activations of}$$

operation op_j belonging to the basic region BR_i; φ_{op_j} is the weight associated to that operation.

The *relative computation load* (*RCL*) of a block of statements, loop, or a subprogram is the computation load of the respective basic region divided by the computation load of the process the region belongs to. The relative computation load of a process is the computation load of that process divided by the total computation load of the system.

2. *Communication intensity* (*CI*) on a channel connecting two processes is expressed as the total number of send operations executed on the respective channel.

Hardware/software partitioning is performed in four successive steps, which are illustrated in Figure 8.1:

1. *Extraction of basic regions*: During the first partitioning step processes are examined individually to identify performance critical regions which are extracted as separate processes.

2. *Process graph generation*: During the second step an internal structure, called the *process graph*, is generated.

3. *Partitioning of the process graph*: We formulate hardware/software partitioning as a graph partitioning problem performed on the process graph.

4. *Process merging*: some processes which have been assigned to the same partition can be, optionally, merged together. Such a typical situation is with basic regions which, during the first step, have been extracted as child processes from a parent process but have been finally assigned to the same partition with their parent.

The first three steps will be presented in more detail in the next sections.

8.2   Extraction of Basic Regions

During the first partitioning step VHDL processes are examined individually to identify regions that are responsible for most of the execution time spent inside a process (regions with a large CL). Candidate regions are typically loops and subprograms, but can also be blocks of statements with a high CL. The designer guides identification and extraction of the regions and decides implicitly on the granularity of further partitioning steps in two ways:

1. By identifying a certain region to be extracted (regardless of its CL) and assigning it to the hardware or software partition. This, for instance, can be the case for some statement sequence on which a hard time constraint has been imposed which can be satisfied only by hardware implementation.

2. By imposing two boundary values:
   • a threshold X on the RCL of processes that are examined for extraction of

```
P1: process                           P2: process
   . . .                                 . . .
   LOOP_1: while X<K loop                procedure P(A: in INTEGER;
      . . .                                           B: out INTEGER) is
      X := ... C + K ... ;               begin
      . . .                                 . . .
   end loop LOOP_1;                         B := ... A ... ;
   . . .                                    . . .
end process P1;                          end P;
                                      begin
                                         . . .
                                         P(7, Z);
                                         . . .
                                      end process P2;
```

Figure 8.2: VHDL example before extraction of basic regions.

basic regions;
- a threshold Y on the RCL of a block, loop, or subprogram to be considered for extraction.

The search for candidate regions in processes with RCL greater than X is performed bottom-up, starting from the inner blocks, loops, and the subprograms that are not containing other basic regions. When a region has been identified for extraction, a new process is built. It has the functionality of the original block, loop, or subprogram and communication channels are established to the *parent* process. These channels are built using the message passing mechanism defined by the SR style. It is essential that synchronization using this process interaction mechanism affects not all processes but only the respective parent and child process involved in the specific communication. Due to this strictly local effect of the transformations performed for interface generation, the original semantics of the VHDL specification is preserved.

Example 8.1: We illustrate the generation of new processes during extraction of basic regions with the VHDL example in Figure 8.2. Given that the RCLs of the loop and the subprogram in the two processes are greater than a threshold, two new processes, P1_LOOP_1 and P2_PROC_P, are generated to execute the loop LOOP_1 and procedure P respectively. Communication channels to and from the new processes are established according to the data dependence relationship. In the above example, signals S_P1_C, S_P1_K, S_P1_X_TO, and S_P1_X_FROM are introduced for communication between P1 and P1_LOOP_1, and signals S_P2_A, and S_P2_B for communication between P2 and P2_PROC_P. The new VHDL code, after process extraction, is given in Figure 8.3.

□

signal S_P1_C,S_P1_K,S_P1_X_TO,S_P1_X_FROM,S_P2_A,S_P2_B: INTEGER;

P1: **process**   P2: **process**

  . . .   . . .

   send(S_P1_C,C,S_P1_X_TO,   send(S_P2_A,7);

      X,S_P1_K,K) ;   . . . *--additional parallelism*

   . . . *--additional parallelism*   receive(S_P2_B);

   receive(S_P1_X_FROM);   Z := S_P2_B;

   X := S_P1_X_FROM;   . . .

  . . .   **end process** P2;

end process P1;

P1_LOOP_1: **process**   P2_PROC_P: **process**

  variable X : INTEGER;   **variable** B: INTEGER;

begin   **begin**

  receive(S_P1_C,S_P1_X_TO,   receive(S_P2_A);

      S_P1_K);   . . .

  X := S_P1_X_TO;   B := ... S_P2_A ... ;

  LOOP_1: **while** X < S_P1_K **loop**   . . .

   . . .   send(S_P2_B,B);

    X := ... S_P1_C+S_P1_K ... ; **end process** P2_PROC_P;

   . . .

  end loop LOOP_1;

  send(S_P1_X_FROM,X);

end process P1_LOOP_1;

Figure 8.3: VHDL example after extraction of basic regions.

As stated, one of the objectives considered during partitioning is to increase parallelism within the resulting system. At process generation we follow this goal by introducing additional parallelism, as far as data dependence allows, between parent and child process. This is achieved by moving statements of the parent process into the sequence between the *send* and the *receive* commands used for communication with the child.

The final decision if processes generated during this step are kept as separate modules or will eventually be merged back, depends on the two subsequent partitioning steps. An important criterion for this decision will be the intensity of communication between parent and child processes.

8.3   The Process Graph

The input to the second partitioning step is a set of interacting processes. Some of them are originally specified by the designer, others are generated during extraction of basic regions. The data structure on which hardware/software partitioning is performed is the *process graph*, and it is generated during the second partitioning step. Each node in this graph corresponds to a process and

port(IP1,IP2:**in** INTEGER; OP1,OP2:**out** INTEGER);

. . .

signal S1,S2,S3,S4,S5,S6:INTEGER;

P1:process	**P3:process**	**P5:process**
. . .	. . .	. . .
receive(IP1);	receive(S4);	receive(S1,S5);
. . .	. . .	. . .
send(S1,...);	send(S2,...);	send(S4,...);
. . .	. . .	. . .
send(S3,...);	**end process P3**;	**end process P5**;
. . .		
receive(S6);		
. . .		
end process P1;		
P2:process	**P4:process**	**P6:process**
. . .	. . .	. . .
receive(IP2);	receive(S3);	receive(S2);
. . .	. . .	. . .
receive(S1);	send(S5,...,S6,...);	send(OP1,...);
. . .	. . .	. . .
send(OP2,...);	**end process P4**;	**end process P6**;
. . .		
end process P2;		

Figure 8.4: VHDL processes interacting through message passing.

an edge connects two nodes if and only if there exists at least one direct communication channel between the corresponding processes.

Example 8.2: Let us consider the VHDL processes described in Figure 8.4, as they resulted from the first partitioning step. The corresponding process graph is depicted in Figure 8.5.

□

The algorithm which has to perform partitioning of the process graph takes into account weights associated to each node and edge. Node weights reflect the degree of suitability for hardware implementation of the corresponding process. Edge weights measure communication and mutual synchronization

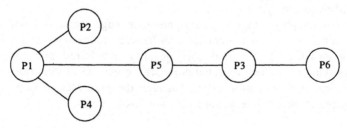

Figure 8.5: Process graph corresponding to example in Figure 8.4.

between processes. The weights capture simulation statistics (CL, RCL, and CI) and information extracted from static analysis of the system specification or of the internal representation resulting after its compilation. The following data extracted from static analysis are captured:

Nr_op_i: total number of operations in the dataflow graph of process i;

$Nr\_kind\_op_i$: number of different operations in process i;

L_path_i: length of the critical path (in terms of data dependency) through process i.

The weight assigned to process node i, has two components. The first one, WI_i^N, is equal to the CL of the respective process. The second one is calculated by the following formula:

$$W2_i^N = M^{CL} \cdot K_i^{CL} + M^U \cdot K_i^U + M^P \cdot K_i^P - M^{SO} \cdot K_i^{SO} \quad ;$$

where:

K_i^{CL} is equal to the RCL of process i, and thus is a measure of the computation load;

$K_i^U = \dfrac{Nr\_op_i}{Nr\_kind\_op_i}$; K_i^U is a measure of the uniformity of operations in process i;

$K_i^P = \dfrac{Nr\_op_i}{L\_path_i}$; K_i^P is a measure of the potential parallelism inside process i;

$K_i^{SO} = \dfrac{\displaystyle\sum_{op_j \in SP_i} w_{op_j}}{Nr\_op_i}$; K_i^{SO} captures the suitability of operations of process i

for software implementation. SP_i is the set of such operations (floating point computation, file access, pointer operations, recursive subprogram call, etc.) in process i and w_{op_j} is a weight associated to operation op_j, measuring the degree to which the operation has to be implemented in software; a large weight associated to such an operation dictates software implementation for the given process, regardless of other criteria.

The relation between the above-named coefficients K^{CL}, K^U, K^P, K^{SO} is regulated by four different weight-multipliers: M^{CL}, M^U, M^P, and M^{SO}, controlled by the designer.

Both components of the weight assigned to an edge connecting nodes i and j depend on the amount of communication between processes i and j. The first one is a measure of the total data quantity transferred between the two processes. The second one does not consider the number of bits transferred but only the degree of synchronization between the processes, expressed in the total number of mutual interactions they are involved in:

$$W1^E_{ij} = \sum_{c_k \in Ch_{ij}} wd_{c_k} \cdot CI_{c_k} ; \qquad\qquad W2^E_{ij} = \sum_{c_k \in Ch_{ij}} CI_{c_k} ;$$

where Ch_{ij} is the set of channels used for communication between processes i and j; wd_{c_k} is the width (number of transported bits) of channel c_k in bits; CI_{c_k} is the communication intensity on channel c_k

8.4 Process Graph Partitioning

After the process graph has been generated hardware/software partitioning can be performed as a graph partitioning task. The partitioning information is captured as weights associated to the nodes and edges of the graph. These weights have to be combined into a cost function which guides the partitioning algorithm towards the desired objective and which can be incrementally updated quickly after each iteration. From the partitioning algorithm we expect to converge fast towards a value of the cost function which is hopefully close to the global minimum.

8.4.1 The Cost Function and the Constraints

The hardware/software partitioning heuristics which we present here are guided by the following cost function which is to be minimized:

$$C(Hw, Sw) = Q1 \cdot \sum_{(ij) \in cut} W1^E_{ij} + Q2 \cdot \frac{\sum_{(i) \in Hw} \frac{\sum_{\exists(ij)} W2^E_{ij}}{W1^N_i}}{N_H}$$

$$-Q3 \cdot \left(\frac{\sum_{i \in Hw} W2^N_i}{N_H} - \frac{\sum_{i \in Sw} W2^N_i}{N_S} \right) \quad ;$$

where Hw and Sw are sets representing the hardware and the software partition respectively; N_H and N_S are the cardinality of the two sets; *cut* is the set of edges connecting the two partitions; *(ij)* is the edge connecting nodes i and j; and *(i)* represents node i.

The three terms of this apparently sophisticated function directly reflect the three objectives of the partitioning process, as stated in Section 8.1. Thus, minimizing the value of these terms guides the partitioning algorithm towards high performance implementations:

• The *first term* represents the total amount of communication between the hardware and the software partition. Decreasing this component of the cost

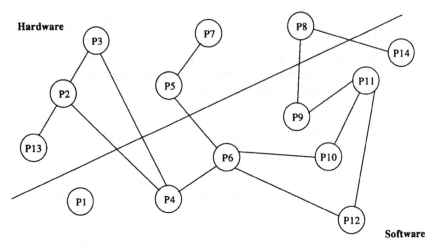

Figure 8.6: Process graph.

function reduces the total amount of communication between partitions, which is a primary design goal. As a second effect it also improves parallelism between processes in the hardware partition and those implemented in software. This is an implicit result of reduced interaction between the two partitions.

The *second term* stimulates placement into hardware of processes which have a reduced amount of interaction with the rest of the system relative to their computation load and, thus, are active most of the time. This strategy improves parallelism between processes inside the hardware partition where physical resources are allocated for real parallel execution. For a given process

i, $\left(\sum_{\exists(ij)} W2_{ij}^{E} \right) \bigg/ W1_{i}^{N}$ is the total amount of interaction the process is involved

in, relative to its computation load. The whole term represents the average of this value over the nodes in the hardware partition.

With the process graph depicted in Figure 8.6 we illustrate the effect of minimizing this second term of the cost function. The 14 processes are distributed between the two partitions according to an intermediate solution which has to be improved during the next partitioning steps. For simplicity let us consider that all edges have the same weight $W2_{ij}^{E}$ and the weight $W1_{i}^{N}$ is also the same for all nodes. By minimizing the second term of the cost function the movement of node *P1* into the hardware partition will be encouraged because it has the lowest amount of interaction with the rest of the system and, thus, provides a high degree of parallelism inside the hardware domain. Similarly, the movement of node *P14* into hardware will be stimulated more than that of any other node except *P1*, according to the objective represented by this term. The same is true, for example, for node

P9 compared to *P11*, *P6*, or *P4*.

- The *third term* in the cost function pushes processes with a high node weight into the hardware partition and those with a low node weight into the software one, by increasing the difference between the average weight of nodes in the two partitions. This is a basic objective of partitioning as it places time critical regions into hardware and also exploits the potential of hardware to implement internal parallelism of processes.

The criteria combined in the cost function are not orthogonal, and sometimes compete with each other. Moving a process with a high node weight into hardware, for instance, can produce an increase in the communication between partitions. This competition between partitioning objectives is controlled by the designer through the cost multipliers *Q1*, *Q2*, and *Q3* which regulate the relative influence of the different metrics.

Minimization of the cost function has to be performed in the context of certain constraints. Thus, our heuristics have to produce a partitioning with a minimum for *C(Hw, Sw)*, so that the total hardware and software cost is within some user specified limits:

$$\sum_{(i)\,\in\,Hw} H\_cost_i \leq Max^H \; ; \qquad\qquad \sum_{(i)\,\in\,Sw} S\_cost_i \leq Max^S \; .$$

A preassignment of processes which are definitively assigned to one of the partitions can be performed optionally by the designer. Often this preassignment is formulated in terms of the weights: nodes with a weight smaller than a given limit have to go into software and those with a weight greater than a certain limit should be assigned to hardware:

$$W2_i^N \geq Lim1 \Rightarrow (i) \in Hw \; ; \qquad\qquad W2_i^N \leq Lim2 \Rightarrow (i) \in Sw \; .$$

Cost estimation has to be performed before graph partitioning, for both the hardware and software implementation alternatives of the processes. The fact that in this approach cost estimation is performed a single time, outside the optimization cycle, makes it not to be critical from the point of view of execution time. Thus, high-level synthesis can be performed for each process in order to get hardware cost estimations in terms of design area. Software cost, in terms of memory size, can be estimated for each process after compilation from VHDL to C.

8.4.2 Iterative Improvement Heuristics

As a final step of the hardware/software partitioning process the weighted graph is to be partitioned into two subgraphs. The partitions containing nodes assigned to hardware and software respectively, are generated so that design constraints are satisfied and the cost function is minimal.

Construct initial configuration $x^{now}:=(Hw_0, Sw_0)$
Initialize Temperature $T:=TI$
repeat
    for $i:=1$ **to** TL **do**
        Generate randomly a neighboring solution $x' \in N(x^{now})$
        Compute change of cost function $\Delta C:=C(x')-C(x^{now})$
        if $\Delta C \leq 0$ **then**
            $x^{now}:=x'$
        else
            Generate $q:=random(0,1)$
            if $q < e^{-\Delta C/T}$ **then** $x^{now}:=x'$ **end if**
        end if
    end for
    Set new temperature $T:=\alpha * T$
until stopping criterion is met
return solution corresponding to the minimum cost function

Figure 8.7: Partitioning with simulated annealing.

Hardware/software partitioning, formulated as a graph partitioning problem, is NP complete. In order to efficiently explore the solution space, heuristics have to be developed which hopefully converge towards an optimal or near-optimal solution. We will present here two such algorithms which both are transformation based. One is using simulated annealing (SA) and the other tabu search (TS). Both the SA and TS approach to optimization were discussed in Chapter 5. They perform neighborhood search and, to avoid being trapped by a local optimum, they allow, at certain stages of the search, uphill moves performed in a controlled manner. In this section we show how SA and TS can be practically adapted to solve graph partitioning for hardware/software co-design purposes and we compare the performances produced by the two approaches. For details concerning the experimental procedure which led to the conclusions presented here we refer the reader to [EPKD97].

Partitioning with Simulated Annealing

Simulated annealing based heuristics have been discussed in general in Chapter 5. The short description of the simulated annealing based hardware/software partitioning algorithm, given in Figure 8.7, practically reproduces the general strategy characteristic for the SA approach. This is typical for simulated annealing, where adaptation to any particular problem is easy and only a few problem specific aspects have to be taken in consideration. With x we denote one solution consisting of the two sets Hw and Sw. x^{now} represents the current solution and $N(x^{now})$ denotes the neighborhood of x^{now} in the solution space.

For implementation of the algorithm, parameters TI (initial temperature), TL (temperature length), and α (cooling ratio) have to be determined. Such a

Generate $k:=random(1, Nr\_of\_nodes)$ until moving node k does not violate constraints
Generate solution x' from x^{now} by moving node k
Generate $q:=random(0,1)$
if $q \geq p$ **then**
    return x'
end if
for each node k' which was direct neighbor of node k and in the same partition with node k **do**
    if moving node k' does not violate constraints and produces $\Delta C \leq 0$ **then**
        Generate solution x' from current x' by moving node k'
    end if
end for
return x'

Figure 8.8: Generation of a new solution with improved move.

tuning of the algorithm usually needs long and laborious experiments [EPKD97].

For the stopping criterion it was assumed that the system is frozen if for three consecutive temperatures no new solution has been accepted.

A problem specific component of the SA algorithm is the generation of a new solution x' by transforming the current one x^{now}. Two transformation strategies have been implemented for solution generation: the *simple move* and the *improved move*.

For the *simple move* a node is randomly selected for being moved to the other partition. The configuration resulting after this move becomes the candidate solution x'. Random node selection is repeated if transfer of the selected node violates some design constraints.

The *improved move* accelerates convergence by moving together with the randomly selected node also some of its direct neighbors (nodes which are in the same partition and are directly connected to it). A direct neighbor is moved together with the selected node if this movement improves the cost function and does not violate any constraint. This strategy stimulates transfer of connected node groups instead of individual nodes. Experiments revealed a negative side effect of this strategy: the repeated move of the same or similar node groups from one partition to the other, which resulted in a reduction of the spectrum of visited solutions. To produce an optimal exploration of the solution space movements of node groups have to be alternated with those of individual nodes: this means that movement of node groups has to be performed not for each iteration but only with a certain probability p. After analysis of experimental results the value for p was fixed at 0.75. In Figure 8.8 we present the algorithm for generation of a candidate solution according to the improved move.

The improved move has been developed as result of a problem specific design of the neighborhood structure. The influence of this solution on the

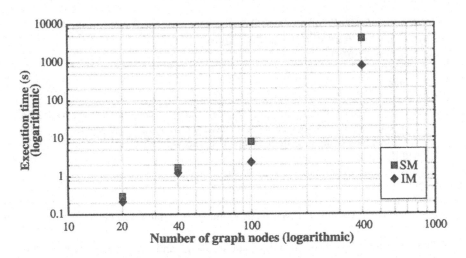

Figure 8.9: Partitioning times with SA: simple moves (SM) and
improved moves (IM).

partitioning time is visible from Figure 8.9. There we show the times needed
for partitioning of process graphs with different number of nodes, using the SA
based algorithm with simple and improved moves respectively. For the same
near-optimal partitioning results, the speedup produced by the improved moves
is between 22% (for graphs with 20 nodes) and 425% (for graphs with 400
nodes) [EPKD97].

Figure 8.10 and Figure 8.11 illustrate the strategy of solution space explo-

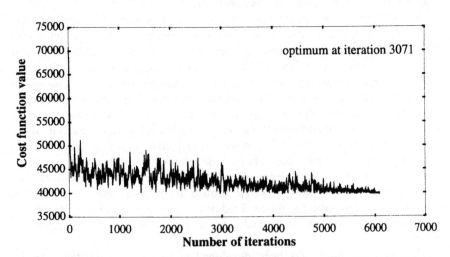

Figure 8.10: Variation of cost function during simulated annealing with simple
moves (100-node graph).

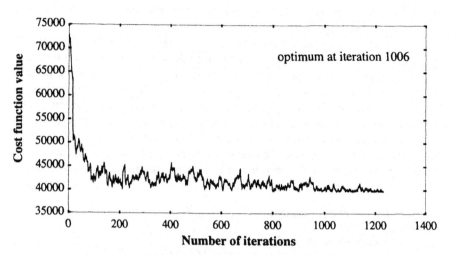

Figure 8.11: Variation of cost function during simulated annealing with improved moves (100-node graph).

ration followed by the SA based algorithm during partitioning of a 100-node graph, using the simple move and the improved move respectively. A comparison of the two curves demonstrates the much faster convergence provided by the improved moves

Partitioning with Tabu Search

In contrast to simulated annealing, tabu search controls uphill moves not purely randomly but in an intelligent way. As it has been discussed in Chapter 5, the tabu search approach accepts uphill moves and stimulates convergence toward a global optimum by creating and exploiting data structures to take advantage of the search history at selection of the next move. The two key elements of the TS algorithm are the data structures called short and long term memory. Short term memory stores information relative to the most recent history of the search. It is used in order to avoid cycling that could occur if a certain move returns to a recently visited solution. Long term memory, on the other side, stores information on the global evolution of the algorithm. These are typically frequency measures relative to the occurrence of a certain event. They can be applied to perform *diversification* which is meant to improve exploration of the solution space by broadening the spectrum of visited solutions.

In Figure 8.12 we give a brief description of our implementation of the TS based partitioning algorithm. In a first attempt an improving move is tried. If no such move exists (or it is tabu and not aspirated) frequency based penalties are applied to the cost function and the best possible non tabu move is performed;

this move can be an uphill step. Finally, in a last attempt, the move which is closest to leave the tabu state is executed.

We consider as a candidate solution x_k the configuration obtained from x^{now} by moving node k from its current partition to the other one, if this movement does not violate any constraints. In the *tabu list* we store the list of the reverse moves of the last τ moves performed, which are considered as being forbidden (tabu). The size τ of this list (the *tabu tenure*) is an essential parameter of the algorithm, which has to be carefully tuned through experiments [EPKD97].

Under certain circumstances it can be useful to ignore the tabu character of a move (the tabu is *aspirated*). The tabu status should be overridden when the selected move improves the search and it does not produce any cycling in the

> Construct initial configuration $x^{now}:=(Hw_0, Sw_0)$
>
> *start*:
>   **for** each solution $x_k \in N(x^{now})$ **do**
>   Compute change of cost function $\Delta C_k := C(x_k) - C(x^{now})$
>   **end for**
>   **for** each $\Delta C_k < 0$, in increasing order of ΔC_k **do**
>   **if** not *tabu(x_k)* or *tabu_aspirated(x_k)* **then**
> $x^{now}:=x_k$
>   **goto** *accept*
>   **end if**
>   **end for**
>   **for** each solution $x_k \in N(x^{now})$ **do**
>   Compute $\Delta C'_k := \Delta C_k + penalty(x_k)$
>   **end for**
>   **for** each $\Delta C'_k$ in increasing order of $\Delta C'_k$ **do**
>   **if** not *tabu(x_k)* **then**
> $x^{now}:=x_k$
>   **goto** *accept*
>   **end if**
>   **end for**
>   Generate x^{now} by performing the least tabu move
> *accept*:
>   **if** iterations since previous best solution $< Nr\_f\_b$ **then**
>   **goto** *start*
>   **end if**
>   **if** restarts $< Nr\_r$ **then**
>   Generate initial configuration x^{now} considering frequencies
>   **goto** *start*
>   **end if**
>   **return** solution corresponding to the minimum cost function

Figure 8.12: Partitioning with tabu search.

exploration of the design space. We ignore the tabu status of a move if the solution produced is better than the best obtained so far.

For diversification purpose we store (in the long term memory structure) the number of iterations each node has spent in the hardware partition. Three means of improving the search strategy by diversification have been implemented:

1. For the second attempt to generate a new configuration (Figure 8.12) moves are ordered according to a penalized cost function which favors the transfer of nodes that have spent a long time in their current partition:

$$\Delta C'_k = \Delta C_k + \frac{\sum_i |\Delta C_i|}{Nr\_of\_nodes} \cdot pen(k)$$

where

$$pen(k) = \begin{cases} -C_H \cdot \dfrac{Node\_in\_Hw_k}{N_{iter}} & \text{if node } k \in Hw \\[4mm] -C_S \cdot \left(1 - \dfrac{Node\_in\_Hw_k}{N_{iter}}\right) & \text{if node } k \in Sw \end{cases}$$

$Node\_in\_Hw_k$ is the number of iterations node k spent in the hardware partition; N_{iter} is the total number of iterations; Nr_of_nodes is the total number of nodes; Coefficients have been experimentally set to $C_H=0.4$ and $C_S=0.15$.

2. We consider a move as forbidden (tabu) if the frequency of occurrences of the node in its current partition is smaller than a certain threshold; thus, a move of node k can be accepted if:

$$\frac{Node\_in\_Hw_k}{N_{iter}} > T_H \qquad\qquad \text{if node } k \in Hw$$

$$\left(1 - \frac{Node\_in\_Hw_k}{N_{iter}}\right) > T_S \qquad\qquad \text{if node } k \in Sw$$

The thresholds have been experimentally set to $T_H=0.2$ and $T_S=0.4$.

3. If the system is frozen (more than Nr_f_b iterations have passed since the current best solution was found) a new search can be started from an initial configuration which is different from those encountered previously.

Figure 8.13 and Figure 8.14 illustrate the strategy of design space exploration for the TS algorithm applied to a 400-node and 100-node graph respectively.

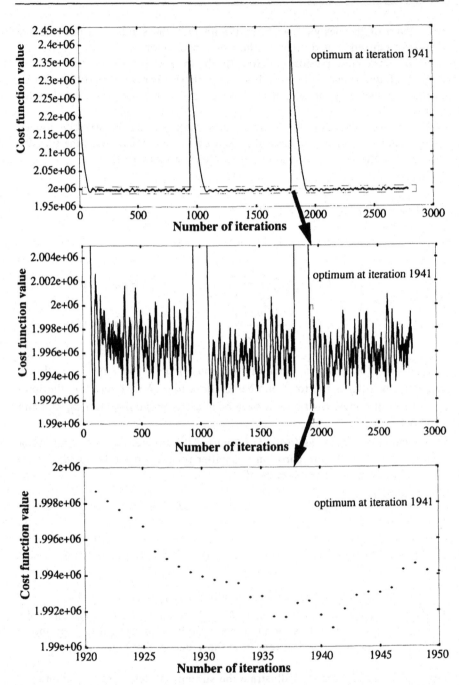

Figure 8.13: Variation of cost function during tabu search partitioning (400 nodes).

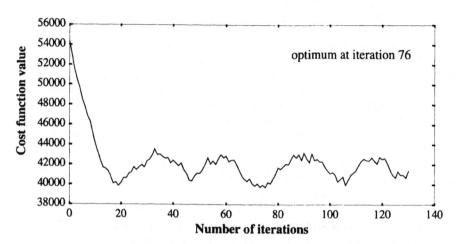

Figure 8.14: Variation of cost function during tabu search partitioning (100 nodes).

Figure 8.14 shows a very fast convergence for 100 nodes, without restarting tours. Partitioning of the 400-node graph needed two restarting tours for diversification, which is shown in Figure 8.13. The two additional detail pictures given in this figure focus successively on the area close to the optimum, in order to illustrate how the algorithm guides the exploration of the solution space towards an optimum by a succession of diversification moves, uphill and improving steps.

Discussion

A wide range of experiments performed with graphs of different dimensions and structure lead to the following conclusions concerning hardware/software partitioning based on SA and TS [EPKD97]:
1. Near-optimal[1] partitioning can be produced both by the SA and TS based algorithm.
2. SA is based on a random exploration of the neighborhood while TS is

---

1. It has to be clarified what we call *near-optimal* in the context of this discussion and of the underlying experiments [EPKD97]. For relatively small graphs (20 or fewer nodes) it is possible to run exhaustive search in order to get the *real* optimum which is then used as a reference value for the experiments. For each of the other graphs very long runs were performed in preparation to the experiments, using both SA and TS. For these runs, very long cooling schedules, for SA, and a high number of restarting tours, for TS, were fixed. The runs have been performed starting from different initial configurations and finally the *best ever* solution produced for each graph (in terms of minimal value of the cost function) has been considered as the reference "near-optimal" value to be produced.

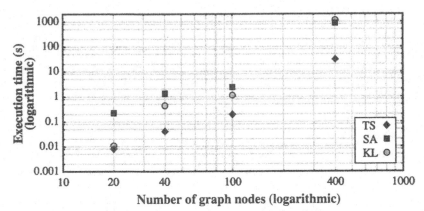

Figure 8.15: Partitioning times with SA, TS, and KL.

completely deterministic. The deterministic nature of TS makes experimental tuning of the algorithm and setting of the parameters less laborious than for SA. At the same time adaptation of the SA strategy for a particular problem is relatively easy and can be performed without a deep study of domain specific aspects. Although, specific improvements can result in large gains of performance (as we have shown with the improved moves). On the contrary, development of a TS algorithm is more complex and has to consider particular aspects of the given problem.

3. Performances obtained with TS are excellent and are definitely superior in comparison to those given by SA (on average more than 20 times faster), as shown in Figure 8.15. There we can compare the times needed for partitioning of process graphs with different number of nodes, using the SA based algorithm with improved moves and the one based on TS. For the same near-optimal partitioning results [EPKD97] the algorithm based on TS is on average more than 20 times faster.

It is also interesting to compare the SA and TS-based heuristics with partitioning based on a classical iterative-improvement approach, the Kernighan-Lin (KL) algorithm, with has been discussed in Subsection 4.4.2. Given the relatively limited capacity of the KL-based algorithm to escape from local minima and its sensitivity to the initial configuration, the partitioning algorithm has to perform several runs for each graph, with randomly selected starting configurations. The number of necessary restarting tours differs depending on the graph dimension. It has been fixed experimentally so that all graphs of a given dimension are near-optimally partitioned with a sufficiently high probability [EPKD97]. As shown in Figure 8.15, partitioning times with KL are slightly better than those with SA for small and medium graphs. For the 400-node graphs SA outperforms the KL-based algorithm. TS is on average 10 times faster than KL for 40 and 100-node graphs, and 30 times faster for graphs with 400 nodes.

8.5 Partitioning Examples

We present now two experiments on real-life models, which have validated our conclusions presented above. The two models are the *Ethernet network coprocessor* and the *OAM block of an ATM switch*. Both were described at system level in VHDL using the SR specification style. After simulation, basic regions were extracted and the annotated process graph has been generated. Partitioning was performed using both the SA based and the TS algorithm, with the cost function presented in Subsection 8.4.1 and a constraint on the hardware cost representing 30% of the cost of a pure hardware implementation.

The *Ethernet network coprocessor* is given in [NVG90] as an example for system specification in SpecCharts and has been used, in a HardwareC version, in [GuDM92] and [Gup95]. We have rewritten it in VHDL, as a model consisting of 10 cooperating processes (730 lines of code). These processes are depicted as rectangles in Figure 8.16. The coprocessor transmits and receives data frames over a network under CSMA/CD (Carrier Sense Multiple Access with Collision Detection) protocol. Its purpose is to off-load the host CPU from managing communication activities. The host CPU programs the coprocessor for specific operations by means of eight instructions. Processes *rcv-comm*, *buffer-comm*, and *exec-unit* are dealing with enqueuing and decoding/executing these instructions.

Transmission to the network is performed in cooperation by three processes. Process *DMA-xmit* gets a memory address from the host CPU and accesses directly the memory in order to read the data. This data is forwarded successively to a second process (*xmit-frame*) which packages it in frames according to a prescribed standard. Frames are then sent as a series of bytes to process *xmit-bit* which outputs them on the serial network line. If a collision is detected normal transmission is suspended, a number of jam bytes are generated and after waiting a certain time the frame will be retransmitted. After a successful transmission, the unit waits for a period of time required between frames before attempting another transmission.

In parallel to the above processes, other four processes deal with reception from the network. Process *rcvd-bit* continuously reads bits from the serial line of the network and sends a succession of bytes to the buffering process *rcvd-buffer*. Process *rcvd-frame* receives the data from *rcvd-buffer* and filters out those frames which have as destination the host system. It first waits to recognize a start-of-frame pattern and then compares the following two bytes with the address of the host. If the addresses are equal the rest of the bytes belonging to the frame are read and sent to process *DMA-rcvd* which writes them to a local memory.

After the first partitioning step, extraction of performance critical blocks,

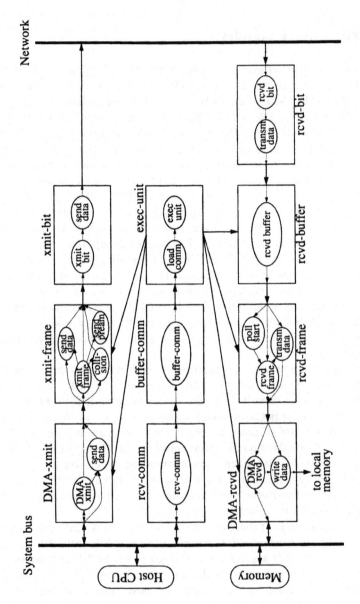

Figure 8.16: The Ethernet network coprocessor.

loops, and subprograms, we got a VHDL specification consisting of 20 processes, represented as ovals in Figure 8.16. Process graph generation and graph partitioning produced a hardware partition with 14 processes and a software partition with 6 processes. Processes implemented in software are shaded in the figure. The most time critical part of those processes that are handling transmission and reception of data on the ethernet line as well as processes which are strongly connected to them have been assigned to hardware and the rest belong to the software partition. This corresponds to the results reported in [GuDM92] and [Gup95], which have been obtained following a different approach.

Our second example implements the *operation and maintenance (OAM) functions corresponding to the F4 level of the ATM protocol layer* [DePr93]. This level handles OAM functionality concerning fault management, performance monitoring, fault localization, and activation/deactivation of functions.

ATM (asynchronous transfer mode) is based on a fixed-size virtual circuit-oriented packet switching methodology. All ATM traffic is broken into a succession of cells. A cell consists of five bytes of header information and a 48-byte information field. The header field contains control information of the cell (identification, cell loss priority, routing and switching information). Of particular interest in the header are the virtual path identifier (VPI) and the virtual channel identifier (VCI). They are used to determine which cells belong to a given connection.

The OAM functions in the network are performed on five hierarchical levels associated with the ATM and Physical layers (PhL) of the protocol reference model [ITU93]. The two highest levels, F4 and F5, are part of the ATM protocol layer. The F4 level handles the OAM functionality concerning virtual paths (VP) [Bellc93]:

- Fault management: when the appearance of a fault is reported to the F4 block, special OAM cells will be generated and sent on all affected connections; if the fault persists, the management system (MS) should be notified.
- Performance monitoring: normal operation of the network is monitored by continuous or periodic checking of cell transmission.
- Fault localization: when a fault occurs it might be necessary to localize it; for this purpose special loop back OAM cells are used.
- Activation/Deactivation: a special protocol for activation and deactivation of OAM functions that require active participation of several F4 blocks, e.g. performance monitoring, has to be implemented.

To perform these functions, in addition to normal *user cells*, specially marked ATM cells are used. They are the *OAM cells*: activation/deactivation cells, performance management cells, and fault management cells (FMC).

We specified functionality of the F4 block of an ATM switch as a VHDL model consisting of 19 interacting processes (1321 lines of code). These

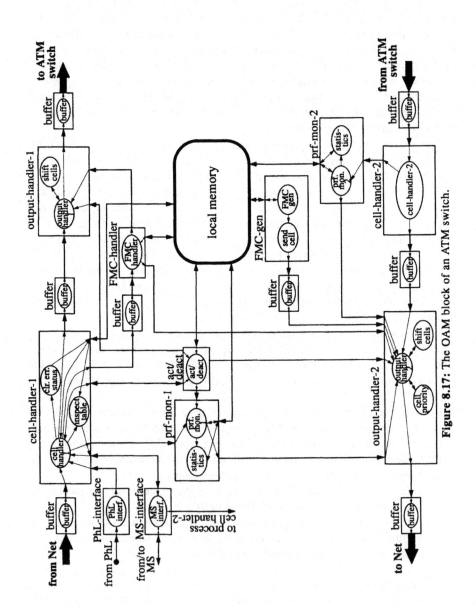

Figure 8.17: The OAM block of an ATM switch.

processes are depicted as rectangles in Figure 8.17. The model resulting after extraction of basic regions has 27 processes, represented as ovals in Figure 8.17. The resulting process graph has been partitioned into 14 processes assigned to hardware and 13 to software. Processes implemented in software are shaded in the figure. Processes performing the filtering of input cells and those handling user cells (which constitute, by far, the majority of received cells) were assigned to hardware. Processes handling exclusively OAM cells (which are arriving at a very low rate), and those assigned to functions which are executed at a low rate and without a hard time constraint (like *inspect-table*, or *clear-error-status*, for example) were assigned to software.

Our experiments with the ethernet coprocessor and the OAM block confirmed the conclusions presented in the previous section. Running times for near-optimal[1] partitioning of the two examples also confirmed the clear superiority of the TS algorithm over that based on SA, with more than an order of magnitude [EPKD97].

---

1. For the ethernet network coprocessor we verified optimallity of the solution by running exhaustive search.

PART III ADVANCED ISSUES

9

Test Synthesis

Testing is an activity which aims at finding design errors and physical faults. *Design errors* are introduced by the designer during the product development process. The testing of design errors, sometimes called design verification, tries to find discrepancies between the design specification and the current implementation. *Physical faults* comprise fabrication errors, fabrication defects and physical failures [ABF90]. Fabrication errors and defects result from an imperfect manufacturing process. A functionally correct design can function incorrectly due to such faults. In this chapter, we will concentrate on hardware testing of production faults. The place of verification and product testing in a general product development flow is depicted in Figure 9.1.

Typical fabrication defects are short circuits, open circuits and bridging faults. They are "injected" into a circuit during the manufacturing process. Physical failures occur also during the life-time of a system due to component wear-out and/or environmental factors. The physical faults have to be detected to sort out faulty chips. This is achieved by performing production tests which can include, for example, acceptance testing, burn-in and quality-assurance testing. Production tests are performed after the product has been manufactured as depicted in Figure 9.1.

Production tests have become more difficult and expensive because of the increasing complexity of designs. The main reasons are that there are a large number of components on a chip and that a relatively small number of input/output ports imposes strict limits on the accessibility to different parts of the chip. To overcome this problem, many people propose that testing should be considered already during the synthesis process. Special synthesis methods

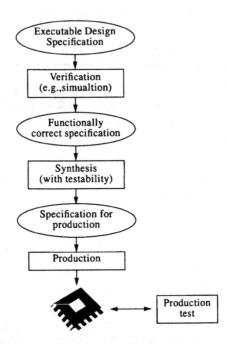

Figure 9.1: Testing activities in the design process.

which take testability of the final product into account have been proposed. These are usually called *test synthesis* methods.

Test synthesis is an emerging design automation technique which is motivated by the high complexity of current designs and large testing costs. The test related activities, such as test generation and test application, usually stand for a relatively big share of the total design and development cost. In some cases this cost can be as high as 50% of the total cost. Thus, the main idea of test synthesis is to improve testability of the design during early synthesis steps, which is expected to reduce the testing costs.

Production testing has basically two phases, test pattern generation and test application, as depicted in Figure 9.2. In the test pattern generation phase test vectors are generated while in the test application phase the test vectors are applied to the circuit. A fault in the circuit is detected when its response is different from the expected one. The fault can be further localized by diagnosis methods, if necessary. The test pattern generation task can be carried out manually by the designer or automatically. Automatic test pattern generation (ATPG) is either done by an ATPG tool or by a built-in circuitry. The later technique is called *built-in self-test* (BIST). In the final test application phase specialized automatic test equipment or the BIST structure of the circuit are used.

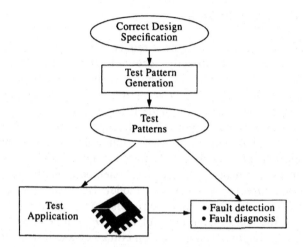

Figure 9.2: Production testing activities.

An ATPG tool is usually implemented as an algorithm which applies a test generation procedure, defined in Subsection 9.1.1, to every fault defined for a design. It has been proven that the test generation problem is *NP*-complete for combinational circuits [IS75, FT82]. Sequential circuits basically require iterative application of the combinational test pattern generation algorithm. Thus for large and complex designs, the ATPG tools require large amount of CPU time while often providing very low fault coverage (see Subsection 9.1.1).

The generally accepted approach for improving the test cost and quality is to use design for testability methods. These methods usually enforce a special design style which reduces the test pattern generation efforts. Several testability improvement methods have been proposed. For example, *scan techniques* are used to convert the sequential test pattern generation problem into a combinational one, which is easier to solve. BIST design styles do not require any external test pattern generation procedures because they provide test patterns and perform analysis of the testing results automatically in hardware.

The main idea of test synthesis is to optimize a circuit, in first place, for testability. Performance and area are still very important constraints which have to be fulfilled but they play only a secondary role. Test synthesis can be performed on different levels. In this chapter, we discuss different aspects of high-level test synthesis, which combines high-level synthesis with design for testability considerations.

9.1   Digital System Testing

Digital systems are tested by applying a sequence of tests to their inputs and observing their outputs. The directly accessible input and output ports are called *primary inputs/outputs* to distinguish them from component inputs/ outputs which are usually not directly accessible by the test equipment. Based on the output signals the decision is made if the circuit is correct or not. The basic test procedure depends very much on the applied method. For example, I_{ddq} testing does not require observation of output ports; instead it monitors the power supply current. BIST methods compress the outputs and provide only the test outcome in the form of a signature. All methods require however input patterns generated based on a given fault model. Here, we consider only the stuck-at fault model [ABF90] which captures also many other faults.

As mentioned, the test generation problem is *NP*-complete even for combinational circuits. To improve test generation efficiency and test coverage factor a number of design for testability methods is used in practice. They provide means to modify a design so that test generation and/or test application become easier and higher test coverage can be achieved.

Test generation and basic design for testability methods will be discussed in the next two sections.

9.1.1   Test Pattern Generation

Test pattern generation is usually based on a fault model. The most commonly used model is the single *stuck-at fault model*. It assumes that only a single line in a circuit is faulty due to that it is permanently connected to 0 (stuck-at-0 fault) or 1 (stuck-at-1 fault). This is of course a simplified assumption but it has been shown that if we can detect a large percentage of the single stuck-at faults, we will usually get a test set that detects a correspondingly high percentage of all failures [WiBr81].

Test pattern generation tries to generate as small as possible set of test vectors while maintaining a high test quality. A measure of test quality is the *fault coverage* which is defined as the number of faults detected by a given test set divided by the total number of faults. If the fault coverage is high this means that the test set is able to detect most of the faults in the circuit. In practice, we would like to achieve a fault coverage which is close to 100%. In some cases, a *test efficiency* measure is also used. It is defined as the number of detected faults plus the number of untestable faults divided by the total number of faults.

The test pattern generation procedure for combinational circuits with the single stuck-at fault model is divided into two phases [KiMe88]: *fault sensitization* and *fault propagation*. The fault sensitization phase finds an input pattern that produces a value 1 (0) at the faulty line in case of stuck-at-0 (stuck-at-1) fault. The fault is sensitized since a correct circuit and the faulty circuit

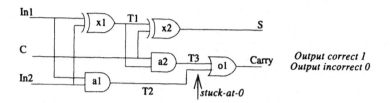

Figure 9.3: Test generation example for one-bit adder.

will have different values on the faulty line. This difference on the faulty line is then propagated to the primary outputs by the fault propagation procedure. This procedure has to find an input pattern which will produce different values at the primary outputs for the correct and faulty circuits. In this way, the fault is observed at the primary output and therefore detected.

Example 9.1: Consider the one-bit adder depicted in Figure 9.3. Assume that a test for the stuck-at-0 at line T2 has to be generated. The fault sensitization procedure should generate a test pattern which sets line T2 to 1. Therefore both inputs to the and gate a1, In1 and In2, are set to 1. The fault propagation needs to propagate the faults through the gate o1. This can be done by setting the line T3 to 0 since in this case we get value 0 at the primary output Carry for the faulty circuit and 1 for the correct one. To set line T3 to zero we need to set one of the inputs of the and gate a2 to 0 which is already done because the xor gate x1 produces 0 as output. Thus the generated test pattern is In1=1, In2=1. In this case, the value of primary input C does not influence the test and can be set to either 0 or 1.

□

In general, the test generation procedure is complicated because of conflicts caused by reconvergent fanouts. A reconvergent fanout refers to different paths started from the same line reconverging at the same component. In Example 9.1 the fanout point at input line In2 reconverges at gate o1 but it does not cause a conflict in our example. A conflict can force an ATPG algorithm to backtrack and reexamine untried alternatives. It can use another line for the propagation phase (for example, C) and/or a repetition of the previous sensitization phase with new values which may not cause conflicts.

Example 9.2: Consider the circuit depicted in Figure 9.4. To test the stuck-at-1 fault at the input B of the gate a1 we need to set this input to 0 for sensitization phase. The propagation phase requires, however, that both inputs of the gate o2 are set to 0 and thus both inputs to the gate o1 have to be set to 0 as well. This can not be achieved due to the conflict caused by the input B which in the propagation phase has to be set to 1 while during the sensitization phase to 0. In this example, there is no

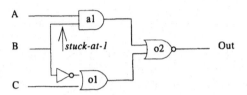

Figure 9.4: A circuit with a reconvergent fanout which causes a conflict during test generation.

alternative solution to set the output of gate o1 to 0 and thus the conflict can not be resolved. The fault is undetectable.

□

Several algorithms have been developed along the line of the above described method for test generation for combinational circuits [KiMe88]. For test generation of synchronous sequential circuits, called sequential test pattern generation, an extension is needed. This extension can be based on a modeling technique transforming a synchronous sequential circuit into an iterative combinational array [ABF90]. Each cell of this array is called a time frame. The combinational test generation algorithm can then be applied to the consecutive time frames and produces a sequence of input test vectors to detect faults.

The above described test generation methods assume a systematic way of test generation for a given fault model. Other methods are based on random test pattern generation. They simply generate random test patterns and then perform fault simulation to find which faults are covered by the generated test vectors. Since these methods can not achieve high fault coverage, they are sometimes combined with the systematic methods. For example, a random test pattern generation can be done first and later tests for the remaining faults are generated using the previously described test generation methods.

9.1.2   Design for Testability

Design for testability means design activities which modify the design in such a way that its final implementation is testable. A circuit is testable if the related test vectors can be generated in a reasonable time and the fault coverage is high. There exist a collection of techniques, rules and guidelines which are employed to ensure that a circuit is testable. They are grouped into two categories. The first category comprises *ad hoc* techniques. These techniques are usually applicable to solve only some specific problems of a design. Examples of techniques belonging to this category are test points and partitioning. The second category of DFT methods are *structured* techniques which aim to solve general testing problems with a systematic design methodology. Rules defined for such techniques are generally applicable. Scan path and

Built-In Self-Test (BIST) techniques are examples of structured methods.

The main idea of DFT techniques is to enhance the *controllability* and *observability* of selected lines in a design. The controllability indicates the relative difficulty of setting a particular line to a specific value from the primary inputs and observability indicates the relative difficulty of propagating a value assigned to a particular line to a primary output. Thus, they reflect the easiness of the two phases of test generation, namely fault sensitization and fault propagation. Methods have been developed for estimation of the controllability and observability at gate level (see for example, [Gol79]) and RT level (see for example, [KuPe90]). These estimations can directly be used by test synthesis tools to guide the selection of design parts to be transformed by selected testability enhancement methods. The detailed discussion of testability analysis at RT level is presented in Subsection 9.2.1 while Subsection 9.2.2 presents a method for the selection of testability enhancement transformations in a transformational approach to high-level synthesis.

In the rest of this section, several well known ad hoc and structured testability enhancement methods will be introduced.

Test Points

The test point technique enhances testability of a line by making it accessible from the primary inputs and/or outputs. Figure 9.5 depicts the method for improving both controllability and observability of a line between components C_1 and C_2. It uses one additional output and two input ports as well as two gates; one and and one or gate. In a normal mode $CP0=1$ and $CP1=0$. The combination $CP0=0$ and $CP1=0$ enforces 0 on the input to component C_2 and the $CP1=1$ will set input of component C_2 to 1. OP is an observation point.

The method is very simple. However, if the number of internal lines which need to be controlled and/or observed grows the overhead in the number of input/output pins can be prohibitive for this method. This difficulty can be alleviated a little bit by using addressing of test points and multiplexing observation points [ABF90]. The advantage of this method, compared to register scan techniques, is that it can be used to control/observe any line of a circuit.

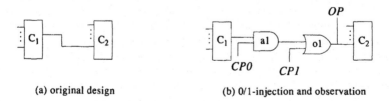

(a) original design   (b) 0/1-injection and observation

Figure 9.5: An example of the test point technique.

Scan Path

The main problem of test pattern generation for sequential logic is to set and observe the state of registers. DFT methods which employ scan registers solve this problem by reducing the sequential test pattern generation problem to a combinational one. This is done by using scan registers which operate in two basic modes: a normal functional mode and a test mode [WiPa83, McC86]. In the normal functional mode, a scan register acts as a normal register while in the test mode all scan registers are interconnected to form one shift register. This makes it possible to shift in and shift out test data. Since one shift register can be created from all registers of the design the technique requires only one extra input and output pin plus some additional control pins. An example of the scan path technique is depicted schematically in Figure 9.6. Registers R0, R1, R2, and R3 which are the only registers of the design depicted in Figure 9.6a are included in the scan path in Figure 9.6b forming a full scan path for the design.

The test procedure with scan registers requires first setting a circuit into a test mode and shifting test patterns into the scan registers. Then the circuit is set into normal mode for, normally, one clock cycle, during which the combinational logic produces output data which are stored into the registers. The output data can then be shifted out from the circuit by setting the circuit into the test mode again.

There exist many types of scan path designs depending on their organization and design of internal flip-flops or latches. Instead of using a single scan path one can, for example, use a multiple scan path or a random access scan path [ABF90]. These scan organizations offer additional capabilities for addressing registers as well as different scan paths. A standard for boundary scan paths [IEEE90] has been developed with the primary goal for interconnection testing at the board level. Designers can use a full scan to include all registers into the

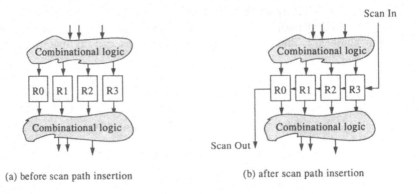

(a) before scan path insertion   (b) after scan path insertion

Figure 9.6: An example of the scan path technique.

scan path or they can use a partial scan which includes only selected registers into the scan path.

BIST

BIST is a DFT technique which requires neither ATPG tool nor external test equipment [AKS93a, AKS93b]. Test pattern generation and test application are accomplished by the built-in hardware and thus this technique is efficient. With BIST, a circuit can also be tested at its own speed which is not possible with, for example, the scan path technique.

The basic BIST architecture does not require expensive test equipment or large test generation efforts but it requires additional hardware blocks to be included in the design. First of all, a *test generator* is needed. It can be implemented as a linear feed-back shift register (LFSR). This register works as an ordinary register in the normal mode and in the test mode it generates test patterns to be fed to the logic connected to its outputs. Second, a *response analyzer* has to be selected. It receives test responses from the tested logic and compacts them into a short signature. The response analyzer usually uses a multiple input signature register (MISR) which is a slight modification of an LFSR. Finally, a *controller* has to be implemented. It activates the test and analyzes the test results.

Typical testing using BIST architecture is quite simple. The controller activates the test generator and the generated tests are fed into the circuit under test. After a number of execution cycles an output is produced. It is compacted by a response analyzer and a *signature* based on a set of outputs is created. The signature, which is a single word, is then compared to the expected signature and a decision, whether the circuit is considered to be correct or not, is made. It has to be noted that there is no known method which can ensure that the signatures for a correct and incorrect circuit are always different. A correct signature can still be produced even for a faulty circuit. This situation is called *error masking* and a faulty response is said to be an *alias* of the correct response. Techniques to alleviate error masking are currently being developed by many research groups.

9.2 High-Level Test Synthesis

The primary goal of high-level test synthesis is to improve the testability of a design while keeping performance and area within given constraints. The synthesis is usually carried out, in this case, by DFT specific transformations together with traditional high-level synthesis methods. This differs from traditional DFT methods where special test hardware is added or a test architecture is selected after that the design has been synthesized.

Test synthesis uses testability measures as one factor of the cost function to guide the synthesis process. The synthesis tries to find a good trade-off among the design testability, performance and area. The testability measures are usually estimated by heuristic algorithms. Heuristics are used since accurate methods to calculate testability are not only too expensive but in many cases even impossible to define when the final implementation at the gate level is not yet known.

The above discussion addresses three main aspects of a high-level test synthesis systems when a transformational approach is used: testability analysis, DFT transformations, and the general framework for transformation selection based on optimization of a cost function which takes testability into account. While the optimization of the cost function and the selection of trans-formations are similar to the corresponding activities in high-level synthesis, the testability analysis and DFT transformations are new.

In this chapter, we will present a high-level test synthesis approach using the methods developed in [GKP91, GKP92, GKP94, GKP95, Gu 96]. Readers can refer, for example, to [AvMc94] to see other high-level test synthesis systems. We will first introduce a testability analysis method which is based on quanti-tative testability measures and a testability calculation algorithm. Having the testability measures defined it is possible to identify hard-to-test spots in a design and improve their testability by applying well known DFT techniques.

An important characteristic of this approach is that testability is estimated early in the design flow (at system or RT level) and decisions regarding its improvement are made. These decisions concern usually a small part of the design which have been identified as hard-to test.

In the next subsections, we will introduce definitions of the testability measures and algorithms for their calculation. Then the testability improvement transformations will be discussed.

9.2.1   Testability Analysis

The testability definition assumes that the stuck-at fault model and a random and/or deterministic ATPG tool are used. These assumptions are justified, since the stuck-at fault model is the mostly used fault model and many ATPGs start by using random test generation to cover as many faults as possible and then switch to deterministic test generation to generate tests for the remaining faults.

Testability Measures

The testability is defined by the measures of controllability and observability. As in other approaches, controllability measures the cost of setting up any specific value on a line. Observability, on the other hand, measures the cost of observing any specific value on a line. The term cost reflects:

1. the costs of the CPU time used to find input vectors to set up a value on an internal line or to distinguishably observe a value on an internal line through primary outputs;
2. the cost of achieving high fault coverage; and
3. the time to control or observe this fault during test application.

Both the controllability and the observability are further defined by two factors: the combinational factor and the sequential factor. The combinational factor is used to measure the cost of generating a test and achieving a reasonable fault coverage. The sequential factor is used to measure the complexity of repeatedly using a combinational test generation algorithm to a sequential circuit, and the cost (time and memory) of executing a test. As a result, we have four measures for testability; combinational controllability (CC), sequential controllability (SC), combinational observability (CO), and sequential observability (SO).

The combinational factors CC and CO range from 0 to 1, where value 1 represents the best combinational controllability (or observability). The CCs (COs) of all primary inputs (outputs) are equal to 1. The value 0 of CC (CO) represents that we are unable to set up (observe) that line. The sequential factors SC and SO are natural numbers which represent 1) the estimated number of steps that a combinational test pattern generation algorithm needs to be repeatedly applied to a sequential circuit, and 2) the number of clock cycles to control or observe a line in the circuit under test. The SC's (SO's) of all primary inputs (outputs) are equal to 0.

A set of heuristics can be developed to find the controllability (observability) at a functional unit's output (input) based on the controllability (observability) at its inputs (outputs). The relationship between the controllability (observability) at a unit's input and output depends on the controllability transfer factor CTF (observability transfer factor OTF) of a unit.

The CTF reflects the probability of setting a value at the output of a unit by randomly exercising its inputs. The OTF reflects the probability of observing the inputs of a unit by randomly exciting its other inputs and observing its outputs. These two factors reflect not only the difficulties of random test pattern generation but also the difficulties of deterministic test pattern generation.

Deterministic ATPG tools usually deal with the computational complexity of the test pattern generation algorithms by using two parameters controlled by the designer. The first one is the maximum CPU time spent by the algorithm to find a test pattern for a given fault. If the test is not found for this fault in the defined CPU time the fault is classified as untestable. The second parameter defines a maximum number of backtrack steps for each vector. This controls the number of alternative paths which are checked during the test generation procedure before the fault is classified as untestable. Faults that are hard to detect will more often cause conflicts between the vectors to control them and the vectors to observe them. Therefore, backtrack steps are necessary, which means that the probability of detecting the faults becomes smaller within a limited CPU time.

Both *CTF* and *OTF* are in the range from 0 to 1, where 1 represents the best controllability and observability transfer factors of a unit. Different types of units have different *CTF*s and *OTF*s. For each type of unit, the *CTF* and *OTF* are calculated only once and stored in a library. In the following, the *CTF* and *OTF* are formally defined.

Definition 9.1 The *Controllability Transfer Factor* for a unit U (CTF_U) is defined by the following formula:

$$CTF_U = \frac{\frac{NI}{NO} - \left(\sum_k \left(\frac{NI}{NO} - NI(k)\right)\right) \cdot \frac{e}{NO} - e_1}{\frac{NI}{NO}} \cdot e_2^{NIL-1} \tag{1}$$

where

- *NI* is the number of possible input values of the component,
- *NO* is the number of possible output values of the component,
- *NI(k)* is the number of alternative input values controlling an output value k,
- *k* is an output value that satisfies $0 < NI(k) < \frac{NI}{NO}$,
- *NIL* is the total number of input lines (in bits),
- *e* (e>0) is a constant which makes the sum proportional to other parts of the expression,
- e_1 is a constant used to reflect the probability of setting a value on an output ($e_1 > 0$),
- e_2 is a constant used to reflect the cost of ATPG in searching for setting values for lines during sensitization and the cost of backtracking when conflict occurs.

In Definition 9.1 *NI/NO* represents the average number of input values which can control the same output value. The sum part is the deviation of the number of input values controlling an output value *k* from the average number of input values controlling the same output value. Only those values whose number is less than the average number but greater than zero are considered. This factor reflects the limitation of a component of controlling some outputs from its inputs. The fraction $e_1/(NI/NO)$ represents the probability of setting an output value by randomly exercise the inputs of a component. The more alternative inputs controlling the same output, the larger the probability of controlling that output and the value *NI/NO* is larger while the fraction $e_1/(NI/NO)$ becomes smaller. Thus, the CTF_U becomes larger since the fraction $e_1/(NI/NO)$ is subtracted from the other terms in the formula (1). The factor e_2^{NIL-1} reflects the cost of test generation in searching several options of setting a line. The more input bits that need to be set up to control an output, the more lines have to be set when sensitizing a fault during ATPG and also more backtracking is needed if a preliminary setting causes a conflict.

Intuitively CTF_U measures the difficulty of setting any value on the output by applying values on the inputs. Since every output value can usually be set by applying several possible values on its inputs, the sum part of the formula reflects a distribution of input/output controllability. If the number of patterns which control every output is the same for all output values than the sum is the lowest and the CTF_U is highest. The basic formula is then multiplied by a factor which is used to represent the difficulty of finding a test by applying random test pattern generation and the cost of searching for a test during ATPG.

Definition 9.2 The *Observability Transfer Factor (OTF)* for an input X_i of a unit is defined by the following formula:

$$OTF_{X_i} = \frac{NAO_{X_i} - \left(\sum_k (NAO_{X_i} - NAO_{X_i}(k))\right) \cdot \frac{e}{NI_{X_i}} - e_1}{NAO_{X_i}} \cdot e_2^{NLC_{X_i} - 1} \cdot e_3^{NLO_{X_i} - 1} \quad (2)$$

where

- NAO_{X_i} is the average number of alternatives to observe a value on input line X_i distinguishably from the faulty one through the outputs of the component,
- $NAO_{X_i}(k)$ is the total number of alternatives to observe a value k on input line X_i distinguishably from the faulty one through the outputs of the component,
- k is an input value on line X_i that satisfies $0 < NAO_{X_i}(k) < NAO_{X_i}$,
- NI_{X_i} is the a number of input values that might appear on line X_i,
- e ($e>0$) is a constant which makes the sum proportional to other parts of the expression,
- e_1 ($e_1>0$) is a constant used to reflect the probability of observing a value on an input,
- e_2 ($e_2>0$) is a constant used to reflect the cost of ATPG in searching for setting values for lines during the sensitization phase and the cost of backtracking when a conflict occurs,
- e_3 ($e_3>0$) is a constant used to reflect the cost of ATPG in searching for setting values for lines during the propagation phase,
- NLC_{X_i} is the total number of input lines (in bits) that needs to be controlled for the observation,
- NLO_{X_i} is the number of output lines (in bits) that need to be observed.

The intuition behind this formula is similar to that behind the controllability transfer factor, discussed above. If the distribution of observable values at the output of the component is uniform the observability transfer factor is higher, i.e., all values are equally easy to be observed. The basic formula is then multiplied by two factors representing the difficulty of ATPG.

The example below explains how the *CTF* and *OTF* are calculated.

Example 9.3: Consider a 4-bit adder with two 4-bit inputs and one 5-bit output and assume that $e=0.1$, $e_1=0.1$, $e_2=0.99$, and $e_3=0.99$.

For calculation of CTF we have the following values: $NIL=8$ since there are two 4-bit inputs; $NI=2^8$ since there are together 8 input bits; the output values are in the range 0..30 and thus $NO=31$; and the number of all alternative input values controlling an output k ($0 < NI(k) < NI/NO$) is given below.

$NI(0)=1$ (0+0)
$NI(1)=2$ (0+1, and 1+0)
$NI(2)=3$ (0+2, 2+0, and 1+1)
$NI(3)=4$ (0+3, 3+0, 1+2, and 2+1)
$NI(4)=5$ (0+4, 4+0, 1+3, 3+1, and 2+2)
$NI(5)=6$ (0+5, 5+0, 1+4, 4+1, 2+3, and 3+2)
$NI(6)=7$ (0+6, 6+0, 1+5, 5+1, 2+4, 4+2, and 3+3)
$NI(7)=8$ (0+7, 7+0, 1+6, 6+1, 2+5, 5+2, 3+4, and 4+3)
$NI(23)=8$ (8+15, 15+8, 9+14, 14+9, 10+13, 13+10, 11+12, and 12+11)
$NI(24)=7$ (9+15, 15+9, 10+14, 14+10, 11+13, 13+11, and 12+12)
$NI(25)=6$ (10+15, 15+10, 11+14, 14+11, 12+13, and 13+12)
$NI(26)=5$ (11+15, 15+11, 12+14, 14+12, and 13+13)
$NI(27)=4$ (12+15, 15+12, 13+14, and 14+13)
$NI(28)=3$ (13+15, 15+13 and 14+14)
$NI(29)=2$ (14+15, and 15+14)
$NI(30)=1$ (15+15)

Using the formula (1) from the Definition 9.1 $CTF=0.8989$.

OTF can be calculated either for input A or input B of the adder, the calculations are however the same and thus we present them only for one input. To observe the value on the input line A it is needed to control the value on line B and observe the value on the output line. Therefore we have $NLC_A=4$ and $NLO_A=5$. The total number of input values that might appear on line A is 2^4. The number of alternatives to effectively observe any value on input line A is 16, since it is possible to distinguishably observe a value on this line with a determined value on another input line B. Thus we have $NAO_A=16$ and the sum part in the formula (2) is 0. The observability transfer factor for this adder is $OTF_A=0.9262$.

□

In the following, we will describe the heuristics for calculating controllability and observability measures for designs represented in ETPN. The controllability of a device is computed by propagating a related measure from its input ports to its output ports while the observability is computed by propagating the measure in the reverse direction. Both controllability and observability measures are computed differently for functional components and registers (sequential units in ETPN). Functional components are usually combinational units which execute basic operations and do not store data while registers do not execute any operation and store data only. Thus, a definition of controllability and observability will be done separately for functional units and sequential units. The combinational and sequential factors (CC, CO, SC, and SO) are computed for both component types.

For combinational units, data transfer from inputs to outputs is finished in the same control state, and combinational ATPG can directly be applied.

Definition 9.3 The *controllability of a combinational unit* which has inputs X_1, ..., X_p with n_1, ..., n_p bits respectively, and outputs Y_1, ..., Y_q with m_1, ..., m_q bits respectively, at each output Y_k is defined as

$$CC_{Y_k} = CC_{in} \cdot CTF_U \tag{3}$$

$$SC_{Y_k} = SC_{in} + \text{clk}(S_i), \qquad (1 \le k \le q) \tag{4}$$

where

$$CC_{in} = \frac{\sum_{i=1}^{p}(CC_{X_i} \cdot n_i)}{\sum_{i=1}^{p} n_i}, \tag{5}$$

$$SC_{in} = \frac{\sum_{i=1}^{p}(SC_{X_i} \cdot n_i)}{\sum_{i=1}^{p} n_i} \tag{6}$$

clk(S_i) is the number of clock cycles needed for data operation at a functional unit.

The formula defines controlability of the functional unit as the average CC at unit inputs (CC_{in}) reduced by the factor CTF_U. This reduction is caused by the data transfer through the unit. SC is, on the other hand, increased because of the delay for data operation. The delay is equal to the number of clock cycles required for the unit to finish its operation.

For sequential units, such as registers, we assume that S_i controls the input X and S_j the output Y of the register in the ETPN notation. The data transfers from input X to output Y are thus determined by the related state transitions (from S_i to S_j) in the control part. In this case, the computation of CC needs to involve a controllability estimation of the condition nodes responsible for the state transitions. The CTF of a sequential unit is one since the register does not change the value from its inputs to outputs.

Definition 9.4 The *controllability of a sequential unit* which has an input X and an output Y, at the output Y is defined as

$$CC_Y = CC_X \cdot CC_{cond(S_i, S_j)} \tag{7}$$

$$SC_Y = SC_X + clk(S_i, S_j) \tag{8}$$

where

$CC_{cond(Si, Sj)}$ is the product of the CCs of all condition nodes for state transitions from S_i to S_j. This reflects the cost of an ATPG to find an input that makes these conditions true.

clk(S_i, S_j) is the sum of clock cycles needed to make a state transitions from S_i to S_j.

If there are several paths from S_i to S_j, the shortest one is selected.

The combinational controllability of a sequential unit becomes worse because of the factor $CC_{cond(S_i, S_j)}$. It represents the difficulty of setting conditions which are needed for state transition from S_i to S_j and value transfer from the input X to the output Y. The sequential controllability is increased by the number of clock cycles needed for state transition from S_i to S_j.

The definition of observability has to take into account that to observe an input of a unit through an output line it is needed to control other input lines of this unit.

Definition 9.5 The *observability of a combinational unit* which, has inputs X_1, ..., X_p with n_1, ..., n_p bits respectively, and outputs Y_1, ..., Y_q with m_1, ..., m_q bits respectively, at each input line X_k is defined as

$$CO_{X_k} = CO_{out} \cdot OTF_U \cdot CC_{in} \tag{9}$$

$$SO_{X_k} = SO_{out} + clk(S_i), \qquad (1 \leq k \leq q) \tag{10}$$

where

$$CO_{out} = \frac{\sum\limits_{i=1}^{q} (CO_{Y_i} \cdot m_i)}{\sum\limits_{i=1}^{q} m_i}, \tag{11}$$

$$SO_{out} = \frac{\sum\limits_{i=1}^{q} (SO_{Y_i} \cdot m_i)}{\sum\limits_{i=1}^{q} m_i} \tag{12}$$

CC_{in} is the average controllability (without CC_{Xk}) as defined in Definition 9.3.

The combinational observability is made worse because of two reasons. First, it is reduced by the observability transfer factor of the component and second it becomes lower because of the combinational controllability of setting other inputs of the unit. The setting of the other inputs is necessary since to observe a value on a given input requires setting of the other inputs to an appropriate value (fault propagation). The sequential observability is similar to the corresponding controllability factor.

The observability calculation of sequential units is similar to controllability calculation for these units. Controllabilities of the condition nodes are responsible for the state transition between S_i and S_j. They are taken into account as the CC factor of the formula. The OTF of a sequential unit is one.

Definition 9.6 The *observability of a sequential unit* which has an input X and an output Y, at the output Y is defined as

$$CO_X = CO_Y \cdot CC_{cond(S_i, S_j)} \tag{13}$$

$$SO_X = SO_Y + \text{clk}(S_i, S_j) \tag{14}$$

<u>Testability Analysis Algorithm</u>

The testability analysis algorithm, to be presented, is defined at RT level for designs modeled using the ETPN design representation (see Section 6.1). It calculates the testability $\{CC, SC, CO, SO\}$ for each line in a design. The controllability calculation algorithm first assigns ones to CCs and zeros to SCs for all primary inputs in the data path of the ETPN. These values are then propagated until the primary outputs are reached. A similar approach is used in calculating observability, but it starts from primary outputs towards inputs. A simplified version of the algorithm for CC and SC calculation is presented in the Figure 9.7.

In the algorithm, $\text{clk}(S_i)$, $\text{clk}(S_i, S_j)$ and $CC_{cond(Si, Sj)}$ are the same as defined

for all primary inputs **do**
   $CC_{prim\_in} = 1; SC_{prim\_in} = 0;$
repeat
   select a next unit U;
   -- calculate controllability at a unit U's outputs by
   -- using the average controllability calculated at its inputs
   if U is a combinational unit **then**
      $CC_{out} = CC_{in} * CTF_U; SC_{out} = SC_{in} + \text{clk}(S_i);$
   if U is a sequential unit **then**
      $CC_{out} = CC_{in} * CC_{cond(Si, Sj)}; SC_{out} = SC_{in} + \text{clk}(S_i, S_j);$
   if U is involved in a feedback loop **then**
      $CC_{out} = CC_{out} * CC_{loop}; SC_{out} = SC_{out} + SC_{loop};$
until all primary outputs are reached;

Figure 9.7: Controllability calculation algorithm.

previously. CC_{loop} and SC_{loop} are the combinational controllability and sequential controllability at the condition node that controls the termination of a feedback loop.

It can be noted that controllability decreases with the increase of depth, in the data path, from primary inputs and observability decreases with the increase of depths from primary outputs.

Traditional testability calculation for feedback loops in a data path usually requires long iterative computations [GoTh80]. To simplify this problem, the controllability at the condition node that controls the termination of a loop is used (CC_{loop} and SC_{loop}) in the approach described here.

Generally speaking, the controllability of all fanout branches X_1, ..., X_n are the same as the controllability of an original line X, and the observability of X should be the best of X_1, ..., X_n. However, if there is a redundant fault caused by a reconvergent fanout, some fanout branches cannot be controlled and observed. Therefore, checking for redundant faults is required at fanout points. To reduce the computational complexity which is *NP*-complete, heuristics can be used as, for example, in [GKP91].

Example 9.4: Consider the ETPN design represented in Figure 9.8. The testability analysis results are presented in Table 9.1. The *CTF* and *OTF* have been calculated for the 8-bit adder and greater-or-equal comparator using $e=0.1$, $e_1=0.1$, $e_2=0.99$ and $e_3=0.99$. Their values are the following:

$$CTF_+ = 0.8380 \qquad\qquad OTF_+ = 0.8684$$
$$CTF_\geq = 0.8599 \qquad\qquad OTF_\geq = 0.6425$$

The controllability of lines L_1, L_2 and L_{11} is not calculated since these lines are connected to constants and are not controllable. The best controllability (both $CC=1.00$ and $SC=0$) has the line L_4 which is

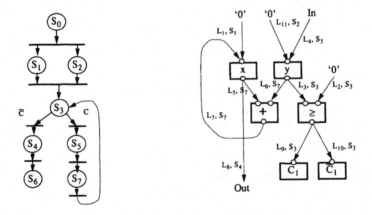

Figure 9.8: Testability analysis example.

directly connected to an input *In*. When the distance from the primary input gets longer the controllability becomes worse. A special case is the loop with line L_5 which has the worst controllability (CC=0.62 and SC=6). Similar reasoning can be applied to observability measures; the best observability have lines which are close to primary outputs, such as line L_8.

□

Table 9.1: Testability analysis results for the Example 9.4.

Line	CC	SC	CO	SO
L_1	-	-	0.86	2
L_2	-	-	0.55	1
L_3	1.00	2	0.55	1
L_4	1.00	0	0.60	3
L_5	0.62	6	0.86	2
L_6	1.00	1	0.60	2
L_7	0.72	3	0.86	2
L_8	0.62	5	1.00	0
L_9	0.86	2	0.86	1
L_{10}	0.86	2	0.86	1
L_{11}	-	-	0.52	5

9.2.2   Testability Improvement Transformations

The described testability analysis algorithm identifies lines of a design which are difficult to test. These lines can then be associated with parts of the design which should be selected for transformation by different testability improvement techniques during high-level or RT level synthesis. The selection of parts to be transformed and appropriate transformations can be performed by a designer after examining the testability analysis results. This can also be incorporated into synthesis tools where transformation selection is based on optimization criteria. In this section, three transformation techniques will be presented. They are based on the previously presented DFT methods: scan paths, test points and BIST.

Partial scan insertion

A popular and relatively simple method to improve testability of a design is to transform all registers into scan registers, which leads to the full scan path. This, in many situations, introduces unnecessary overhead in terms of design area. The problem can be solved by partial scan path which selects only a subset of all registers to be included in the scan path. The register selection is, however, not straightforward and methods for determining an optimal subset of registers are needed. These methods can directly make use of the testability measures, defined in the previous section, and some heuristics.

The testability analysis algorithm described above assigns a 4-tuple $\{CC, SC, CO, SO\}$ to every line in the design. A line may have a very good controllability but poor observability and thus it is difficult to select a register for partial scan inclusion directly on these measures. There is a need to create a single testability measure for a register which can later be used for selection of candidates for the partial scan. Such a single measure is defined in the following definition.

Definition 9.7   The *testability evaluation for a register* R_i is defined by the following formula:

$$E_{R_i} = \frac{C_{R_i} - \overline{C}}{\overline{C}} + k \cdot \frac{\overline{S} - S_{R_i}}{\overline{S}} \tag{15}$$

where
- $C_{R_i} = CC_{R_i} + CO_{R_i}$, where CC_{R_i} is the combinational controllability at the output of the register R_i and CO_{R_i} the combinational observability at the input of the register R_i,
- $S_{R_i} = SC_{R_i} + SO_{R_i}$, where SC_{R_i} is the sequential controllability at the output of the register R_i and SO_{R_i} the sequential observability at the input of the register R_i,
- $\overline{C}$ is the average of all C_{R_i} in the design,
- $\overline{S}$ is the average of all S_{R_i} in the design,
- $k = \overline{C}/\overline{S}$, which scales the combinational and sequential factors to the same level.

The testability of a register is good if the testability evaluation of a register is large. Thus, to improve testability, the registers with the lowest E_{R_i} should be selected first.

The register selection strategy for a scan path is defined by the following three steps:
1. select registers which are in a data path loop; the selected registers must be in different loops,
2. if the testability still does not satisfy the given requirements select a number of registers with the worst testability using formula (15),
3. re-calculate testability measures and repeat step 2 if necessary.

The first step of the selection procedure is based on a heuristic which assumes that the registers located in data path loops are responsible for test generation problems. This is well known and similar rules are very often used by engineers. Then the testability measures are used to select registers with the worst testability factors. This step is repeated until acceptable results are obtained. After insertion of registers into a scan path, the testability measures are re-calculated, taking into account different testability factors for scan registers [Gu96].

The result of the register selection procedure is a partial scan chain which contains all the selected registers. If there exist no more registers to be selected the testability can still be further improved by the test point insertion strategy described below.

Test point insertion

To insert test points into a design to improve its testability, in an efficient way, a very small set of lines of the design should be selected. The line selection strategy can use a formula to evaluate relative difficulty of the lines and select the worse. The formula is similar to (15), introduced in Definition 9.7 for registers. We only need to change C_{Ri} and S_{Ri} for registers in the above formula to those for lines. This strategy suggests the insertion of a special kind of a cell, called T-cell, in the line with the worst testability. The T-cell is a scan-like element designed to enhance controllability and observability at any line in a design [Rac93]. This strategy selects only one line for transformation each time because of testability dependencies between relevant lines. Testabilities at these relevant lines must be re-calculated after each transformation.

Partitioning

The general idea of a testability guided circuit partitioning is to use DFT techniques to transform some hard-to-test registers and/or lines detected by the testability analysis algorithm to boundary components [GKP95, Gu96]. These components act as ordinary registers and lines in normal mode and serve as partitioning boundaries in test mode. Therefore, a design is partitioned into several disjoint sub-circuits and each of them can be tested independently through the boundary components including I/O ports.

The advantage of the testability guided partitioning approach is that it can accurately find hard-to-test parts and make them directly accessible. The complexity of the partitioning algorithm can also be significantly reduced due to the direct use of testability analysis results. Since each partition can be controlled independently in the test mode, it is, therefore, possible to apply different test strategies, such as scan path for deterministic test and BIST for random test, to the different partitions.

The main objective of data path partitioning is to use testability analysis results and other heuristics to find partitioning boundaries and isolate data communication among partitions in the test mode. This is done in two steps:
• selecting partitioning boundaries, and
• identifying partitions by clustering a set of components surrounded by the

partitioning boundaries.

Partitioning boundaries are formed by three types of components: transformed registers, test modules (transformed lines), and I/O ports.

Selection of registers and/or lines to be boundary components is similar to the scan register selection and test point insertion transformations, and is based on two heuristics. The first one tries to break feedback loops identified by the testability analysis algorithm. This is done by selection and transformation of the registers with the worst testability evaluation involved in feedback loops. The second heuristic selects registers or lines with the worst testability.

The next step, component clustering, actually performs the preliminary partitioning. It clusters directly interconnected components excluding boundary components. Figure 9.9 presents the basic structure of this part of the partitioning algorithm

begin

    call testability analysis algorithm;

    identify all boundary components $B_1, ..., B_s$ by applying the following rules:

        1) all primary inputs/outputs are boundary components;

        2) for each feedback loop in the data path select a register with the worst testability and classify it as a boundary component;

        3) select registers with the worst testability to boundary components;

    initialize input boundary component sets $B_1, ..., B_s$ by

        $B_i := \{B_i\};$

    initialize partitions $P_1, ..., P_s$ by assigning components C_k to partition P_i iff there exist a direct connection from B_i to C_k;

    for all $\{P_1, ..., P_s\}$ **do**

      repeat

        if component $C_i \in P_i$ has a direct connection to component C_k **then**

          begin

            $P_i := P_i \cup \{C_k\};$

            if $C_k \in P_j$ **then**

              begin

                $P_i := P_i \cup P_j;$

                $B_i := B_i \cup B_j;$

              end;

          end;

      until all $C_i \in P_i$ are connected to $C_k \in P_i$ or to a boundary component;

    end for;

end;

Figure 9.9: Preliminary partitioning algorithm - component clustering.

Three heuristics are then used to improve the preliminary partitioning. One idea is to transform a design into one which can use BIST techniques, i.e., boundary registers can be transformed into test generators and/or response analyzers. The first heuristic identifies a boundary component which is used as both the input and the output of the same partition. This partition needs to be further partitioned into two such that the boundary component is only used as the input of one partition and as the output of the other. The algorithm for this heuristic is presented in Figure 9.10.

Two other heuristics are used to improve test efficiency and quality. One tries to eliminate fanout points since they can cause redundant faults. This is done by further partitioning a design using an approach similar to the previously presented algorithm. The other one is specific for the BIST implementation and tries to further partition clusters which are random pattern resistant since this kind of resistant logic reduces fault coverage when pseudorandom patterns are used.

After applying the described partitioning procedure it is guaranteed that each partition is acyclic. Moreover, the size and depth of each partition are usually suitable for the ATPG tool to be used. And when BIST technique is used, we ensure that the partitioned design is random pattern resistant free.

> **Example 9.5:** Consider a design depicted in Figure 9.11. It contains two inputs, In1 and In2, and three outputs Out1, Out2 and Out3. They are part of the partitioning boundaries. In addition, the heuristic for selecting partitioning boundaries selects registers R1 and R2 to form the first partitioning. This results in two partitions P1 and P2. In this partitioning, the register R2 is still both an output and an input of partition P1. Thus the heuristic further partitions the design into three partitions (not shown in the figure) by inserting a test point T1. Finally, the rest of the partition P1 is partitioned into P1b and P1c by the heuristic which removes fanouts. The resulting circuit has four partitions of a similar size as depicted in Figure 9.11b.
>
> □

The data transfers and operations specified in the data path of an ETPN

begin

    create the set of boundary components, B_{inout}, which are both inputs and outputs
    to components in the same partition;

    for all $B_k \in B_{inout} \neq \emptyset$ **do**

        split partition P_i, which uses $B_k \in B_{inout}$, to P_{i1} and P_{i2} such that B_k is an output from
        one partition and an input to the other one;

    end for;

end;

Figure 9.10: Partitioning algorithm- the BIST heuristic.

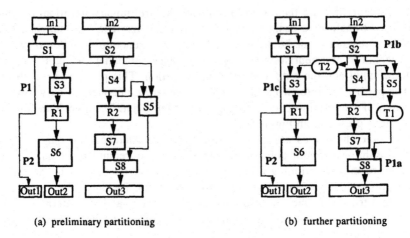

(a) preliminary partitioning   (b) further partitioning

Figure 9.11: Data path partitioning example.

representation are controlled by the control part of the ETPN. The data path partitioning only provides the facility to isolate data communications between partitions. We also need to modify the control part so that it works in two modes, *normal mode* and *test mode*. In the normal mode, it controls the whole design to perform the initially specified functions. In the test mode, it controls the execution of tests for each partition independently.

Assume that Z_i is a set of states in the control part controlling the data transfers in partition P_i. The objectives of control part modification are to: 1) construct a token-supplier to provide token(s) to initialize, in the test mode, the part that controls P_i, and 2) construct a token-consumer to terminate the control of P_i. If there is only one state in Z_i which needs to be initialized the token-supplier can simply be built by adding an additional transition controlled by $T \wedge Clk$, where T is a test signal and Clk is a clock signal. This situation is indicated in Figure 9.12. When the test signal T is set, the token-supplier will provide a token to the control part Z_i at the rising edge of the clock signal Clk.

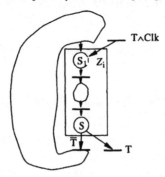

Figure 9.12: Control part modification for data path partitioning.

The token-consumer is created as an additional transition from the last state in Z_i controlled by the condition T. A complementary condition T provides a token for normal execution of the controller. If there are several states in Z_i which need to be initialized or there are branches, the token-supplier and token-consumer can be constructed using similar methods (see [GKP95]).

Low-Power Synthesis

Traditionally the main concern of a designer was performance and silicon area of the final product. Accordingly, research and development in the design automation community have focused on methods and tools which mainly optimize performance and/or area. Recently, power consumption has become an important design parameter. This shift in design emphasis is motivated by the increasingly important area of portable electronics and wireless communication. These systems are battery-operated and the goal is to keep the battery life long, while maintaining a reasonable weight. Since advanced battery technologies currently yield around 65 Wh/kilogram [Ped96], the only way to provide an acceptable operation time between recharges is to produce low power devices which can operate on the same batteries for a longer time.

There also exist other reasons for optimizing designs for low power. While current technologies make it possible to design more and more complex chips, some having tens of millions of transistors and running at several hundred MHz already, their power consumption is also increasing all the time. For example, the Alpha processor consumes 72 W at 600 MHz. The trend seems to continue. This kind of systems produces heat and requires additional cooling which is also power consuming. High temperature increases also the number of failures on a chip and should therefore be avoided because of reliability reasons. Finally, hot chips require expensive packaging techniques.

As discussed earlier, synthesis is usually defined as an optimization problem. For example, we might want a design to provide a given throughput while minimizing area or to maximize throughput while meeting certain area constraints. This formulation is typical both for high-level synthesis and for logic synthesis. Unfortunately it does not take the power consumption into account and very often leads to power hungry designs. New methods and tools need to be developed to produce low power designs.

While doing synthesis for low power a number of assumptions is usually made. First, it is assumed that certain performance specifications are given. For example, in digital signal processing applications a certain throughput is imposed. Different synthesis trade-offs can be made as long as this throughput is guaranteed. Second, it is generally accepted that low power design may require a larger silicon area. How big an increase in the silicon area can be accepted depends very much on the specific application and constraints imposed by the market. Finally, low power synthesis usually requires additional information on the switching activity in the circuit, since lowering the switching activity is an efficient way to decrease power consumption. This information is traditionally not considered during synthesis.

10.1 Sources of Power Consumption and Its Reduction

There are three major sources of the power consumption in CMOS circuits: switching, short-circuits and leakage. While short-circuits and leakage can be made negligible if proper design techniques are used, switching is the main component responsible for power consumption. Switching is the effect of charging and discharging of the node capacitances in the circuit and it is given by the following formula [Ped96]:

$$P_{switching} = \frac{1}{2}C_L V_{dd}^2 E_{sw} f_{clk} \qquad (1)$$

where C_L is the physical capacitance at the output of the node, V_{dd} is the supply voltage, E_{sw} is the average number of output transitions per clock cycle (also called switching activities), and f_{clk} is the clock frequency.

According to the above formula reduction of one or several factors of the formula will result in a lower power consumption of the design. Obviously, the reduction of supply voltage (V_{dd}) will be the most attractive because it is in a quadratic relation to power. This has, however, a negative impact on the speed of the design because the reduction of the supply voltage increases the delay of the components and thus reduces the throughput of the design. This is a very undesirable effect since our general goal is to keep the predefined throughput unchanged while reducing power consumption.

There exist several methods to compensate the above described reduction of speed. First, it is possible to modify not only the supply voltage but also the threshold voltage of the device [CPRMB95b, LiSv93]. Reducing threshold voltage of the device together with the reduced supply voltage allows the device to be run at the same speed. Second, in the cases when the larger delay of the devices does not exceed the clock cycle limits it is possible to tolerate it. Moreover it has been shown that while power consumption decreases quadratically the speed decreases linearly [CPRMB95b, MaKn95]. Thus the throughput penalty can be relatively small comparing to power savings. Finally, the reduced speed can be, in some cases, compensated by architectural transformations; an efficient use of parallelism and pipelining can speed up the design to meet the original throughput requirements. For example, duplication of a module makes it possible to run it at half speed while retaining the same throughput. The first two solutions represent typical power consumption minimization methods which are technology dependent while the last one combines a technology dependent method of supply voltage reduction with an architectural solution that increases the parallelism of the design.

The reduction of the clock frequency (f_{clk}), without other transformations, directly produces a slower design. This however can be combined, in some cases, with an architectural transformation which duplicates a module and makes it possible to run the design at half speed (more discussion on module duplication can be found on page 339). Thus it is possible to reduce the clock frequency by half without any impact on the throughput of the design which gives the possibility to reduce the V_{dd}.

The physical capacitance offers another opportunity to reduce power consumption. It can be achieved at all levels of the design in the physical and, to some extend, the structural domain. In the physical domain, there are known methods of transistor sizing and minimizing interconnect routing length which minimize physical capacitance. It is also well known that the logic style and the circuit topology have an impact on physical capacitance and thus on the power consumption. In the structural domain at the logic level, it has been observed that the choice of a particular implementation for a given function can lead to different values of the physical capacitance. For example, an adder can be implemented using either ripple-carry or carry-lookahead approach. The carry-lookahead adder is the best solution if speed-capacitance trade-offs [CPMRB95] are taken into account.

Finally, reducing the switching activity offers a very attractive way to lower the power consumption. The switching activity is caused by two components: static and dynamic. The static activity is a function of the circuit's structure and signal statistics. It does not depend on the timing behavior of the circuit. The dynamic component, on the other hand, depends on the timing behavior of the circuit and accounts for spurious transitions, also called glitches. The glitches are caused by timing skew between signals which means that a module can have several signal transitions before setting to a correct stable value.

In principle, the static switching activity of the device depends on the consecutive patterns applied to the device inputs. The power consumption will be higher if more bits in the input pattern switch between two consecutive clock cycles. Thus, the general optimization rule for reducing the switching activity in the circuit is to reduce the Hamming distance between the two input patterns. This can be achieved in different ways. Selection of the number representation

Chapter 10: Low-Power Synthesis

is a well known method. It has been shown, for example, that the two's complement representation of integer numbers works worse than sign-magnitude representation in respect to power consumption. This is caused by the negative numbers representation in two's complement numbers that use sign-extension which produces unnecessary switching activity in the situation when the values oscillate around zero.

Switching activity can also be reduced when proper decisions are made on resource sharing. For example, the decision on sharing a register should be made based on careful statistical analysis of input data patterns for both registers [MaKn95, ChPe95a]. The selection of variables which can be assigned to the same register should be made based on predicted minimal switching activity for this register. The same method can be applied for functional units assignment.

Spurious transitions (glitches) cause extra power consumption over that strictly required to perform a computation. The number of glitches is a function of input patterns, internal state assignment in the logic design, delay skew, and logic depth. Careful logic design can eliminate these transitions. At the high-level design it is possible to reduce this factor by reducing the depth of the combinational circuits, for example.

Power consumption can be optimized at different levels as well as in different design domains as indicated by the above discussion. In this book, we concentrate on the system and high level transformations and thus we will not discuss physical or logic level methods to reduce power consumption. We assume that basic module optimizations are done separately during component design and that a number of implementation solutions is available in the component library. Different modules, implementing the same functionality, are characterized by their power consumption figures in addition to traditionally provided data on area and delay. A detailed discussion on different aspects of high-level synthesis for low power can be found in Section 10.3.

10.2 Power Estimation

Power estimation is a crucial part of any low power design activities since it makes it possible to predict the power consumption early in the design process. Having this information, it is possible to optimize power consumption of the design, either manually or automatically. Power estimation can be done on the circuit, logic or RT level [Ped96, Naj94].

There exists an important difference between power estimation and many other design estimations. The power consumption formula (1) contains the switching activity factor. This factor can not be estimated based on the design characteristics, as in the case of other factors of the formula. The switching activity depends on the input data patterns and the correlation between them. Other factors can be estimated based on the technology data or the circuit architecture. Thus, the main efforts at power estimation are usually directed toward efficient techniques to predict switching activities of the design. For this purpose, two main approaches can be used: *simulation based* and *non-simulation (analysis) based*.

Simulation-Based Techniques

The earliest methods for power estimation used simulation of the circuits while monitoring power consumption. The selection of input data patterns influences both the simulation and estimation time as well as the accuracy of this technique. Usually patterns are selected to represent the typical applications run on the circuit so that the estimation can be quite accurate. The main drawback is a long execution time of the simulation since a long simulation with a long sequence of input data patterns is usually needed.

To overcome the main drawback of this classical approach two solutions can be used. First, we can use a higher-level simulator. For example, instead of using a circuit-level simulator we can use a logic or RT level simulator. This however suffers from reduced accuracy of the estimation. The other method is to use shorter but representative input data pattern sequences and this is called the *statistical sampling* technique. This method is applicable, in practice, only to combinational circuits.

In the statistical sampling technique, the circuit is simulated using randomly generated input data patterns and the power consumption is monitored. In general, the power consumption converges to an average value when the number of patterns is infinite. In this method, the simulation is topped earlier to obtain an estimation of the average power consumption which is close enough to the true average value. This is based on statistical mean estimation techniques, essentially the Monté Carlo method. In practice, a sample size of 30-50 input patterns is enough for well-behaved combinational circuits [Ped96]. The method has also been extended to finite state machines [NGH95].

Non-Simulation Based Techniques

The non-simulation based techniques are essentially analytical techniques. The simplest but most convenient method is based on *parametrized module libraries*. In this method, the power consumption characteristics of a single module are stored in the library and the power consumption of the whole design is computed based on these characteristics. The power consumption figures are usually parametrized by, for example, bit length. Other analytical power consumption estimation techniques usually make use of stochastic and information-theoretic models to evaluate switching activity.

Non-simulation based techniques have been developed for both behavioral and logic level specifications. Thus these techniques can be directly used as estimators for high level or logic optimization.

The method based on a parametrized module library assumes that in addition to other parameters stored in the library for every module there is also the power consumption figure. This figure is obtained from other power consumption estimation methods, for example, using simulation based techniques on random input data patterns. To get good estimation accuracy the power consumption figure is usually parametrized or a formula is given to calculate the actual power consumption. In this case, a module may be parametrized by its world length, actual input switching activity or correlation between input pattern bits. The parametric model is presented, for example, in [SvLi94]. The work presented in [LaRa93], on the other hand, takes into account the correlation of data in the data path and requires input switching activity values.

The popular methods which use *stochastic models* are probabilistic techniques which calculate signal probabilities in the circuit based on the provided signal probabilities at primary inputs. In these methods, the signal probability is defined as the average fraction of clock cycles in which the steady state value of a given signal is high. The input signal probabilities are propagated into the circuit to obtain probabilities at every node. To perform this, special models of the basic modules are required. They must be developed and stored in the module library. The signal probabilities have to be supplied by the user. For input signals this can be done based on the knowledge of an application. Since the methods are mainly applicable to combinational circuits, the signal probabilities for inner latches outputs have to be provided as well. This is more complex and requires knowledge of the circuit.

An *information theoretic* approach relies on the entropy measure of a circuit's activity [MMP95, NeNa96]. Entropy is used in information theory as a measure of information-carrying capacity. If x is a random variable with probability p of being one then the entropy is defined as

$$H(x) = p\log_2\frac{1}{p} + (1-p)\log_2\frac{1}{(1-p)} \qquad (2)$$

A plot of $H(x)$ is depicted in Figure 10.1. Intuitively, the switching activity is maximum when the probability of a variable being 1 is 0.5 since in this case the probability of it being 0 is also 0.5. When this probability decreases or increases the switching activity decreases as well. Thus it can be concluded that the switching activity is related to the entropy of the signal. Indeed, it is shown that, under the temporal independence assumption, the average switching activity of a bit is upper-bounded by one half of its entropy.

The general power estimation method based on entropy calculation can be described as follows. First, the entropy for output signals is calculated based on the input signal entropy. Based on the calculated entropies an average entropy per signal is calculated and used to estimate the average switching activity of the circuit which is later used to estimate the power consumption of the circuit.

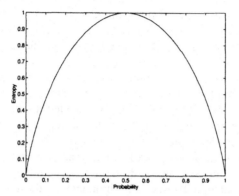

Figure 10.1: The entropy, $H(x)$, of a Boolean variable.

The method is directly applicable to combinational circuits. For other circuits, it has to be combined with estimations of latches and clock power [NeNa96].

10.3 High-Level Power Optimization

The general structure of a low power synthesis system based on a transformational approach is depicted in Figure 10.2. The design specification given in, for example, VHDL is first compiled into an internal design representation, such as ETPN or CDFG. Then the design is transformed by a high-level synthesis system into a final RT level design. This is done in the same way as in classical high-level synthesis but the main optimization emphasis is on power, not on area and performance. An optimization algorithm selects transformations based on their estimated effect on the design. This effect is provided by an

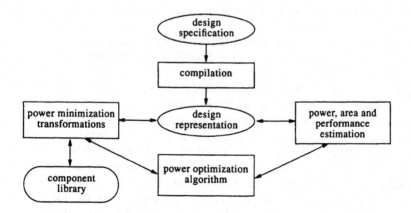

Figure 10.2: The general structure of a low power synthesis system.

estimator which predicts the power consumption of the design as well as its area and performance. The system optimizes power consumption while fulfilling performance and area constraints. In some cases, it optimizes several parameters by using an objective function which combines power consumption with other parameters.

The component library is an important part of such a design system. It contains different implementation solutions for storage elements, functional units, and buses. The library modules are needed to be characterized, in addition to performance and area, by their power consumption. There are usually several possible implementations of every module. For example, an addition can be implemented by a carry-lookahead or a ripple-carry adder. The selection can also be made between diferent supply voltages. These selections are made based on the optimization algorithm and the design criteria to be met.

10.3.1   Estimation

High-level power consumption estimation is an important part of high-level synthesis for low power. To minimize power consumption during synthesis, the optimization algorithm needs an estimation of the power consumption of the design in order to make optimization decisions. The optimization decision usually concerns the selection of a transformation which effects power consumption as well as performance and area. In this case, the power consumption estimation does not need to be very accurate but it needs to be consistent; the decrease/increase of the power consumption in the design has to be reflected by the proportional decrease/increase of the estimated power consumption value. The power consumption estimation has to be fast since it will be performed very often during the optimization. In principle, every time we apply a transformation to the design we have to estimate the power consumption. For a typical iterative improvement optimization algorithm, such as simulated annealing, genetic algorithms or tabu search, such estimations will be done thousands of times.

In Section 10.2, we discussed different methods for power consumption estimation. These methods can be used directly or indirectly here. The general approach is to use the parametrized module library together with other heuristics or analytical methods to get the power consumption for the whole design [CPMRB95, MeRa94, KKRV95, MaKn95].

A power consumption of a module, or more precisely switched capacitance, is obtained using previously discussed power estimation methods. Usually a simulation-based method is used and a module is simulated using a circuit level simulator like SPICE [MaKn95]. Since the power consumption depends on the switching activity on inputs of the module, the estimation can be different for different applications. It is normally assumed that a uniform white noise is used for all inputs for the simulation. More accurate models can characterize a power consumption of a module using signal statistics [LaRa93]. The switched capacitance is stored in the module library.

The power estimation for the whole design is made based on the internal design representation (such as ETPN or CDFG) and the module library only. The power consumption formula (1) can be rewritten to the following form which is more convenient at RT level [MeRa94]:

$$P_{switching} = N_{access}C_{average}V_{dd}^2 f_{clk} \tag{3}$$

where N_{access} is the number of accesses to the modules over the period of computation, $C_{average}$ is the average capacitance switched per one access, V_{dd} is the supply voltage and f_{clk} is the clock frequency. Using this formula the power consumption can be directly estimated if the number of accesses (N_{access}) to the different modules is known since the average capacitance ($C_{average}$) is available for every module in the library. The number of accesses can be obtained by, for example, profiling [KKRV95].

The average access parameter is estimated differently for different types of modules. Usually it is done separately for data path units, interconnections, controllers and the clock [MeRa94, KKRV95]. Furthermore, data path units are divided into functional units and registers as well as, for some architectures, bus buffers.

For example, the capacitance switched by the functional units (the factor $N_{access}C_{average}$) is estimated by multiplying the number of times a given functional unit is used during the period of computation by its capacitance as given by the following formula:

$$C_{fu} = \sum_{i=1}^{units} N_i C_i \tag{4}$$

where N_i is the number of times the functional unit i is accessed during the period of computation and C_i is the average capacitance of the functional unit i obtained from the library. Similar computations have to be made for other data path units.

The estimation of interconnections is more complex since the layout is not available and the average length of wires can not be determined. The known parameters which can be used for estimation are the number of interconnection buses and their width as well as needed multiplexers. Different estimations for interconnections can be found in [MeRa94, KKRV95].

Finally, the power consumption of the controller should be estimated. A simple method proposed in [CPRB95] estimates controller power consumption based on the number of states. The switched capacitance is proportional to this number. A more advanced method which takes into account other factors is presented in [MeRa94]. A detailed estimation of the controller power consumption is presented in [KKRV95]. The authors assume PLA implementation of the controller and estimate power consumption as a sum of power for the input plane, the output plane and the I/O buffers.

The reported average error of power estimation methods for high-level synthesis is usually around 10-20% [MeRa94, KKRV95].

10.3.2   Optimization

The main goal of a high-level power optimization system is to produce an RT level design which has a minimum power consumption while a required throughput is still achieved. In addition, area has to be minimized as well or, at least a certain area constraint can not be violated. The computational complexity of this power optimization problem is high (many problems which has to be solved during the power optimization, such as operation scheduling or module allocation and binding, are proved to be *NP*-complete) and optimization heuristic algorithms have to be used. Different approaches are described in the literature which use either specific heuristics or meta-heuristic algorithms. For example, the Leiserson-Saxe algorithm for critical path minimization using retiming has been used in [CPRB95] as one of the optimization algorithms and the genetic algorithm was used in [MaKn95]. Basically, any optimization algorithms presented in Chapter 5 can be used for power minimization. In addition, several existing specialized algorithms can be used for single transformations or parts of the optimization. For example, a Leiserson-Saxe algorithm can be used for critical path minimization using retiming.

10.3.3   High-level Synthesis Transformations for Low Power

High-level synthesis transformations for low power do not differ fundamentally from classical transformations used in high-level synthesis systems. Many high-level transformations can be directly applied here. The selection of transformations is however based on power minimization criteria. There are also specific power optimization transformations which are not present in classical high-level synthesis systems. For example, change of the supply voltage or voltages is such a transformation. Another example is a module binding transformation which takes into account switching statistics on the inputs of the modules. It decides on the sharing of the module based on the correlation between variables on the inputs of the considered modules.

The transformation selection does not necessarily mean that, at a certain step, the transformation with the best power estimation for the current design always has to be chosen. To avoid local minima and enable other subsequent power reduction transformations, in some cases, it is required to make the current design worse in terms of power consumption.

In the following sections, we will present typical transformations which are proposed for low power synthesis. This presentation is based on recent research in the area of high-level synthesis for low power (see for example [CPRMB95, ChBr95, MaKn95, ChPe95a, ChPe95b, Ped96]).

Architecture-driven voltage scaling

Power consumption reduction can be achieved efficiently by reducing the supply voltage. Reducing the voltage from 5 V to 3.3 V, for example, can achieve 60% of power consumption reduction. This technology-based approach can not be used alone, in most cases, since the reduction of supply voltage increases the delay of basic components, as discussed earlier. This reduction in speed has to be compensated by architectural solutions. Several solutions are presented in [CPMRB95b]. The main idea behind these solutions is to increase the parallelism of the design. Below we will discuss selected transformations which compensate speed reduction and make it possible to reduce the supply voltage.

Parallelism can be introduced in the data path by duplication of its resources. The computation can then be scheduled on the two identical paths each of which needs twice the original delay. Since there exist two parallel paths, the results are still produced with the original data rate.

Example 10.1: Consider the data path represented in Figure 10.3a [CPMRB95]. Inputs A and B are added and compared with an input C. The computation is performed in one clock cycle using clock frequency *Clk*. If the supply voltage is reduced the delay through the adder and the comparator is too high and one clock cycle is not long enough for the completion. The duplication of the data path modules together with an additional multiplexer makes it possible to run both modules at half clock frequency (*Clk/2*) providing two times longer clock cycle, as

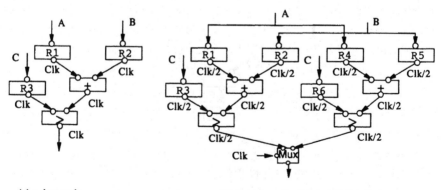

(a) a data path   (b) the data path after parallelization

Figure 10.3: Parallelisation of the data path.

indicated in Figure 10.3b.

□

Pipelining is another way to introduce parallelism into the design and provide the same throughput when the delay of single modules is longer. In this case, the data path modules are not duplicated but the throughput of the design can still be maintained at the same level. As it has been indicated in Section 6.5 pipelining is a very cost efficient implementation.

> **Example 10.2:** Consider the simple data path from Example 10.1. Figure 10.4 represents a pipelined solution for this data path which is run at half the speed.

□

There also exists a clear way to obtain a given throughput by the *reduction of the number of control steps* required to perform a given computation. Since the throughput is constant, the execution time is constant. If the execution time is represented by t_{alg}, the clock cycle time is T_{clock} and n is the number of clock cycles need for the algorithm execution, the following formula represents the algorithm's execution time:

$$t_{alg} = T_{clock} \cdot n \tag{5}$$

To keep t_{alg} constant, we can reduce the number of clock cycles n while increasing the clock cycle time T_{clock} (making the clock slower). This can be done by using previously introduced operation-scheduling transformations. This however is not always possible due to data dependencies between operations. In such situations, scheduling transformations have to be combined with compiler-oriented transformations, such as algebraic, constant propagation or loop unrolling transformations [CPMRB95b].

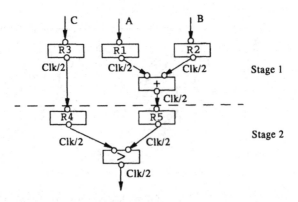

Figure 10.4: Pipelined implementation of the data path from Figure 10.3a.

Low power module selection

One obvious way to reduce power consumption is to select modules from the library which provide required throughput of the design but consume less power. This is, in principle, a binding decision. The synthesis system has to select an optimal binding of modules to provide a given throughput and minimize the power consumption. This transformation can be combined with voltage scaling by selecting modules with lower supply voltage, for example 3.3 V instead of 5 V. Such transformations can lead, in certain situations, to the decision of using two supply voltages for the same design together with dc/dc convertors.

Low power module sharing

In classical high-level synthesis module sharing is mainly used to reduce the area of a design. The synthesis system makes decisions on registers and functional units sharing to optimize the area and performance of a design. For low power synthesis, the decision on sharing is made to reduce the switching activity in the circuit. This can be achieved by selecting for sharing those modules whose switching activity on pairs of data is minimal. Module sharing for low power can be applied to both registers and functional units, however the formulation of these two problems is slightly different [ChPe95a, ChPe95b].

Since the idea of module sharing is well known the main problem of low-power module sharing is the calculation of joint switching activities between different pairs of signals. It has to be calculated for all possible candidates for sharing. A method for calculation of the switching activities which is based on statistical methods has been proposed in [ChPe95a]. The method assumes that the joint probability density function of the primary input random variables is known. Similar formulae are used for calculation of the switching activities for registers [ChPe95a] and functional units [ChPe95b].

Number representation selection

In most applications two's complement is used to represent numbers. This is justified because arithmetic operations of two's complement numbers are easy to perform. This representation is however not appropriate when low power design is considered, specially in the case when data values oscillate around zero. For arithmetical operations, a more preferable representation is, in this case, sign-magnitude. When sign-magnitude number system is used only one bit toggles if the value switches sign. For a sequence of consecutive numbers transmitted, for example, on a bus, Gray code is a good solution since only one bit toggles when a new data arrives.

The number representation has impact on three main components of a design: buses, storage (register and memories), and functional units. Buses usually consume quite much power and thus the number representation selection is very important. In many applications when a sequence of consecutive values is transmitted, the Gray code is a good solution because it reduces switching activities on the bus. It applies, for example, to instruction-address bus of a microprocessor where the probability of fetching the instruction at the following address is quite high.

Selection of the number representation for registers and memories can be reasoned in the same way as for buses. In addition, the mapping of data into different memories can influence switching activity and this will be discussed later in this section.

Functional units are different. They can provide power savings when using the sign-magnitude number representation. A careful design analysis is however needed because some operations are more complex in this representation. For example, subtraction is more complex in sign-magnitude than in two's complement representation. There are also new innovative designs of functional units, such as those presented in [CPMRB95b].

Algebraic transformations for low power

Algebraic transformations have been discussed in Subsection 6.4.1. The main purpose of using these transformations in high-level synthesis is to reduce design area. It is achieved by reducing the number of operators and thus the number of functional units. This goal is still valid in low power synthesis since reducing the number of operations reduces power consumption. Thus, the proposed transformations which are based on algebraic laws of associativity, commutativity and distributivity are directly applicable here. Additionally new transformations can be used for reducing switching activities by ordering of input signals and minimizing glitches.

The *ordering of input signals* can be changed by applying associativity and commutativity laws to the expressions. In this way, by carefully studying the characteristics of the input signals, we can reduce the switching activity.

Example 10.3: Consider the following expression $(\text{In}+\text{In}*2^7)+\text{In}*2^8$ [ChBr95] and its dataflow graph depicted in Figure 10.5a. The multiplication by 2^7 and 2^8 represents a shift operation by 7 and 8 position respectively. The resulting switching activity of the first data path will be quite high if we assume that signal In has a large variance and occupies the whole bit-width. In the second data path, depicted in Figure 10.5b, the signal In shifted by 7 or 8 positions to the left has zeros on these positions and adding two such signals does not create any switching activity at the lower 7 bit positions and thus this adder consumes less power. In this example, the expression is transformed first into $\text{In}+(\text{In}*2^7+\text{In}*2^8)$ and then to $(\text{In}*2^7+\text{In}*2^8)+\text{In}$ using commuta-

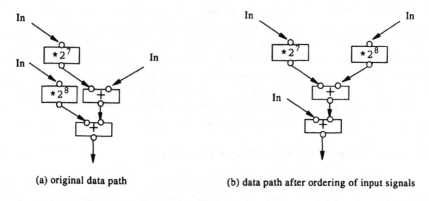

(a) original data path   (b) data path after ordering of input signals

Figure 10.5: The input signal ordering transformation for switching activity reduction.

tivity and associativity respectively.

□

The *glitching activity* is a function of several factors as indicated in Section 10.1. In high-level synthesis it is possible to reduce one of the factors, namely the depth of combinational logic. This can be done using a tree-height reduction transformation described in Subsection 6.4.1. The reduction of the depth of the combinational logic leads also to a decrease of the total delay of the logic unit.

Storage allocation and data mapping

Memories are very often responsible for a major part of the power consumption. The decisions on memory selection and mapping of data to one or several memory modules can influence the power consumption very much. There are several ways to deal with this problem during high-level synthesis which is motivated by the following two facts. First, a larger memory requires switching of a greater capacitance during every memory access. Second, the switching on buses between the memory and the rest of the design depends on the correlation between the transferred data/addresses. Thus the synthesis can influence power consumption caused by memories basically by optimizing the number of memory modules and their size as well as deciding on the data mapping to the selected memory modules.

The selection of the number of memory modules for a given algorithm is the first decision which can be made during high-level synthesis. A number of memory modules can be selected based on the access patterns to the data. The most frequently used data can be mapped into internal registers or a small on-chip memory while the other data can be mapped into one or several bigger memory modules placed off the chip.

Access patterns influence the switching activities on the memory data and address buses. To reduce the activity on the address bus, the data should be located in consecutive memory location. This will result in minimal switches on the address bus if combined together with a Gray code addressing method. This approach is possible for regular data structures, such as arrays. An array mapping algorithm has been presented in [PaDu96] and applied to a number of image processing benchmarks. The reported results indicate up to 63% reduction of the switching activities on the address bus.

10.3.4   Summary of Low Power Transformations

In this section, we present a summery of transformations used for low power synthesis. First, in Table 10.1, evaluation of the previously introduced high-level synthesis transformations is given from the point of view of low power design. Later, in Table 10.2, specific low power transformations proposed in the literature and discussed in this chapter are presented. In the tables, we try to present the effect of every transformation on power consumption, delay (speed or throughput) and area of the design. The delay presented in the table is the total time required to perform the computation. This is the module delay in the case of combinational circuits and the execution time for sequential circuits. The execution time is computed as a product of the clock cycle time and the number of steps required for the computation. In some cases, a specific transformation does not reduce the power consumption directly but makes changes in the design that enable further low power transformations. This is indicated in the comments. The comments also explain the sources of reduction or increase in power, delay, and area.

In Table 10.1, the transformations, discussed in Section 6.4, are presented. It can be noted that the compiler-oriented transformations are well suited for power consumption reduction. Algebraic transformations, common subexpression elimination, dead code elimination, and constant and variable propagation directly reduce power consumption and reduce or, at least do not increase, other design parameters. In-line procedure expansion and loop unrolling do not reduce power consumption directly but make it possible to apply other power reduction transformations.

Because of the complexity of the power optimization problem, it can not be concluded directly from the table that some transformations are not applicable to low power synthesis. For example, pipelining increases both power consumption and area but decreases the delay. This can be later used to slow down the whole design by, for example, reducing the supply voltage, which decreases the power consumption, as discussed earlier.

Table 10.1: The impact of some high-level synthesis transformations on power consumption, delay, and area.

	Transformation	Power	Delay	Area	Comments
Compiler oriented	algebraic	−	−	−	mainly based on the use of commutativity, associativity and distributivity laws
	common subexpression elimination	−	−	−	
	dead code elimination	−	≈	−	
	code motion	−	−	≈	mainly moving loop invariants out of a loop
	constant and variable propagation	−	−	−	
	in-line procedure expansion	≈	−	+	can allow other transformations, such as resource sharing
	loop unrolling	≈	≈	+	used to allow parallelization and pipelining transformations
Operation scheduling oriented	place-stretch	≈	+	−	
	parallelization	≈	−	+	
	place-splitting	+	−	+	a small increase in power and area is caused by introduction of a new register in the data path
	place-folding	−	+	−	
	rescheduling	≈	≈	−	area is reduced because of possible sharing of functional units
data path oriented	vertex merger	≈	≈	−	delay is slightly increased by adding a multiplexer
	vertex splitter	≈	≈	+	delay is slightly reduced by removing a multiplexer
control oriented	control merger	≈	≈	≈	
	control->data path moving of conditional statements	+	−	+	allows easier pipelining
	data path->control moving of conditional statements	−	+	−	

Table 10.1: The impact of some high-level synthesis transformations on power consumption, delay, and area. *(Continued)*

	Transformation	Power	Delay	Area	Comments
pipeline oriented	functional pipelining	+	–	+	the power consumption increase can later be compensated by reducing supply voltage and/or slowing down the clock
	algorithmic pipelining	+	–	+	the power consumption increase can later be compensated by reducing supply voltage and/or slowing down clock

+: increase of the parameter, -: decrease of the parameter, ≈: no direct impact on the parameter.

All transformation presented in Table 10.2 reduce power consumption. They can however not all be used alone, which is indicated in some comments. For example, reducing supply voltage requires some kind of compensation of the throughput which becomes lower.

Table 10.2: The impact of low power specific transformations on power consumption, delay, and area.

Transformation	Power	Delay	Area	Comments
lower V_{dd}	–	+	≈	the reduced speed can be later compensated by introduction of *parallelism, pipelining* or *reduction of the number of control steps.*
low power module selection	–	+	≈	
low power module sharing	–	≈	≈	low power module sharing uses vertex-merger transformation
number representation	–	≈	≈, +	buses and registers do not require bigger area; arithmetical units can be bigger
Storage allocation and data mapping	–	≈	≈	

+: increase of the parameter, -: decrease of the parameter, ≈: no direct impact on the parameter.

10.4 Low-power Synthesis Systems and Algorithms

High-level synthesis for low power is a new research area which currently develops very fast. There are several research groups working on design automation tools for low power synthesis. In this chapter, we will briefly present some selected tools developed by these groups.

10.4.1   HYPER-LP

HYPER-LP is a high-level synthesis system for minimization of power consumption in application-specific data-path intensive CMOS circuits (mainly filters) using architectural and algebraic transformations [CPMRB95]. The main idea behind this system is to maintain throughput at reduced supply voltages through hardware duplication or pipelining. The basic principles are described in Subsection 10.3.3. In addition compiler-oriented and retiming transformations are used.

Minimization of power consumption is defined as an optimization problem in a similar way as depicted in Figure 10.2. The objective function is defined as the total power consumption:

$$P_{total} = C_{total} \cdot V^2 \cdot f_{sampling} \tag{5}$$

where V is the supply voltage, $f_{sampling}$ is the clock frequency and C_{total} is the total switched capacitance. Since V and $f_{sampling}$ are known the main parameter which has to be estimated is the total switched capacitance. It depends on four components and can be further defined as:

$$C_{total} = C_{exe} + C_{registers} + C_{interconnect} + C_{control} \tag{6}$$

where C_{exe}, $C_{registers}$, $C_{interconnect}$ and $C_{control}$ are the capacitance switched by the functional units, registers, interconnections and controller respectively. The estimations for these four parameters are done separately using methods similar to those described in Subsection 10.3.1.

The power consumption optimization algorithm used by HYPER-LP selects appropriate transformations in order to reduce the objective function (5). The transformations are divided into global and local ones. Global transformations optimize the whole design represented by CDFG graph while the local ones are applicable only to one or very few nodes of the CDFG graph. The following global transformations are used: i) retiming and pipelining for critical path reduction, ii) associativity, iii) constant elimination, and iv) loop unrolling. Three local transformations have been implemented: i) associativity, ii) commutativity, and iii) local retiming.

The power optimization algorithm uses both heuristic and probabilistic optimization algorithms. The heuristic part of the optimization algorithm uses global transformations. It is used to minimize the critical path using polynomial complexity algorithm which produces optimal results. The modified Leiserson-Saxe algorithm for critical path minimization using retiming is used (for the original algorithm see [LeSa91]). The associativity transformations are selected based on the minimization of critical path using dynamic programming. Loop unrolling does not involve any optimization and it is selected when it reveals a large amount of parallelism for other transformations. The probabilistic optimization algorithm is simulated annealing and it is used to select local transformations.

The power optimization is done in several steps. First, constant propagation is applied to reduce both the number of operations and the critical path. This is followed by the modified Leiserson-Saxe algorithm and the dynamic programming algorithm. Finally, the simulated annealing optimization algorithm is used to improve the current solution using local transformations.

The authors have reported very good results [CPMRB95]. For example, the synthesis of the Wavelet filter provided 18 times power reduction (from 107 mW to 5.8 mW) while the area increased only 7.4 times (from 8.5 mm^2 to 62.9 mm^2).

10.4.2   Profile-Driven Synthesis System (PDSS)

The PDSS system makes low power synthesis of a design by minimizing estimated switching activity [KKRV95]. The switching activity is estimated early in the design process and the results are used to predict the aggregate (total) switching activity (ASA) in the whole circuit.

According to the PDSS strategy, the design process is started with profiling. Probes (counters) are inserted into behavioral VHDL specifications by the profiler to collect event activity data. These data are later used during high-level synthesis to estimate the ASA of the design. The following data are gathered by the profiler and assigned to nodes and edges of a CDFG: i) the event activity of the operation, ii) the transaction activity of the edge, and iii) the event activity of the edge. The annotated graph is submitted for synthesis.

The synthesis consist of three steps:
- *Scheduling and performance estimation* - this step provides a minimal length schedule fulfilling resource constraints. It is done by applying force-directed scheduling to a number of valid bags[1] of library components and selecting the schedule which meets both area and clock period constraints while having the least estimated ASA.
- *Register allocation* - allocates and assigns registers into the design while optimizing their number. It is based on a clique-partitioning heuristic.
- *Interconnect optimization* - a Min-max algorithm that prefers shared buses

---

1. A bag, like a set, is a collection of elements over some domain but, unlike a set, bags allow multiple occurrences of elements.

before multiplexer-based connections wherever possible is used.
- *Controller generation* - the FSM is generated.

The ASA is estimated during the scheduling and performance estimation phase based on the profiled, scheduled and operator-bound data-flow graph. The estimation is done separately for operators, registers, interconnections and controllers. It is assumed that controllers are implemented using PLAs.

The PDSS system mainly performs trade-offs between different schedules and power consumption. It does not make use of behavioral transformations, such as algebraic transformations or algorithmic pipelining.

10.4.3   Power-Profiler

The main idea of the Power-Profiler high-level synthesis tool is to make trade-offs in module allocation and binding [MaKn94, MaKn95]. The system has a component library of predefined modules. In the current implementation 32-bit adders and multipliers are implemented. There are two types of adders: ripple-carry and carry-lookahead, and four types of multipliers: Booth, Quasi Bit-Serial, Array, and a 16-bit Digit-Serial Array. Every component is implemented for 5 and 3.3 volts. The component library contains the related area, delay and power consumption for every module.

The scheduling, allocation and binding of data path modules are optimization problems. They are combined in this system and are implemented by a single genetic algorithm. The algorithm is able to perform the following optimizations:
- minimize either average or peak power with area and/or delay constraints,
- minimize delay with area and/or power constraints,
- minimize area with delay and/or power constraints,
- minimize any weighted combination of area, average and peak power with multiple constraints.

10.4.4   Module and Register Allocation and Binding

The main idea behind the system presented in [ChPe95a, ChPe95b] is to reduce switching activities within a circuit by an appropriate decisions regarding module sharing. These principles are discussed in Subsection 10.3.3.

The main assumption of this approach is that the joint probability density function of the primary input random variables is known or that a sufficiently large number of input vectors has been given. Allocation and binding is then separately formulated for registers and functional units. The register assignment problem for minimum power consumption is formulated as a minimum cost clique covering problem and is solved using a max-cost flow algorithm [ChPe95a]. This is possible since the compatibility graph is transitively oriented. The similar problem for assigning functional units for pipeline computation has not this property and thus is solved using multi-commodity flow formulation [ChPe95b].

Bibliography

[ABF90]   M. Abramovic, M. A. Breuer and A. D. Friedman, *Digital System Testing and Testable Design*, IEEE Press, 1990.

[AdTh96]   J. K. Adams and D. E. Thomas, "The Design of Mixed Hardware/Software Systems", *Proc. 33rd ACM/IEEE Design Automation Conference*, 1996, pp. 515-520.

[AKS93a]   V. D. Agrawal, C. R. Kime and K. K. Saluja, "A Tutorial on Build-in Self-Test, Part 1: Principles", *IEEE Design and test of Computers*, March 1993.

[AKS93b]   V. D. Agrawal, C. R. Kime and K. K. Saluja, "A Tutorial on Build-in Self-Test, Part 2: Applications", *IEEE Design and Test of Computers*, June 1993.

[AlKa95]   C. J. Alpert and A. B. Kahng, "Recent Directions in Netlist Partitioning: a Survey", *INTEGRATION, the VLSI journal*, 19, 1995, pp. 1-81.

[AHU83]   A. V. Aho, J. E. Hopcroft and J. D. Ullman, *Data Structures and Algorithms*, Adison-Wesley, Reading, MA, 1983.

[ASU88]   A. V. Aho, R. Sethi and J. D. Ullman, *Compilers: Principles, Techniques and Tools*, Adison-Wesley, Reading, MA, 1988.

[AvMc94]   L. J. Avra and E. J. McCluskey, "High-Level Synthesis of Testable Designs: An Overview of University Systems", International Test Conference, ITC'94, *Test Synthesis Seminar, Digest of Papers*, Washington, D. C., USA, October 4, 1994.

[BaBo69]   J, Baer and D. Bovet, "Compilation of Arithmetic Expressions for Parallel Computations", *Proc. IFIP*, 1969, pp. 340-346.

[BAJ94]   T. Ben Ismail, M. Abid and A. Jerraya, "COSMOS: A CoDesign Approach for Communicating Systems", *Proc. 3rd International Workshop on Hardware-Software Co-design*, Grenoble 1994, pp. 17-24.

[BCO96]   G. Boriello, P. Chou and R. Ortega, "Embedded System Co-Design: Towards Portability and Rapid Integration", in G. De Micheli and M. G. Sami, eds: *Hardware/Software Co-Design*, NATO ASI 1995, Kluwer Academic Publisher, 1996.

[BeKu93]   R. A. Bergamaschi and A. Kuehlmann, "A System for Production Use of High-Level Synthesis", *IEEE Transactions on Very Large Scale Integration (VLSI) Systems*, vol. 1, no. 3, September 1993, pp. 233-243.

[Bellc93]   *Generic Requirements for Operations of Broadband Switching Systems*, Bellcore TANWT-001248 issue 2, October 1993.

[Ben96]   A. Bender, "Design of an Optimal Loosely Coupled Heterogeneous Multiprocessor System", *Proc. European Design&Test Conference*, Paris 1996, pp. 275-281.

[BFMR93]   J. M. Bergé, A. Fonkoua, S. Maginot and J. Rouillard, *VHDL'92, The New Features of the VHDL Hardware Description Language*, Kluwer Academic Publisher, 1993.

[BhLe90]   J. Bhasker and Huan-Chih Lee, "An Optimizer for Hardware Synthesis", *IEEE Design and Test of Computers*, October 1990.

[BHS91]   F. Belina, D. Hogrefe and A. Sarma, *SDL with Applications from Protocol Specification*, Prentice Hall, 1991.

[Bie93]   J. Biesenack et al., "The Siemens High-Level Synthesis System CALLAS", *IEEE Transactions on Very Large Scale Integration (VLSI) Systems*, vol. 1, no. 3, September 1993, pp. 244-253.

[BIJe95]   T. Ben Ismail and A. A. Jerraya, "Synthesis Steps and Design Models for Codesign", *IEEE Computer*, vol. 28, no. 2, February 1995, pp.44-52.

[BOJ94]   T. Ben Ismail, K. O'Brian and A. A. Jerraya, "Interactive System-Level Partitioning with PARTIF", *Proc. European Design&Test Conference*, Paris 1994, pp. 464-468.

[BuRo95]   K. Buchenrieder and J. W. Rozenblit, "Codesign: An Overview", in J. W. Rozenblit and K. Buchenrieder, eds: *Codesign - Computer-Aided Software/ Hardware Engineering*, IEEE Press, 1995.

[Cam90]   R. Camposano, "Behavior-preserving transformations for high-level synthesis," in *Hardware Specification, Verification and Synthesis: Mathematical Aspects*, M. Leeser and G. Brown, Eds. Berlin: Springer-Verlag, pp.106-128, 1990.

[CaKu86]   R. Camposano and A. Kunzmann, "Considering Timing Constraints in Synthesis from a Behavioral Description", *Proc. IEEE International Conference on Computer Design*, 1986, pp. 6-9.

[CaWi96]   R. Camposano and J. Wilberg, "Embedded System Design", *Design Automation for Embedded Systems*, vol. 1, no. 1-2, January 1996, pp. 5-50.

[ChBo94]   P. Chou and G. Boriello, "Software Scheduling in the Co-Synthesis of Reactive Real-Time Systems", *Proc. 31st ACM/IEEE Design Automation Conference*, 1994, pp. 1-4.

[ChBo95]   P. Chou and G. Boriello, "Interval Scheduling: Fine-Grained Code Scheduling for Embedded Systems", *Proc. 32nd ACM/IEEE Design Automation Conference*, 1995, pp. 462-467.

[ChBr95]   A. P. Chandrakasan and R. W. Brodersen, "Minimizing Power Consumption in Digital CMOS Circuits", *Proceedings of the IEEE*, vol. 83, no. 4, April 1995.

[ChPe95a]   Jui-Ming Chang and M. Pedram, "Register Allocation and Binding for Low Power", *Proceedings of the 32nd Design Automation Conference*, San Francisco, USA, June 1995.

[ChPe95b]   Jui-Ming Chang and M. Pedram, "Power Efficient Module Allocation and Binding", Technical Report CENG 95-16, University of Southern California.

[CKL96]   W. T. Chang, A. Kalavade and E. A. Lee, "Effective Heterogeneous Design and

Co-simulation", in G. De Micheli and M. G. Sami, eds: *Hardware/Software Co-Design*, NATO ASI 1995, Kluwer Academic Publisher, 1996.

[COB92]   P. H. Chou, R. B. Ortega and G. Boriello, "Synthesis of the Hardware/Software Interface in Microcontroller-Based Systems", *Proc. IEEE International Conference on Computer-Aided Design*, 1992, pp. 488-495.

[COB95a] P. H. Chou, R. B. Ortega and G. Boriello, "The Chinook Hardware/Software Co-Synthesis System", *Proc. 8th International Symposium on System Synthesis*, September 1995, pp. 22-27.

[COB95b] P. Chou, R. B. Ortega and G. Boriello, "Interface Co-Synthesis Techniques for Embedded Systems", *Proc. IEEE International Conference on Computer-Aided Design*, 1995, pp. 280-287.

[CPMRB95] A. P. Chandrakasan, M. Potkonjak, R. Mehra, J. Rabaey, and R. W. Brodersen, "Optimizing Power Using Transformations", *IEEE Transactions on Computer-Aided Design of Integrated Circuits and Systems*, vol. 14, no. 1, January 1995.

[CST91]   R. Camposano, L. F. Saunders and R. M. Tabet, "VHDL as Input for High-Level Synthesis", *IEEE Design and Test of Computers*, March 1991, pp. 43-49.

[CWB94]   P. Chou, E. A. Walkup and G. Boriello, "Scheduling for Reactive Real-Time Systems", *IEEE Micro*, August 1994, pp. 37-47.

[DBJ95]   J. M. Daveau, T. Ben Ismail and A. A. Jerraya, "Synthesis of System-Level Communication by an Allocation-Based Approach", *Proc. 8th International Symposium on System Synthesis*, September 1995, pp. 150-155.

[DeMi92] G. De Micheli, "High-Level Synthesis of Digital Circuits,", *Advamces in Computers*, Vol. 37, 1993, pp.207-283.

[DeMi94a] G. De Micheli, *Synthesis and Optimization of Digital Circuits*, McGraw-Hill 1994.

[DeMi94b] G. De Micheli, "Computer-Aided Hardware-Software Codesign", *IEEE Micro*, August 1994, pp. 10-16.

[DeMi96] G. De Micheli, "Hardware-Software Co-Design: Application Domains and Design Technologies", in G. De Micheli and M. G. Sami, eds: *Hardware/Software Co-Design*, NATO ASI 1995, Kluwer Academic Publisher, 1996.

[DePr93] M. De Prycker, *Asynchronous Transfer Mode: Solution for Broadband ISDN*, Ellis Horwood, 1993.

[DKMT90] G. De Micheli, D. C. Ku, F. Mailhot and T Truong, "The Olympus Synthesis System for Digital Design", *IEEE Design and Test of Computers*, October 1990, pp. 37-53.

[Dutt93] S. Dutt, "New Faster Kernighan-Lin Type Graph-Partitioning Algorithms", *Proc. IEEE International Conference on Computer-Aided Design*, 1993, pp. 370-377.

[EKPM92] P. Eles, K. Kuchcinski, Z. Peng and M. Minea, "Compiling VHDL into a High-Level Synthesis Design Representation", *Proc. EURO-DAC with EURO-VHDL*, September 1992, Hamburg, pp. 604-609.

[EKPM93a] P. Eles, K. Kuchcinski, Z. Peng and M. Minea, "Synthesis of VHDL Subprograms and Processes in the CAMAD System", *Proc. Workshop on Design Methodologies for Microelectronics and Signal Processing*, Gliwice-Cracow, 1993, pp. 359-367.

[EKPM93b] P. Eles, K. Kuchcinski, Z. Peng and M. Minea, "Two Methods for Synthesizing

VHDL Concurrent Processes", Research Report, LiTH-IDA-R-93-22, Linköping University, 1993.

[EKPM94] P. Eles, K. Kuchcinski, Z. Peng and M. Minea, "Synthesis of VHDL Concurrent Processes", *Proc. EURO-DAC with EURO-VHDL*, September 1994, Grenoble, pp. 540-545.

[EKPD95] P. Eles, K. Kuchcinski, Z. Peng and A. Doboli, "Timing Constraints Specification and Synthesis in Behavioral VHDL", *Proc. EURO-DAC with EURO-VHDL*, September 1995, Brighton, pp. 452-457.

[EPKD96a] P. Eles, Z. Peng, K. Kuchcinski and A. Doboli, "Hardware/Software Partitioning of VHDL System Specifications", *Proc. EURO-DAC with EURO-VHDL*, September 1996, Geneva, pp. 434-439.

[EPKD96b] P. Eles, Z. Peng, K. Kuchcinski and A. Doboli, "Hardware/Software Partitioning with Iterative Improvement Heuristics", *Proc. 9th International Symposium on System Synthesis*, November 1996, pp. 71-76.

[EKP96] P. Eles, K. Kuchcinski and Z. Peng, "Synthesis of Systems Specified as Interacting VHDL Processes", *INTEGRATION, the VLSI journal*, 21, 1996, pp. 113-138.

[EKPD97] P. Eles, K. Kuchcinski, Z. Peng and A. Doboli, "Post-Synthesis Back-Annotation of Timing Information in Behavioral VHDL", *Journal of Systems Architecture*, vol 42, no. 9-10, February 1997, pp. 725-741.

[EPKD97] P. Eles, Z. Peng, K. Kuchcinski and A. Doboli, "System Level Hardware/Software Partitioning Based on Simulated Annealing and Tabu Search", *Design Automation for Embedded Systems*, vol. 2, no. 1, January 1997, pp. 5-32.

[EHB93] R. Ernst, J. Henkel and T. Benner, "Hardware-Software Cosynthesis for Microcontrollers", *IEEE Design and Test of Computers*, December 1993, pp. 64-75.

[Fje92] Björn Fjellborg, *Pipeline Extraction for VLSI Data Path Synthesis*, PhD Dissertation, Dept. of Computer and Information Science, Linköping University, 1992, No. 273.

[Fou81] L. R. Foulds, *Optimization Techniques*, Springer-Verlag, New York, 1981.

[FKCD93] D. Filo, D. C. Ku, C. N. Coelho and G. De Micheli, "Interface Optimization for Concurrent Systems under Timing Constraints", *IEEE Transactions on Very Large Scale Integration (VLSI) Systems*, vol. 1, no. 3, September 1993, pp. 268-281.

[FT82] H. Fujiwara and S. Toida, "The Complexity of Fault Detedtion Problems for Combinational Logic Circuits", *IEEE Transactions on Computers*, vol. C-31, June 1982, pp. 555-560.

[GaGl96] M. Gasteier and M. Glesner, "Bus-Based Communication Synthesis on System-Level", *Proc. 9th International Symposium on System Synthesis*, November 1996, pp. 65-70.

[Gaj96] D. D. Gajski et al., "System Design Methodologies: Aiming at the 100 h Design cycle", *IEEE Transactions on Very Large Scale Integration (VLSI) Systems*, vol. 4, no. 1, March 1996, pp. 70-82.

[GaJo75] M.R. Garey and D.S. Johnson "Complexity Results for Multiprocessor Scheduling under Resource Constraints," *SIAM J. Computing*, vol. 4, 1975, pp. 397-411.

[GaKu83] D. Gajski and R. Kuhn, "Guest Editors' Introduction: New VLSI Tools," *IEEE*

Computer, vol. 6, no. 12, pp.11-14, December 1983.

[GaVa95] D. D. Gajski and F. Vahid, "Specification and Design of Embedded Hardware-Software Systems", *IEEE Design and Test of Computers*, Spring 1995, pp. 53-67.

[GCD92] R. K. Gupta, C. N. Coelho Jr. and G. De Micheli, "Synthesis and Simulation of Digital Systems Containing Interacting Hardware and Software Components", *Proc. 29th ACM/IEEE Design Automation Conference*, 1992, pp. 225-230.

[GCD94] R. K. Gupta, C. N. Coelho Jr. and G. De Micheli, "Program Implementation Schemes for Hardware-Software Systems", *IEEE Computer*, vol. 27, no. 1, February 1994, pp.48-55.

[GDWL92] D. Gajski, N. Dutt, A. Wu and S. Lin, *High-Level Synthesis, Introduction to Chip and System Design*, Kluwer Academic Publisher, 1992.

[GEKP96] P. Grün, P. Eles, K. Kuchcinski and Z. Peng, "Automatic Parallelization of a Petri Net-Based Design Representation for High-Level Synthesis", *Proc. 22nd EUROMICRO Conference*, September 1996, pp. 185-192.

[GGB96] J. Gong, D. D. Gajski and S. Bakshi, "Model Refinement for Hardware-Software Codesign", *Proc. European Design&Test Conference*, Paris 1996, pp. 270-274.

[GGN95] J. Gong, D. D. Gajski and S. Narayan, "Software Estimation Using A Generic-Processor Model", *Proc. European Design&Test Conference*, Paris 1995.

[GKP91] X. Gu, K. Kuchcinski and Z. Peng, "Testability Measure with Reconvergent Fanout Analysis and Its Applications", *Microprocessing and Microprogramming, The Euromicro Journal*, vol. 32, no. 1-5, August 1991.

[GKP92] X. Gu, K. Kuchcinski and Z. Peng, "An Approach to Testability Analysis and Improvement for VLSI Systems", *Microprocessing and Microprogramming, The Euromicro Journal*, vol. 35, no. 1-5, September 1992.

[GKP94] X. Gu, K. Kuchcinski and Z. Peng, "Testability Analysis and Improvement from VHDL Behavioral Specifications", *Proc. EURO-DAC with EURO-VHDL*, September 1994, Grenoble, France.

[GKP95] X. Gu, K. Kuchcinski and Z. Peng, "An Efficient and Economic Partitioning Approach for Testability", *Proc. International Test Conference*, ITC'95, Washington D.C., October 23-25, 1995.

[Gol79] L. H. Goldstein, "Controllability/Observability Analysis of Digital Systems", *IEEE Transactions on Circuits and Systems*, vol. CAS-26, no.9, September 1979.

[Goo96] G. Goossens et al., "Programmable Chips in Consumer Electronics and Telecommunications: Architecture and Design Technology", in G. De Micheli and M. G. Sami, eds: *Hardware/Software Co-Design*, NATO ASI 1995, Kluwer Academic Publisher, 1996.

[GoTh80] L. H. Goldstein, and E. L. Thigpen, "SCOAP: SANDIA Controllability/Observability Analysis Program", *Proc. 17th Design Automation Conf.*, 1980.

[GTW93] F. Glover, E. Taillard and D. de Werra, "A User's Guide to Tabu Search", *Annals of Operations Research*", vol. 41, 1993, pp. 3-28.

[Gu96] X. Gu, *RT Level Testability Improvement by Testability Analysis and Transformations*, PhD Dissertation, Dept. of Computer and Information Science,

Linköping University, 1996, No. 414.

[GuDM92] R. K. Gupta and G. De Micheli, "System Synthesis via Hardware-Software Co-Design", Technical Report No. CSL-TR-92-548, Comp. Syst. Lab., Stanford Univ., 1992.

[GuDM93] R. K. Gupta and G. De Micheli, "Hardware-Software Cosynthesis for Digital Systems", *IEEE Design and Test of Computers*, September 1993, pp. 29-41.

[GuDM94] R. K. Gupta and G. De Micheli, "Constrained Software Generation for Hardware-Software Systems", *Proc. 3rd International Workshop on Hardware-Software Co-design*, Grenoble 1994, pp. 56-63.

[GuDM96] R. K. Gupta and G. De Micheli, "A Co-Synthesis Approach to Embedded System Design Automation", *Design Automation for Embedded Systems*, vol. 1, no. 1-2, January 1996, pp. 69-120.

[Gup95] R. K. Gupta, *Co-Synthesis of Hardware and Software for Digital Embedded Systems*, Kluwer Academic Publisher, 1995.

[Gup96a] R. K. Gupta, "A Framework for Interactive Analyses of Timing Constraints in Embedded Systems", *Proc. 4th International Workshop on Hardware-Software Co-design*, Pittsburgh 1996, pp. 44-51.

[Gup96b] R. K. Gupta, "Operation Serializability for Embedded Systems", *Proc. European Design & Test Conference*, Paris 1996, pp. 108-114.

[GVNG94] D. Gajski, F. Vahid, S. Narayan and J. Gong, *Specification and Design of Embedded Systems*, P T R Prentice Hall, 1994.

[GVN94] D. D. Gajski, F. Vahid and S. Narayan, "A System-Design Methodology: Executable- Specification Refinement", *Proc. European Design & Test Conference*, Paris 1994.

[HaPe95] J. Hallberg and Z. Peng, "Synthesis under Local Timing Constraints in the CAMAD High-Level Synthesis System," *Proc. Euromicro'95 Conference on Design of Hardware/Software Systems*, Como, Italy, Sept. 4-7, 1995, pp. 650-656.

[HaPe96] J. Hallberg and Z. Peng, "Multicycle Scheduling under Local Timing Constraints using Genetic Algorithms and Tabu Search," *Short Contributions, Euromicro'96 Conference on Hardware and Software Design Strategies*, Prague, Czech Republic, Sept. 2-5, 1996, pp. 138-143.

[HeEr95] J. Henkel and R. Ernst, "A Path-Based Technique for Estimating Hardware Runtime in Hw/Sw-Cosynthesis", *Proc. 8th International Symposium on System Synthesis*, September 1995, pp. 116-121.

[HEHB94] J. Henkel, R. Ernst, U. Holtmann and T. Benner, "Adaptation of Partitioning and High-Level Synthesis in Hardware/Software Co-Synthesis", *Proc. IEEE International Conference on Computer-Aided Design*, 1994, pp. 96-100.

[HKL90] P. Harper, S. Krolikoski and O. Levia, "Using VHDL as a Synthesis Language in the Honeywell VSYNTH System", in J. A. Darringer and F. J. Rammig, eds: *Computer Hardware Description Languages and their Applications*, North-Holland, Amsterdam, 1990, pp. 315-330.

[Hwa93] Kau Hwang, *Advanced Computer Architecture: Parallelism, Scalability, Programmability*, McGraw-Hill, Inc., 1993.

[IEEE87] IEEE Standard VHDL Language Reference Manual, IEEE Std 1076-1987, March

31, 1988.

[IEEE 90] *IEEE Std 1149.1-1990: IEEE Standard Test Access Port and Boundary-Scan Architecture.* Published by the Institute of Electrical and Electronics engineers, Inc., 345 East 47th Street, New York, NY 10017, USA.

[IEEE93] IEEE Standard VHDL Language Reference Manual, ANSI/IEEE Std 1076-1993 (Revision of IEEE Std 1076-1987), June 6, 1994.

[IS75] O. H. Ibarra and S. K. Sahni, "Polynomially Complete Fault Detection Problems", *IEEE Transactions on Computers*, vol. C-24, March 1975, pp. 242-249.

[ITU93] *B-ISDN Operation and Maintenance Principles and Functions*, ITU-T Recommendation I.610, 1993.

[JeOB95] A. A. Jerraya and K. O'Brian, "Solar: An Intermediate Format for System-Level Modelling and Synthesis", in J. W. Rozenblit and K. Buchenrieder, eds: *Codesign - Computer-Aided Software/Hardware Engineering*, IEEE Press, 1995.

[Joh67] S. C. Johnson, "Hierarchical Clustering Schemes", *Psychometrika*, vol. 32, no. 3, September 1967, pp. 241-254.

[Joh96] F. M. Johannes, "Partitioning of VLSI Circuits and Systems", *Proc. 33rd ACM/ IEEE Design Automation Conference*, 1996, pp. 83-87.

[JGB96] H.-P. Juan, D. D. Gajski and S. Bakshi, "Clock Optimization for High-performance Pipelined Design", *Proc. EURO-DAC with EURO-VHDL*, September 1996, Geneva, pp. 330-335.

[KAJW96] S. Kumar, J. H. Aylor, B. W. Johnson and Wm. A. Wulf, *The Codesign of Embedded Systems: A Unified Hardware/Software Representation*, Kluwer Academic Publisher, 1996.

[KeLi70] B. W. Kernighen and S. Lin, "An Efficient Heuristic Procedure for Partitioning Graphs", *Bell Systems Tech. J.*, vol. 49, no. 2, 1970, pp. 291-307.

[KGP83] S. Kirpatrick, C. D. Gelatt and M. P. Vecchi, "Optimization by Simulated Annealing", *Science*, vol. 220, no. 4598, 1983, pp. 671-680.

[KiMe88] T. Kirkland and M. R. Mercer, "Algorithms for Automatic Test Pattern Generation", *IEEE Design and Test of Computers*, June 1988.

[KLMM95] D. Knapp, T. Ly, D. MacMillen and R. Miller, "Behavioral Synthesis Methodology for HDL-Based Specification and Validation, *Proc. 32nd ACM/IEEE Design Automation Conference*, 1995, pp. 286-291.

[Kuck76] D. Kuck, "Parallel Processing of Ordinary Programs", *Advances in Computers*, Academic Press, vol. 15, 1976, pp. 119-179.

[KuDM92] D. C. Ku and G. De Micheli, *High-Level Synthesis of ASICs Under Timing and Synchronization Constraints*, Kluwer Academic Publisher, 1992.

[KuPe 87] K. Kuchcinski and Z. Peng, "Microprogramming implementation of timed Petri nets," *Integration, the VLSI Journal*, vol.5, pp.133-144, 1987.

[KuPe90] K. Kuchcinski and Z. Peng, "Testability Analysis in a VLSI High-Level Synthesis System", *Microprocessing and Microprogramming, The Euromicro Journal*, vol.28, nrs 1-5, March, 1990, pp. 295-300.

[KKRV95] N. Kumar, S. Katkoori, L. Rader and R. Vemuri, "Profile-Driven Behavioral Synthesis for Low-Power VLSI Systems", *IEEE Design and test of Computers*, Fall 1995.

[LaRa 93]   P. E. Landman and J. M. Rabaey, "Power estimation for High.Level Synthesis", *Proceedings of the European Design Automation Conference 1993*, Paris, February 1993.

[LaRa94]   P. E. Landman and Jan M. Rabaey, "Black-Box Capacitance Models for Architectural Power Analysis", *Proceedings of the 1994 International Workshop on Low-Power Design*, Napa Valley, CA, April 1994.

[LaTh91]   E. D. Lagnese and D. E. Thomas, "Architectural Partitioning for System Level Synthesis of Integrated Circuits", *IEEE Transactions on Computer-Aided Design of Integrated Circuits and Systems*, vol. 10, no. 7, July 1991, pp. 847-860.

[LeWh82]   J. Y. T. Leung and J. Whitehead, "On the Complexity of Fixed-Priority Scheduling of Periodic Fixed Priority Tasks", *Performance Evaluation*, vol. 2, 1982, pp. 237-250.

[LeSa91]   C. E. Leiserson and J. B. Saxe, "Retiming Synchronous Circuitry", *Algorithmica*, vol. 6, 1991, pp. 5-35.

[LiGa89]   J. S. Lis and D. D. Gajski, "VHDL Synthesis Using Structured Modelling", *Proc. 26th ACM/IEEE Design Automation Conference*, 1989, pp. 606-609.

[LiLa73]   C. L. Liu and J. W. Layland, "Scheduling Algorithms for Multiprogramming in a Hard Real-Time Environment", *Journal of the ACM*, vol. 20, no. 1, January 1973, pp. 46-61.

[LiMa95]   Y. S. Li and S. Malik, "Performance Analysis of Embedded Software Using Implicit Path Enumeration", *Proc. 32nd ACM/IEEE Design Automation Conference*, 1995, pp. 456-461.

[LiSv 93]   D. Liu and C. Svensson, "Trading Speed for Low-Power by Choice of Supply and Threshold Voltage", *IEEE Journal of Solid-State Circuits*, vol. 28, pp. 10-17, 1993.

[LSH96]   L. Lavagno, A. Sangiovanni-Vincentelli and H. Hsieh, "Embedded System Co-Design. Synthesis and Verification", in G. De Micheli and M. G. Sami, eds: *Hardware/Software Co-Design*, NATO ASI 1995, Kluwer Academic Publisher, 1996.

[LVM96]   B. Lin, S. Vercauteren and H. De Man, "Embedded Architecture Co-Synthesis and System Integration", *Proc. 4th International Workshop on Hardware-Software Co-design*, Pittsburgh 1996, pp. 2-9.

[McC86]   E. J. McCluskey, "A Survey of Design for Testability Scan Techniques" *Semicustom Design Guide*, Summer 1986.

[MaKn94]   R. San Martin and J. P. Knight, "A Tutorial on Behavioral Synthesis Power Optimization", http://www.doe.carleton.ca/~rsm/Power/tutorial.html.

[MaKn95]   R. San Martin and J. P. Knight, "Power-Profiler: Optimizing ASICs Power Consumption at the Behavioral Level", *Proceedings of the 32nd Design Automation Conference*, San Francisco, USA, June 1995.

[Mal96]   S. Malik, et al., "Performance Analysis of Embedded Systems", in G. De Micheli and M. G. Sami, eds: *Hardware/Software Co-Design*, NATO ASI 1995, Kluwer Academic Publisher, 1996.

[McF83]   M. C. McFarland, "Computer-Aided Partitioning of Behavioral Hardware Descriptions", *Proc. 20th ACM/IEEE Design Automation Conference*, 1983, pp. 472-478.

[McF86] M. C. McFarland, "Using Bottom-Up Design Techniques in the Synthesis of Digital Hardware from Abstract Behavioral Descriptions", *Proc. 23rd ACM/IEEE Design Automation Conference*, 1986, pp. 474-480.

[McKo90] M. C. McFarland and T. J. Kowalski, "Incorporating Bottom-Up Design into Hardware Synthesis", *IEEE Transactions on Computer-Aided Design of Integrated Circuits and Systems*, vol. 9, no. 9, September 1990, pp. 938-950.

[MeRa94] R. Mehra and J. Rabey, "Behavioral Level Power Estimation and Exploration", *Proceedings of the 1994 International Workshop on Low-Power Design*, Napa Valley, CA, April 1994.

[Mic94] Z. Michalewicz, *Genetic Algorithms + Data Structure = Evolution Programs*, Springer-Verlag, 1994.

[MLD92] P. Michel, U. Lauther and P. Duzy, Editors, *The Synthesis Approach to Digital System Design*, Kluwer Academic Publisher, 1992.

[MMP95] R. Marculescu, D.a Marculescu and M. Pedram, "Efficient Power Estimation for Highly Correlated Input Streams", *Proceedings of the 32nd Design Automation Conference*, San Francisco, USA, June 1995.

[MPC90] M. C. McFarland, A. C. Parker and R. Camposano, "The high-level synthesis of digital systems," *Proceedings of the IEEE*, vol. 78, pp.301-318, Feb. 1990.

[Mura71] Y. Muraoka, *Parallelism Exposure and Exploitation in Programs*, PhD dissertation, Univ. of Illinois at Urbana-Champaign, 1971.

[NaGa94a] S. Narayan and D. D. Gajski, "Protocol Generation for Communication Channels", *Proc. 31st ACM/IEEE Design Automation Conference*, 1994, pp. 547-551.

[NaGa94b] S. Narayan and D. D. Gajski, "Synthesis of System-Level Bus Interfaces", *Proc. European Design & Test Conference*, Paris 1994, pp. 395-399.

[NaGa95] S. Narayan and D. D. Gajski, "Interfacing Incompatible Protocols Using Interface Process Generation", *32nd ACM/IEEE Design Automation Conference*, 1995, pp. 468-473.

[NaGa96] S. Narayan and D. D. Gajski, "Rapid Performance Estimation for System Design", *Proc. EURO-DAC with EURO-VHDL*, September 1996, Geneva, pp. 206-211.

[Naj94] F. N. Najm, "A Survey of Power Estimation Techniques in VLSI Circuits", *IEEE Transactions on VLSI*, December 1994.

[NBD92] V. Nagasamy, N. Berry and C. Dangelo, "Specification, Planning, and Synthesis in a VHDL Design Environment", *IEEE Design and Test of Computers*, June 1992, pp. 58-68.

[NeNa96] M. Nemani and F. N. Najm, "Towards a High-Level Power Estimation Capacity", *IEEE Transactions on Computer-Aided Design of Integrated Circuits and Systems*, vol. 15, no. 6, June 1996.

[NGH95] F. N. Najm, S. Goel and I. Hajj, "Power Estimation in Sequential Circuits", *32nd ACM/IEEE Design Automation Conference*, pp. 635-640.

[NiMa96] R. Niemann and P. Marwedel, "Hardware/Software Partitioning Using Integer Programming", *Proc. European Design & Test Conference*, Paris 1996, pp. 473-479.

[NiMa97] R. Niemann and P. Marwedel, "An Algorithm for Hardware/Software Partitioning

Using Mixed Integer Linear Programming", *Design Automation for Embedded Systems*, vol. 2, no. 2, 1997, pp. 165-193.

[NVG90] S. Narayan, F. Vahid and D. D. Gajski, "Modeling with SpecCharts", Technical Report #90-20, Dept. of Inf. and Comp. Science, Univ. of California, Irvine, 1990/1992.

[NVG92] S. Narayan, F. Vahid and D. D. Gajski, "System Specification with the SpecCharts Language", *IEEE Design and Test of Computers*, December 1992, pp. 6-13.

[OKDX95] S. Y. Ohm, F. J. Kurdahi, N. Dutt and M. Xu, "A Comprehensive Estimation Technique for High-Level Synthesis", *Proc. 8th International Symposium on System Synthesis*, September 1995, pp. 122-127.

[OrBo97] R. B. Ortega and G. Boriello, "Communication Synthesis for Embedded Systems with Global Considerations", *Proc. 5th International Workshop on Hardware-Software Co-design*, Braunschweig 1997, pp. 69-73.

[PaDu96] P. R. Panda and N. D. Dutt, "Reducing Address Bus Transitions for Low Power Memory Mapping", *Proc. European Design & Test Conference*, Paris, France, March 11-14, 1996.

[PPM86] A. C. Parker, J. T. Pizarro and M. Mlinar, "MAHA: A program for datapath synthesis", *Proc. 23rd Design Automation Conf.*, pp.461-466, 1986.

[PaKn89] P. Paulin and J. Knight, "Force-directed scheduling for the behavioral synthesis of ASIC's," *IEEE Trans. Computer-Aided Design*, vol. 8, pp.661-679, June 1989.

[Pau96] P. G. Paulin et al., "Trends in Embedded System Technology: an Industrial Perspective", in G. De Micheli and M. G. Sami, eds: *Hardware/Software Co-Design*, NATO ASI 1995, Kluwer Academic Publisher, 1996.

[Ped96] M. Pedram, "Power Minimization in IC Design: Principles and Applications", *ACM Transactions on Design Automation of Electronic Systems*, vol. 1, no. 1, January 1996.

[PeKu94] Z. Peng and K. Kuchcinski, "Automated Transformation of Algorithms into Register-Transfer Level Implementation," *IEEE Transactions on Computer-Aided Design of Integrated Circuits and Systems*, vol. 13, no. 2, Feb. 1994, pp. 150-166.

[Pen86] Z. Peng, "Synthesis of VLSI Systems with the CAMAD Design Aid," *Proc. 23rd ACM/IEEE Design Automation Conf.*, Las Vegas, Jun. 1986, pp.278-284.

[Pen87] Z. Peng, *A Formal Methodology for Automated Synthesis of VLSI Systems*, Ph.D. Dissertation, No. 170, Dept. of Computer and Information Science, Linköping University, 1987.

[Pen88] Z. Peng, "Semantics of a Parallel Computation Model and its Applications in Digital Hardware Design," *Proc. 1988 International Conference on Parallel Processing*, Pennsylvania State University, August 15-19, 1988, pp. 69-73.

[PKL89] Z. Peng, K. Kuchcinski and B. Lyles, "CAMAD: A Unified Data Path/Control Synthesis Environment," in D. A. Edwards (Editor), *Design Methodologies for VLSI and Computer Architecture*, Elsevier Science Publishers, 1989, pp. 53-67.

[Pet81] J. Peterson, *Petri Net Theory and Modeling of Systems*, Englwood Cliffs, NJ: Prentice-Hall, 1981.

[PoRa94] M. Potkonjak and J. Rabaey, "Optimizing Resource Utilization Using Transformations", *IEEE Transactions on Computer-Aided Design of Integrated Circuits and Systems*, vol. 13, no. 3, March 1994.

[Pos91]   A. Postula, "VHDL Specific Issues in High-Level Synthesis", *Proc. EURO-VHDL*, 1991, Stockholm, pp. 70-77.

[PrPa92] S. Prakash and A. Parker, "SOS: Synthesis of Application-Specific Heterogeneous Multiprocessor Systems", *Journal of Parallel and Distributed Computing*, vol. 16, 1992, pp. 338-351.

[Rac93]   *Racal-Redac Intelligen 2.0 Reference Manual*, 1993.

[Ree93]   C. R. Reeves, *Modern Heuristic Techniques for Combinatorial Problems*, Blackwell Scientific Pub., London, 1993.

[RKDV92] J. Roy, N. Kumar, R. Dutta and R. Vemuri, "DSS: A Distributed High-Level Synthesis System", *IEEE Design and Test of Computers*, June 1992, pp. 18-32.

[RNVG93] L. Ramachandran, S. Narayan, F. Vahid and D. D. Gajski, "Synthesis of Functions and Procedures in Behavioral VHDL", *Proc. EURO-DAC with EURO-VHDL*, September 1993, Hamburg, pp. 560-565.

[SrBr95] M. B. Srivastava and R. W. Brodersen, "SIERA: A Unified Framework for Rapid-Prototyping of System-Level Hardware and Software", *IEEE Transactions on Computer-Aided Design of Integrated Circuits and Systems*, vol. 14, no. 6, June 1995, pp. 676-693.

[SrPa94] M. Srinivas and L. M. Patnaik, "Genetic Algorithms: A Survey," *Computer*, Vol. 27, No. 6, 1994, pp.17-26.

[Sto87]   Harold S. Stone, *High-Performance Computer Architecture*, Addison-Wesley Publishing Company, 1987.

[SuSa96] K. Suzuki and A. Sangiovanni-Vincentelli, "Efficient Software Performance Estimation Methods for Hardware/Software Codesign", *Proc. 33rd ACM/IEEE Design Automation Conference*, 1996, pp. 605-610.

[Syn92]   VHDL Compiler Reference Manual. V3.0, Synopsis, Nov. 1992, Chapter 11, VHDL Compiler Directives.

[SvLi94] C. Svensson and D. Liu "Power Consumption in CMOS VLSI Chips", *IEEE Journal of Solid-State Circuits*, vol. 30. pp. 663-670, 1994.

[ThMo91] D. E. Thomas and P. R. Moorby, *The Verilog Hardware Description Language*, Kluwer Academic Publisher, 1991.

[VaGa92] F. Vahid and D. D. Gajski, "Specification Partitioning for System Design", *Proc. 29th ACM/IEEE Design Automation Conference*, 1992, pp. 219-224.

[VaGa95a] F. Vahid and D. D. Gajski, "Incremental Hardware Estimation during Hardware/Software Functional Partitioning", *IEEE Transactions on Very Large Scale Integration (VLSI) Systems*, vol. 3, no. 3, September 1995, pp. 459-464.

[VaGa95b] F. Vahid and D. D. Gajski, "Closeness Metrics for System-Level Functional Partitioning", *Proc. EURO-DAC with EURO-VHDL*, September 1995, Brighton, pp. 328-333.

[VaGa95c] F. Vahid and D. D. Gajski, "Clustering for Improved System-Level Functional Partitioning", *Proc. 8th International Symposium on System Synthesis*, September 1995, pp. 28-33.

[Vah95]   F. Vahid, "Procedure Exlining: A New System-Level Specification Transformation", *Proc. EURO-DAC with EURO-VHDL*, September 1995, Brighton, pp. 508-513.

[VaLe97] F. Vahid and T. Le, "Extending the Kernighan/Lin Heuristic for Hardware and Software Functional Partitioning", *Design Automation for Embedded Systems*, vol. 2, no. 2, 1997, pp. 237-261.

[VGG94] F. Vahid, J. Gong and D. D. Gajski, "A Binary-Constraints Search Algorithm for Minimizing Hardware During Hardware/Software Partitioning", *Proc. EURO-DAC with EURO-VHDL*, September 1994, Grenoble, pp. 214-219.

[VLH96] F. Vahid, T. Le and Y. Hsu, "A Comparison of Functional and Structural Partitioning", *Proc. 9th International Symposium on System Synthesis*, November 1996, pp. 121-126.

[VLM96a] S. Vercauteren, B. Lin and H. De Man, "Constructing Application-Specific Heterogeneous Embedded Architectures from Custom HW/SW Applications", *Proc. 33rd ACM/IEEE Design Automation Conference*, 1996, pp. 521-526.

[VLM96b] S. Vercauteren, B. Lin and H. De Man, "A Strategy for Real-Time Kernel Support in Application Specific HW/SW Embedded Architectures", *Proc. 33rd ACM/IEEE Design Automation Conference*, 1996, pp. 678-683.

[VNG95] F. Vahid, S. Narayan and D. D. Gajski, "SpecCharts: A VHDL Front-End for Embedded Systems", *IEEE Transactions on Computer-Aided Design of Integrated Circuits and Systems*, vol. 14, no. 6, June 1995, pp. 694-706.

[VRBM96] D. Verkest, K. Van Rompaey, I. Bolsens and H. De Man, "CoWare-A Design Environment for Heterogeneous Hardware/Software System", *Design Automation for Embedded Systems*, vol. 1, 1996, pp. 357-386.

[WaCh95] R. A. Walker and S. Chaudhuri, "Introduction to the Scheduling Problem," *IEEE Design and Test of Computers*, vol. 12, no. 2, 1995, pp. 60-69.

[WiBr81] T. W. Williams and N. Brown, "Defect Level as a Function of Fault Coverage", *IEEE Transactions on Computers*, Dec. 1981.

[WiPa83] T. W. Williams and K. P. Parker, "Design for Testability— A Survey", *Proceedings of the IEEE*, vol. 71, no. 1, January 1983.

[Wolf94] W. Wolf, "Hardware-Software Co-Design of Embedded Systems", *Proceedings of the IEEE*, vol. 82, no. 7, July 1994, pp. 967-989.

[WoMa93] W. Wolf and R. Manno, "High-Level Modelling and Synthesis of Communicating Processes Using VHDL", *IEICE Trans. Inform. Systems*, vol. E76-D, no. 9, September 1993, pp. 1039-1046.

[YEBH93] W. Ye, R. Ernst, T. Benner and J. Henkel, "Fast Timing Analysis for Hardware-Software Co-Synthesis", *Proc. IEEE International Conference on Computer Design*, 1993, pp. 452-457.

[YeWo95a] T. Y. Yen and W. Wolf, "Performance Estimation for Real-Time Distributed Embedded Systems", *Proc. IEEE International Conference on Computer Design*, 1995, pp. 64-69.

[YeWo95b] T. Y. Yen and W. Wolf, "Communication Synthesis for Distributed Embedded Systems", *Proc. IEEE International Conference on Computer Aided Design*, 1995, pp. 288-294.

[YeWo95c] T. Y. Yen and W. Wolf, "Sensitivity-Driven Co-Synthesis of Distributed Embedded

Systems", *Proc. 8th International Symposium on System Synthesis*, September 1995, pp. 4-9.

[YeWo97] T. Y. Yen and W. Wolf, *Hardware-Software Co-Synthesis of Distributed Embedded Systems*, Kluwer Academic Publisher, 1997.

INDEX

A

Abstraction level, 4, 21, 60, 104, 105
  circuit, 5, 21
  logic, 4, 21
  RT (register transfer), 4, 21, 262
  system, 4, 21, 60, 61, 99, 105, 120, 121, 243
Allocation, 8, 9, 80–86, 102, 106
  See also Data path allocation
  system components, 101–103, 126–127
APARTY, 113
Architecture-driven voltage scaling, 339
  parallelism, 339
  pipelining, 340
ATM switch, 295, 297
ATPG (automatic test pattern generation), 304
  See also Test pattern generation

B

Binding, 8, 9, 80–86, 106, 120
  clique partitioning, 83
  compatibility graph, 83
  conflict graph, 84, 87
  graph coloring, 84
  integer linear programming, 82–83
  left-edge algorithm, 85–86
  module library, 81
  register assignment, 86
BIST (built-in-self-test), 13, 305, 311
  alias, 311
  controller, 311
  error masking, 311
  response analyzer, 311
  signature, 311
  test generator, 311
Branch-and-bound, 138–140

BUD, 111

C

CALLAS, 264
CAMAD, 78
CDFG (control/data flow graph), 7, 68, 105, 347
Chinook, 133
Closeness function, 108–110, 112–113
Combinatorial optimization, 137
Communication channel, 102, 104, 120–122, 128, 130, 133
Communication protocol, 123
Communication synthesis, 9, 101–102, 104, 120, 132–135
  channel binding, 120
  communication refinement, 120–121
  interface generation, 120, 123, 134
Constraint, 99, 107, 126, 285
  area, 107
  cost, 99, 127, 128, 132, 285, 295
  performance, 102, 107, 124, 132
  pin, 107
  rate, 99, 124, 128, 130, 134
  timing, 127, 128, 130, 131, 134–135, 261–262, 268
Control unit, 235–237, 247, 249–254, 260
Controller synthesis, 87–97
  controller-style selection, 89
  finite state machine (FSM), 88
  generation, 93
  handshaking protocol, 96
  hierarchical controller, 89
  implementation, 94
  parallel controller, 90
  single controller, 89
COSMOS, 132